# Current Trends in Dynamical Systems, Fractal Geometry, and Their Applications

CONTEMPORARY MATHEMATICS

839

# Current Trends in Dynamical Systems, Fractal Geometry, and Their Applications

AMS Special Session
Dynamical Systems: Statistical Properties, Spectral Theory, and Fractal Geometry
September 14–15, 2024
University of Texas at San Antonio, San Antonio, Texas

Sangita Jha
William Ott
Mrinal Kanti Roychowdhury
Editors

AMS AMERICAN MATHEMATICAL SOCIETY

2020 *Mathematics Subject Classification.* Primary 37A25, 37A35, 37D05, 37E05, 37F10, 60F15, 60G15, 60H30, 46N30, 47B32.

---

**Library of Congress Cataloging-in-Publication Data**

Contemporary Mathematics ISSN: 0271-4132 (print); ISSN: 1098-3627 (online)
DOI: https://doi.org/10.1090/conm/839

---



10 9 8 7 6 5 4 3 2 1 31 30 29 28 27 26

We dedicate the volume to our beloved families.

# Contents

# Preface

This volume presents a curated collection of research articles that reflect current trends at the intersection of dynamical systems, fractal geometry, ergodic theory, and symbolic dynamics. The contributions arise from a diverse international cohort of mathematicians and highlight a broad spectrum of theoretical advances and applications in contemporary mathematical science.

The selected papers delve into rich and varied topics. These include novel formulations of relative topological pressure in nonautonomous settings, intricate tiling theories using graph iterated function systems, and statistical frameworks for understanding weak chaos in dynamical systems. Foundational concepts such as hyperbolic laminations and the structure of Julia sets are studied alongside new inequalities for ergodic Hilbert transforms and investigations into oscillatory behavior inspired by Sarnak's conjecture.

Several papers offer innovative methodologies. From reciprocal transformations and Maharam extensions in ergodic theory to symbolic dynamics shaped by concatenations of periodic words and Artin-Mazur zeta function analysis, the breadth is substantial. Other works advance our understanding of dimension theory through dyadic cube constructions, explore homotopical complexity in dispersing billiard systems, and apply operator theory to diffusions governed by self-similar measures. The volume concludes with a modern treatment of fractal interpolation functions using Rakotch contractions to model complex forms with partial self-similarity.

By uniting these contributions, this book aims to foster deeper insights, inspire new research directions, and facilitate cross-disciplinary dialogue among researchers in mathematics and allied fields. We hope the reader finds this volume both intellectually stimulating and a valuable resource in the ongoing exploration of nonlinear dynamics, geometry, and analysis.

We would like to express our sincere gratitude to all those who made this volume possible. First and foremost, we thank the authors for their valuable contributions and the anonymous referees for their careful and thoughtful evaluations of the manuscripts. Their expertise and dedication, even under very tight deadlines, were instrumental in shaping the quality of this collection. Last but not least, we extend our appreciation to Ms. Bridget Manzella for her assistance with the editorial process, and to the American Mathematical Society for their continued support of our proposal. Without the collective efforts of the contributors and reviewers, this volume would not have been conceivable. Our heartfelt thanks to each of you.

Dr. Mrinal Kanti Roychowdhury
(On behalf of the editors)

Contemporary Mathematics
Volume **839**, 2026
https://doi.org/10.1090/conm/839/16722

# Relative topological pressure of nonautonomous dynamical systems

Baimei Shi, Bilel Selmi, and Zhiming Li

ABSTRACT. Based on the study of relative topological pressure for the autonomous dynamical system, we define and study relative topological pressure for the non-autonomous dynamical systems on a non-compact metric space. In this paper, we obtain the basic properties and the invariant with respect to uniformly conjugate of relative topological pressure for the non-autonomous dynamical systems.

## 1. Introduction

Let $(X, d)$ be a metric space and $f : X \to X$ be a continuous self-map. Then we will call $(X, f)$ an autonomous dynamical system.

In the study of autonomous dynamical system, entropy is an important invariant of topological conjugate, which describes the complexity of a system. The notion of topological entropy was first introduced by Adler, Konheim and McAndrew [**1**] in 1965. Later Bowen [**2**] and Dinaburg [**5**] gave a new, but equivalent definition in the case when the space under consideration is metrizable.

Topological pressure as a generalization of topological entropy was first introduced by Ruelle [**10**] in 1973 for an expansive dynamical system and later by Walters [**14, 15**] for the general case. The theories related to the topological pressure, variational principle and equilibrium states play a fundamental role in statistical mechanics, ergodic theory and dynamical systems [**3, 4, 6, 12**]. Since the works of Bowen [**4**] and Ruelle [**11**], the topological pressure has become a basic tool in multi-fractal analysis of hyperbolic and expanding system.

In 1996, S. Kolyada and L. Snoha [**7**] studied topological entropy for the non-autonomous dynamical systems on a compact topological space. In 2006, Tian Chuanjun and Chen Guanrong [**13**] gave the concept of topological (uniformly) conjugate for the non-autonomous dynamical systems. For the recent progress on this direction, we refer to [**16**].

All we mentioned above are mainly focus on the compact system. For non-compact system, in 2008, M.Malziri and M.R.Molaei [**8**] gave an extension of the

---

2020 *Mathematics Subject Classification.* Primary 37A35, 37H99.
*Key words and phrases.* Non-autonomous dynamical system; relative topological pressure.
This research is supported by the Science and Technology Project Foundation of Xi'an (2021XDJH36), Shaanxi Provincial Department of Science and Technology Project (2022JQ-837), Shaanxi Provincial Department of Education Project (21JK0873) and by Analysis, Probability & Fractals Laboratory (No. LR18ES17).

notion of topological entropy via the notion of a base of a map and further in 2010, M.R.Molaei and H.Molaei [**9**] gave an extension of the notion of topological pressure via the notion of base of a map. In this paper, we will define and study topological pressure for a non-autonomous relative topological dynamical system (abbreviated as NRTDS, which is generated by a sequence of maps with a base and an observer). In section 2, we give a precise definition of NRTDS and then give the concepts of relative topological entropy and relative topological pressure by spanning set and separated set for NRTDS. The relationship between the two ways of definitions are also studied. In section 3, we investigate the basic properties of relative topological pressure for NRTDS and prove that under a mild assumption, the relative topological pressure is independent of the choice of base. In section 4, we show that the relative topological pressure for NRTDS is invariant with respect to uniformly topological conjugate.

## 2. Preliminary

Let $(X,d)$ be a metric space (not necessarily compact) and $f_{1,\infty}=\{f_i\}_{i=1}^{\infty}$ be a sequence of continuous maps from X to X. The identity map on X will be denoted by $id_X$ or shortly by $id$. Let $N,R$ be the set of all positive integers, real numbers, respectively. For any $i\in N$, let $f_i^0=f_i^{-0}=id_X$ and for any $i,n\in N$, set $f_i^n=f_{i+(n-1)}\circ\cdots\circ f_{i+1}\circ f_i$ (first apply $f_i$) and $f_i^{-n}=(f_i^n)^{-1}=f_i^{-1}\circ f_{i+1}^{-1}\circ\cdots\circ f_{i+(n-1)}^{-1}$ (the last notations will be applied to sets, we do not assume that the maps $f_i$ are invertible). Finally, denote by $f_{1,\infty}^n$ the sequence of maps $\{f_{(i-1)n+1}^n\}_{i=1}^{\infty}$ and by $f_{1,\infty}^{-1}$ the sequence $\{f_i^{-1}\}_{i=1}^{\infty}$. We denote the cardinal number of a set $S$ by $\sharp S$.

**2.1. Relative topological dynamical system.** In order to overcome the difficulty of non-compactness of the space, we need to introduce the concept of a base of $X$. Then we can give a reasonable definition of topological entropy and topological pressure in a non-compact system.

Definition 2.1. Let $(X,d)$ be a metric space, $f_{1,\infty}=\{f_i\}_{i=1}^{\infty}$ a sequence of continuous maps from X into itself, $K$ a compact subset of $X$. If for all compact set $C\subseteq X$ there exists $m(C)\in N$ such that $f_1^n(K)\supseteq C$ for all $n\geq m(C)$, then we call $(f_{1,\infty},K)$ is **a base of** $X$.

Definition 2.2. Let $(X,d)$ be a metric space, $f_{1,\infty}=\{f_i\}_{i=1}^{\infty}$ a sequence of continuous maps from X into itself, $g_{1,\infty}=\{g_i\}_{i=1}^{\infty}$ a sequence of uniformly continuous maps from X into itself. If $(f_{1,\infty},K)$ is a base of $X$, then we call $(X,d,g_{1,\infty})$ **a relative topological dynamical system with respect to** $(f_{1,\infty},K)$ and denote it by $(g_{1,\infty}|_{(f_{1,\infty},K)},X,d)$.

In the rest of this paper, we always assume that $(g_{1,\infty}|_{(f_{1,\infty},K)},X,d)$ is a relative topological dynamical system.

**2.2. Relative topological entropy.** For each $n\in N$, set

$$d_n(x,y)=\max_{0\leq i\leq n-1} d(g_1^i(x),g_1^i(y)).$$

Definition 2.3. For any $m,n\in N$ and $\varepsilon>0$, let $F\subseteq f_1^m(K)$, if for any $x\in f_1^m(K)$ there exists $y\in F$ such that $d_n(x,y)\leq\varepsilon$, then we call F is **an** $(m,n,g_{1,\infty},\varepsilon)$**-spanning set of** $f_1^m(K)$. Let $E\subseteq f_1^m(K)$, if for any $x,y\in E$ we have $d_n(x,y)>\varepsilon$, then we call E is **an** $(m,n,g_{1,\infty},\varepsilon)$**-separated set of** $f_1^m(K)$.

Let $r(g_{1,\infty}|_{f_1^m(K)}, n, \varepsilon)$ denote the smallest cardinality of any $(m, n, g_{1,\infty}, \varepsilon)$-spanning set of $f_1^m(K)$, $s(g_{1,\infty}|_{f_1^m(K)}, n, \varepsilon)$ denote the largest cardinality of any $(m, n, g_{1,\infty}, \varepsilon)$-separated set of $f_1^m(K)$. By the compactness of $f_1^m(K)$, we have the next lemma.

LEMMA 2.1. *For any $m, n \in N$ and $\varepsilon > 0$, we have*

(1) $r(g_{1,\infty}|_{f_1^m(K)}, n, \varepsilon) < +\infty$, $s(g_{1,\infty}|_{f_1^m(K)}, n, \varepsilon) < +\infty$;

(2) $r(g_{1,\infty}|_{f_1^m(K)}, n, \varepsilon_1) \geq r(g_{1,\infty}|_{f_1^m(K)}, n, \varepsilon_2)$,
$s(g_{1,\infty}|_{f_1^m(K)}, n, \varepsilon_1) \geq s(g_{1,\infty}|_{f_1^m(K)}, n, \varepsilon_2)$ *for* $0 < \varepsilon_1 < \varepsilon_2$;

(3) $r(g_{1,\infty}|_{f_1^m(K)}, n, \varepsilon) \leq s(g_{1,\infty}|_{f_1^m(K)}, n, \varepsilon) \leq r(g_{1,\infty}|_{f_1^m(K)}, n, \frac{\varepsilon}{2})$.

PROOF.

(1) Since $g_{1,\infty} = \{g_i\}_{i=1}^{\infty}$ is a sequence of uniformly continuous maps on $X$, the map $g_1^i$ is uniformly continuous on $X$. Then for any $\varepsilon > 0$ there exists $\delta > 0$ such that for all $x, y \in X$, $d(x, y) < \delta$ implies that $d(g_1^i(x), g_1^i(y)) < \varepsilon$, $i = 1, 2, \cdots, n-1$. Put $\alpha = \{B_d(y, \delta) | y \in X\}$, then $\alpha$ is an open cover of $f_1^m(K)$. By the compactness of $f_1^m(K)$, $\alpha$ has an open subcover which denoted by $\alpha_1 = \{B_d(y_1, \delta), B_d(y_2, \delta), \cdots, B_d(y_l, \delta)\}$. For any $x \in f_1^m(K)$, there exists $1 \leq k \leq l$ such that $x \in B_d(y_k, \delta)$, i.e. $d(x, y_k) < \delta$. By the uniformly continuous of $g_1^i$, we have $d(g_1^i(x), g_1^i(y_k)) < \varepsilon$, $i = 1, 2, \cdots, n-1$, i.e. $d_n(x, y_k) < \varepsilon$. Hence $F_0 = \{y_1, y_2, \cdots, y_l\}$ is an $(m, n, g_{1,\infty}, \varepsilon)$-spanning set of $f_1^m(K)$. Therefore $r(g_{1,\infty}|_{f_1^m(K)}, n, \varepsilon) \leq l < +\infty$.

To obtain the second inequality, we prove

$$s(g_{1,\infty}|_{f_1^m(K)}, n, \varepsilon) \leq r(g_{1,\infty}|_{f_1^m(K)}, n, \frac{\varepsilon}{2}).$$

Let $E$ be an $(m, n, g_{1,\infty}, \varepsilon)$-separated set of $f_1^m(K)$, $F$ be an $(m, n, g_{1,\infty}, \frac{\varepsilon}{2})$-spanning set of $f_1^m(K)$, then for any $x \in f_1^m(K)$ there exists $y \in F$ such that $d_n(x, y) \leq \frac{\varepsilon}{2}$. Particularly, for any $x \in E$ there exists $y_x \in F$ such that $d_n(x, y_x) \leq \frac{\varepsilon}{2}$. Then we have the following claim: if $x_1, x_2$ are two distinct points in $E$, then $y_{x_1}, y_{x_2}$ are two distinct points in $F$. Suppose that there exists $x_1, x_2 \in E$ , $x_1 \neq x_2$ such that $y_{x_1} = y_{x_2}$, then $d_n(x_1, y_{x_1}) \leq \frac{\varepsilon}{2}$ and $d_n(x_2, y_{x_2}) \leq \frac{\varepsilon}{2}$ and hence $d_n(x_1, x_2) \leq d_n(x_1, y_{x_1}) + d_n(y_{x_1}, x_2) = d_n(x_1, y_{x_1}) + d_n(x_2, y_{x_2}) \leq \frac{\varepsilon}{2} + \frac{\varepsilon}{2} = \varepsilon$, which contradicts with the fact that $E$ is an $(m, n, g_{1,\infty}, \varepsilon)$-separated set of $f_1^m(K)$. Hence $y_{x_1} \neq y_{x_2}$. By the above argument, we know that there exists an injective from $E$ to $F$. This means $\sharp E \leq \sharp F$. Hence $s(g_{1,\infty}|_{f_1^m(K)}, n, \varepsilon) \leq \sharp F$. Therefore $s(g_{1,\infty}|_{f_1^m(K)}, n, \varepsilon) \leq r(g_{1,\infty}|_{f_1^m(K)}, n, \frac{\varepsilon}{2}) < +\infty$.

(2) Let $F_1$ be an $(m, n, g_{1,\infty}, \varepsilon_1)$-spanning set of $f_1^m(K)$, then for any $x \in f_1^m(K)$ there exists $y \in F_1$ such that $d_n(x, y) \leq \varepsilon_1$. Since $\varepsilon_1 < \varepsilon_2$, we have $d_n(x, y) < \varepsilon_2$. Hence $F_1$ is an $(m, n, g_{1,\infty}, \varepsilon_2)$-spanning set of $f_1^m(K)$. Therefore $\{\sharp F_1 |\ F_1$ is an $(m, n, g_{1,\infty}, \varepsilon_1)$-spanning set of $F_1\} \subseteq \{\sharp F_2 |\ F_2$ is an $(m, n, g_{1,\infty}, \varepsilon_2)$-spanning set of $f_1^m(K)\}$. It follows that $r(g_{1,\infty}|_{f_1^m(K)}, n, \varepsilon_1) \geq r(g_{1,\infty}|_{f_1^m(K)}, n, \varepsilon_2)$. Let $E_2$ be an $(m, n, g_{1,\infty}, \varepsilon_2)$-separated set of $f_1^m(K)$, then for any $x, y \in E_2$, we have $d_n(x, y) > \varepsilon_2$. Since $\varepsilon_1 < \varepsilon_2$, we have $d_n(x, y) > \varepsilon_1$. Hence $E_2$ is an $(m, n, g_{1,\infty}, \varepsilon_1)$-separated set of $f_1^m(K)$. Therefore $\{\sharp E_2 |\ E_2$ is an $(m, n, g_{1,\infty}, \varepsilon_2)$-separated set of $f_1^m(K)\} \subseteq \{\sharp E_1 |\ E_1$ is an $(m, n, g_{1,\infty}, \varepsilon_1)$-separated set of $f_1^m(K)\}$. It follows that $s(g_{1,\infty}|_{f_1^m(K)}, n, \varepsilon_1) \geq s(g_{1,\infty}|_{f_1^m(K)}, n, \varepsilon_2)$.

(3) Let $E_0$ be an $(m, n, g_{1,\infty}, \varepsilon)$-separated set of $f_1^m(K)$ such that the cardinality of $E_0$ is equal to $s(g_{1,\infty}|_{f_1^m(K)}, n, \varepsilon)$. Then $E_0$ is an $(m, n, g_{1,\infty}, \varepsilon)$-spanning set of $f_1^m(K)$. Otherwise there exists $x_0 \in f_1^m(K)$ such that for all $y \in E_0$ we

have $d_n(x_0, y) > \varepsilon$. Hence $E_0 \bigcup \{x_0\}$ is an $(m, n, g_{1,\infty}, \varepsilon)$-separated set of $f_1^m(K)$, which contradicts with the fact that $\sharp E_0 = s(g_{1,\infty}|_{f_1^m(K)}, n, \varepsilon)$. Thus $E_0$ is an $(m, n, g_{1,\infty}, \varepsilon)$-spanning set of $f_1^m(K)$. Therefore $r(g_{1,\infty}|_{f_1^m(K)}, n, \varepsilon) \leq \sharp E_0 = s(g_{1,\infty}|_{f_1^m(K)}, n, \varepsilon)$. By (1) we obtain $s(g_{1,\infty}|_{f_1^m(K)}, n, \varepsilon) \leq r(g_{1,\infty}|_{f_1^m(K)}, n, \frac{\varepsilon}{2})$. Hence $r(g_{1,\infty}|_{f_1^m(K)}, n, \varepsilon) \leq s(g_{1,\infty}|_{f_1^m(K)}, n, \varepsilon) \leq r(g_{1,\infty}|_{f_1^m(K)}, n, \frac{\varepsilon}{2})$ .

□

Let

$$r(g_{1,\infty}|_{f_1^m(K)}, \varepsilon) = \limsup_{n\to\infty} \frac{1}{n} logr(g_{1,\infty}|_{f_1^m(K)}, n, \varepsilon),$$

$$s(g_{1,\infty}|_{f_1^m(K)}, \varepsilon) = \limsup_{n\to\infty} \frac{1}{n} logs(g_{1,\infty}|_{f_1^m(K)}, n, \varepsilon).$$

Then by lemma 2.1, we have the next corollary.

COROLLARY 2.1. *For any $m \in N$ and $\varepsilon > 0$, we have*

(1) $r(g_{1,\infty}|_{f_1^m(K)}, \varepsilon_1) \geq r(g_{1,\infty}|_{f_1^m(K)}, \varepsilon_2)$,
$s(g_{1,\infty}|_{f_1^m(K)}, \varepsilon_1) \geq s(g_{1,\infty}|_{f_1^m(K)}, \varepsilon_2)$ *for* $0 < \varepsilon_1 < \varepsilon_2$;

(2) $r(g_{1,\infty}|_{f_1^m(K)}, \varepsilon) \leq s(g_{1,\infty}|_{f_1^m(K)}, \varepsilon) \leq r(g_{1,\infty}|_{f_1^m(K)}, \frac{\varepsilon}{2})$.

By corollary 2.1 (2), we have $\lim_{\varepsilon\to 0} r(g_{1,\infty}|_{f_1^m(K)}, \varepsilon) = \lim_{\varepsilon\to 0} s(g_{1,\infty}|_{f_1^m(K)}, \varepsilon)$. This common limit is denoted by $h(g_{1,\infty}|_{f_1^m(K)})$, that is

$$h(g_{1,\infty}|_{f_1^m(K)}) = \lim_{\varepsilon\to 0} r(g_{1,\infty}|_{f_1^m(K)}, \varepsilon) = \lim_{\varepsilon\to 0} s(g_{1,\infty}|_{f_1^m(K)}, \varepsilon).$$

DEFINITION 2.4. Let $(g_{1,\infty}|_{(f_{1,\infty},K)}, X, d)$ be a relative topological dynamical system. Then we call

$$h(g_{1,\infty}|_{(f_{1,\infty},K)}) = \limsup_{m\to\infty} h(g_{1,\infty}|_{f_1^m(K)})$$

the **relative topological entropy for** $(g_{1,\infty}|_{(f_{1,\infty},K)}, X, d)$.

REMARK 2.1. This definition is a natural generalization of the notion of topological entropy for the non-autonomous dynamical systems[8]. Because when $X$ is a compact metric space then $h(g_{1,\infty}|_{(I_{1,\infty},X)})$ will be the topological entropy of $g_{1,\infty}$, where $I_{1,\infty}$ is a sequence of identity maps on $X$.

**2.3. Relative topological pressure.** Now, we will use the spanning set and separated set to define the relative topological pressure in non-autonomous dynamical systems.

Let $(C(f_1^m(K), R); \|\cdot\|)$ be the space of all continuous functions on $f_1^m(K)$, where $\| \varphi \| = \max_{x\in f_1^m(K)} |\varphi(x)|$, $\varphi \in C(f_1^m(K), R)$; $(BC(X, R); \|\cdot\|_B)$ be the space of all bounded continuous functions on $X$, where $\| \psi \|_B = \sup_{x\in X} |\psi(x)|$, $\psi \in BC(X, R)$. For $m, n \in N$, $x \in f_1^m(K)$, $\lambda : X \to [0, 1]$ (we can treat the function $\lambda$ as a one dimensional observer of $X$) and $\varphi \in C(f_1^m(K), R)$, we denote the summation by

$$(S^{\lambda}_{m,n}\varphi)(x) = \sum_{i=0}^{n-1} \delta^{g_1^i(x)}_{f_1^m(K)} \lambda(g_1^i(x))\varphi(g_1^i(x)),$$

where

$$\delta^{g_1^i(x)}_{f_1^m(K)} = \begin{cases} 1, & \text{if } g_1^i(x) \in f_1^m(K) \\ \\ 0, & \text{if } g_1^i(x) \notin f_1^m(K). \end{cases}$$

2.3.1. *Define the relative topological pressure by a spanning set.* For any $\varepsilon > 0$, we put

$$Q^{\lambda}_{m,n,f_{1,\infty}}(g_{1,\infty},\varphi,\varepsilon) = \inf\left\{\sum_{x\in F} e^{(S^{\lambda}_{m,n}\varphi)(x)} \mid F \text{ is an } (m,n,g_{1,\infty},\varepsilon)\text{-spanning set of } f_1^m(K)\right\}.$$

Then we have the next lemma.

LEMMA 2.2. *For any $m,n \in N$, $\varepsilon > 0$ and $\varphi \in C(f_1^m(K),R)$, we have*
(1) $Q^{\lambda}_{m,n,f_{1,\infty}}(g_{1,\infty},\varphi,\varepsilon_1) \geq Q^{\lambda}_{m,n,f_{1,\infty}}(g_{1,\infty},\varphi,\varepsilon_2)$ *for* $0 < \varepsilon_1 < \varepsilon_2$*;*
(2) $0 < Q^{\lambda}_{m,n,f_{1,\infty}}(g_{1,\infty},\varphi,\varepsilon) \leq e^{n\|\varphi\|} r(g_{1,\infty}|_{f_1^m(K)},n,\varepsilon)$*;*
(3) $Q^{\lambda}_{m,n,f_{1,\infty}}(g_{1,\infty},0,\varepsilon) = r(g_{1,\infty}|_{f_1^m(K)},n,\varepsilon)$.

PROOF.

(1) Since $\{F_1 \mid F_1 \text{ is an } (m,n,g_{1,\infty},\varepsilon_1)\text{-spanning set of } f_1^m(K)\}$

$$\subseteq \{F_2 \mid F_2 \text{ is an } (m,n,g_{1,\infty},\varepsilon_2)\text{-spanning set of } f_1^m(K)\},$$

then

$$\left\{\sum_{x\in F_1} e^{(S^{\lambda}_{m,n}\varphi)(x)} \mid F_1 \text{ is an } (m,n,g_{1,\infty},\varepsilon_1)\text{-spanning set of } f_1^m(K)\right\}$$
$$\subseteq \left\{\sum_{x\in F_2} e^{(S^{\lambda}_{m,n}\varphi)(x)} \mid F_2 \text{ is an } (m,n,g_{1,\infty},\varepsilon_2)\text{-spanning set of } f_1^m(K)\right\}.$$

Therefore

$$Q^{\lambda}_{m,n,f_{1,\infty}}(g_{1,\infty},\varphi,\varepsilon_1) \geq Q^{\lambda}_{m,n,f_{1,\infty}}(g_{1,\infty},\varphi,\varepsilon_2).$$

(2) Let $F_0$ be an $(m,n,g_{1,\infty},\varepsilon)$-spanning set of $f_1^m(K)$ such that the cardinality of $F_0$ is equal to $r(g_{1,\infty}|_{f_1^m(K)},n,\varepsilon)$, then by

$$(S^{\lambda}_{m,n}\varphi)(x) \leq |(S^{\lambda}_{m,n}\varphi)(x)| = \left|\sum_{i=0}^{n-1} \delta^{g_1^i(x)}_{f_1^m(K)} \lambda(g_1^i(x))\varphi(g_1^i(x))\right| \leq n \parallel \varphi \parallel,\ \forall x \in f_1^m(K),$$

we have

$$Q^{\lambda}_{m,n,f_{1,\infty}}(g_{1,\infty},\varphi,\varepsilon) \leq \sum_{x\in F_0} e^{(S^{\lambda}_{m,n}\varphi)(x)} \leq e^{n\|\varphi\|}\cdot \sharp F_0 = e^{n\|\varphi\|} r(g_{1,\infty}|_{f_1^m(K)},n,\varepsilon).$$

And since

$$Q^{\lambda}_{m,n,f_{1,\infty}}(g_{1,\infty},\varphi,\varepsilon) > 0,$$

then

$$0 < Q^{\lambda}_{m,n,f_{1,\infty}}(g_{1,\infty},\varphi,\varepsilon) \leq e^{n\|\varphi\|} r(g_{1,\infty}|_{f_1^m(K)},n,\varepsilon).$$

(3) $Q^{\lambda}_{m,n,f_{1,\infty}}(g_{1,\infty},0,\varepsilon) = \inf\{\sharp F \mid$ F is an $(m,n,g_{1,\infty},\varepsilon)$-spanning set of $f_1^m(K)\} = r(g_{1,\infty}|_{f_1^m(K)},n,\varepsilon)$. □

Put

$$Q^{\lambda}(m, f_{1,\infty}, g_{1,\infty}, \varphi, \varepsilon) = \limsup_{n\to\infty} \frac{1}{n} log Q^{\lambda}_{m,n,f_{1,\infty}}(g_{1,\infty}, \varphi, \varepsilon).$$

Then by lemma 2.2, we have the next corollary.

COROLLARY 2.2. *For any $m \in N$, $\varepsilon > 0$ and $\varphi \in C(f_1^m(K), R)$, we have*

(1) $Q^{\lambda}(m, f_{1,\infty}, g_{1,\infty}, \varphi, \varepsilon_1) \geq Q^{\lambda}(m, f_{1,\infty}, g_{1,\infty}, \varphi, \varepsilon_2)$ *for* $0 < \varepsilon_1 < \varepsilon_2$;
(2) $Q^{\lambda}(m, f_{1,\infty}, g_{1,\infty}, \varphi, \varepsilon) \leq \| \varphi \| + r(g_{1,\infty}|_{f_1^m(K)}, \varepsilon)$.

Put

$$P^{\lambda}(m, f_{1,\infty}, g_{1,\infty}, K, \varphi) = \lim_{\varepsilon\to 0} Q^{\lambda}(m, f_{1,\infty}, g_{1,\infty}, \varphi, \varepsilon).$$

Then by corollary 2.2 (1), we know that $P^{\lambda}(m, f_{1,\infty}, g_{1,\infty}, K, \varphi)$ exists but could be $\infty$.

DEFINITION 2.5. Let $(g_{1,\infty}|_{(f_{1,\infty},K)}, X, d)$ be a relative topological dynamical system. Define a map

$$P^{\lambda}(g_{1,\infty}|_{(f_{1,\infty},K)}, \cdot) \ : \ C(X, R) \ \to \ R\bigcup\{\infty\}$$

$$\varphi \mapsto P^{\lambda}(g_{1,\infty}|_{(f_{1,\infty},K)}, \varphi),$$

where

$$P^{\lambda}(g_{1,\infty}|_{(f_{1,\infty},K)}, \varphi) = \limsup_{m\to\infty} P^{\lambda}(m, f_{1,\infty}, g_{1,\infty}, K, \varphi).$$

Then we call $P^{\lambda}(g_{1,\infty}|_{(f_{1,\infty},K)}, \cdot)$ the **relative topological pressure** for

$$(g_{1,\infty}|_{(f_{1,\infty},K)}, X, d).$$

REMARK 2.2. This definition is a natural generalization of the notion of topological pressure for the non-autonomous dynamical systems[6]. Because when $X$ is a compact metric space then $P^{\chi_X}(g_{1,\infty}|_{(I_{1,\infty},X)}, \varphi)$ will be the topological pressure of $g_{1,\infty}$ with respect to $\varphi$, where $I_{1,\infty}$ is a sequence of identity maps on $X$, and $\chi_X$ is the characteristic function of $X$.

2.3.2. *Define the relative topological pressure by a separated set.* For any $\varepsilon > 0$, we put

$$\begin{aligned} &P^{\lambda}_{m,n,f_{1,\infty}}(g_{1,\infty}, \varphi, \varepsilon) \\ &\quad = \sup\left\{ \sum_{x\in E} e^{(S^{\lambda}_{m,n}\varphi)(x)} | E \text{ is an } (m, n, g_{1,\infty}, \varepsilon)\text{-separated set of } f_1^m(K) \right\}. \end{aligned}$$

Then we have the next lemma.

LEMMA 2.3. *For any $m, n \in N$, $\varepsilon > 0$ and $\varphi \in C(f_1^m(K), R)$, we have*

(1) $P^{\lambda}_{m,n,f_{1,\infty}}(g_{1,\infty}, \varphi, \varepsilon_1) \geq P^{\lambda}_{m,n,f_{1,\infty}}(g_{1,\infty}, \varphi, \varepsilon_2)$ *for* $0 < \varepsilon_1 < \varepsilon_2$;
(2) $Q^{\lambda}_{m,n,f_{1,\infty}}(g_{1,\infty}, \varphi, \varepsilon) \leq P^{\lambda}_{m,n,f_{1,\infty}}(g_{1,\infty}, \varphi, \varepsilon)$;
(3) $P^{\lambda}_{m,n,f_{1,\infty}}(g_{1,\infty}, 0, \varepsilon) = s(g_{1,\infty}|_{f_1^m(K)}, n, \varepsilon)$.

PROOF.

(1) Since $\{E_2|\ E_2$ is an $(m,n,g_{1,\infty},\varepsilon_2)$-separated set of $f_1^m(K)\}$

$$\subseteq \{E_1|\ E_1 \text{ is an } (m,n,g_{1,\infty},\varepsilon_1)\text{-separated set of } f_1^m(K)\},$$

then

$$\left\{\sum_{x\in E_2} e^{(S^\lambda_{m,n}\varphi)(x)}|\ E_2 \text{ is an } (m,n,g_{1,\infty},\varepsilon_2)\text{-separated set of } f_1^m(K)\right\}$$
$$\subseteq \left\{\sum_{x\in E_1} e^{(S^\lambda_{m,n}\varphi)(x)}|\ E_1 \text{ is an } (m,n,g_{1,\infty},\varepsilon_1)\text{-separated set of } f_1^m(K)\right\}.$$

Therefore

$$P^\lambda_{m,n,f_{1,\infty}}(g_{1,\infty},\varphi,\varepsilon_1) \geq P^\lambda_{m,n,f_{1,\infty}}(g_{1,\infty},\varphi,\varepsilon_2).$$

(2) Let $E_0$ be an $(m,n,g_{1,\infty},\varepsilon)$-separated set of $f_1^m(K)$ such that the cardinality of $E_0$ is equal to $s(g_{1,\infty}|_{f_1^m(K)},n,\varepsilon)$, then by lemma 2.1 (3), we know that $E_0$ is an $(m,n,g_{1,\infty},\varepsilon)$-spanning set of $f_1^m(K)$. It follows that

$$Q^\lambda_{m,n,f_{1,\infty}}(g_{1,\infty},\varphi,\varepsilon) \leq \sum_{x\in E_0} e^{(S^\lambda_{m,n}\varphi)(x)} \leq P^\lambda_{m,n,f_{1,\infty}}(g_{1,\infty},\varphi,\varepsilon).$$

(3) $P^\lambda_{m,n,f_{1,\infty}}(g_{1,\infty},0,\varepsilon) = \sup\{\sharp E|$ E is an $(m,n,g_{1,\infty},\varepsilon)$-separated set of $f_1^m(K)\} = s(g_{1,\infty}|_{f_1^m(K)},n,\varepsilon)$. □

Put

$$P^\lambda(m,f_{1,\infty},g_{1,\infty},\varphi,\varepsilon) = \limsup_{n\to\infty} \frac{1}{n} log P^\lambda_{m,n,f_{1,\infty}}(g_{1,\infty},\varphi,\varepsilon).$$

Then by lemma 2.3, we have the next corollary.

COROLLARY 2.3. *For any* $m \in N$, $\varepsilon > 0$ *and* $\varphi \in C(f_1^m(K),R)$, *we have*

(1) $P^\lambda(m,f_{1,\infty},g_{1,\infty},\varphi,\varepsilon_1) \geq P^\lambda(m,f_{1,\infty},g_{1,\infty},\varphi,\varepsilon_2)$ *for* $0<\varepsilon_1<\varepsilon_2$;
(2) $Q^\lambda(m,f_{1,\infty},g_{1,\infty},\varphi,\varepsilon) \leq P^\lambda(m,f_{1,\infty},g_{1,\infty},\varphi,\varepsilon)$.
(3) $P^\lambda(m,f_{1,\infty},g_{1,\infty},0,\varepsilon) = s(g_{1,\infty}|_{f_1^m(K)},\varepsilon)$.

Put

$$\widetilde{P^\lambda}(m,f_{1,\infty},g_{1,\infty},K,\varphi) = \lim_{\varepsilon\to 0} P^\lambda(m,f_{1,\infty},g_{1,\infty},\varphi,\varepsilon).$$

Then by corollary 2.3 (1), we know that $\widetilde{P^\lambda}(m,f_{1,\infty},g_{1,\infty},K,\varphi)$ exists but could be $\infty$.

Set

$$\widetilde{P^\lambda}(g_{1,\infty}|_{(f_{1,\infty},K)},\varphi) = \limsup_{m\to\infty} \widetilde{P^\lambda}(m,f_{1,\infty},g_{1,\infty},K,\varphi).$$

This shows that we can also use a separated set to define the relative topological pressure for $(g_{1,\infty}|_{(f_{1,\infty},K)},X,d)$.

2.3.3. *The relationship between $P^\lambda(g_{1,\infty}|_{(f_{1,\infty},K)},\varphi)$ and $\widetilde{P^\lambda}(g_{1,\infty}|_{(f_{1,\infty},K)},\varphi)$.* In the sequel, we will further study the relationship between $P^\lambda(g_{1,\infty}|_{(f_{1,\infty},K)},\varphi)$ and $\widetilde{P^\lambda}(g_{1,\infty}|_{(f_{1,\infty},K)},\varphi)$. We already know that the topological pressure defined by a spanning set and separated set are equivalent on a compact metric space (see[6]). Now, a natural question is whether the relative topological pressure $P^\lambda(g_{1,\infty}|_{(f_{1,\infty},K)},\varphi)$ and $\widetilde{P^\lambda}(g_{1,\infty}|_{(f_{1,\infty},K)},\varphi)$ are equal? We show in the following theorem (see theorem 2.4) that under certain assumptions, the relative topological pressure $P^\lambda(g_{1,\infty}|_{(f_{1,\infty},K)},\varphi)$ and $\widetilde{P^\lambda}(g_{1,\infty}|_{(f_{1,\infty},K)},\varphi)$ are equal.

THEOREM 2.1. *If for any $x,y\in X$ and $\delta>0$, $d(x,y)<\frac{\varepsilon}{2}$, $|\lambda(x)\varphi(x)-\lambda(y)\varphi(y)|<\delta$ and for all $m\geq 1$, $f_1^m(K)$ is invariant for $g_{1,\infty}$, then*

$$P^\lambda(g_{1,\infty}|_{(f_{1,\infty},K)},\varphi)=\widetilde{P^\lambda}(g_{1,\infty}|_{(f_{1,\infty},K)},\varphi).$$

PROOF.

STEP 1. We prove $P^\lambda(m,f_{1,\infty},g_{1,\infty},\varphi,\varepsilon)\leq\delta+Q^\lambda(m,f_{1,\infty},g_{1,\infty},\varphi,\frac{\varepsilon}{2})$. Let $E$ be an $(m,n,g_{1,\infty},\varepsilon)$-separated set of $f_1^m(K)$ and $F$ be an $(m,n,g_{1,\infty},\frac{\varepsilon}{2})$-spanning set of $f_1^m(K)$. Define a map $\phi:E\to F$ by choosing, for each $x\in E$ there exists some point $\phi(x)\in F$ with $d_n(x,\phi(x))\leq\frac{\varepsilon}{2}$. Then $\phi:E\to F$ is injective and

$$\sum_{y\in F}e^{(S^\lambda_{m,n}\varphi)(y)}\geq\sum_{y\in\phi(E)}e^{(S^\lambda_{m,n}\varphi)(y)}=\sum_{x\in E}e^{(S^\lambda_{m,n}\varphi)(\phi(x))}$$
$$\geq\min_{x\in E}e^{(S^\lambda_{m,n}\varphi)(\phi(x))-(S^\lambda_{m,n}\varphi)(x)}\sum_{x\in E}e^{(S^\lambda_{m,n}\varphi)(x)}.$$

Since

$$\begin{aligned}(S^\lambda_{m,n}\varphi)(\phi(x))-(S^\lambda_{m,n}\varphi)(x)\\&=\sum_{i=0}^{n-1}[\lambda(g_1^i(\phi(x)))\varphi(g_1^i(\phi(x)))-\lambda(g_1^i(x))\varphi(g_1^i(x))]\\&\geq\sum_{i=0}^{n-1}(-\delta)=-n\delta,\quad\forall\ x\in E.\end{aligned}$$

Then

$$\sum_{y\in F}e^{(S^\lambda_{m,n}\varphi)(y)}\geq e^{-n\delta}\sum_{x\in E}e^{(S^\lambda_{m,n}\varphi)(x)}.$$

Hence

$$Q^\lambda_{m,n,f_{1,\infty}}(g_{1,\infty},\varphi,\frac{\varepsilon}{2})\geq e^{-n\delta}\sum_{x\in E}e^{(S^\lambda_{m,n}\varphi)(x)}.$$

Thus

$$Q^\lambda_{m,n,f_{1,\infty}}(g_{1,\infty},\varphi,\frac{\varepsilon}{2})\geq e^{-n\delta}P^\lambda_{m,n,f_{1,\infty}}(g_{1,\infty},\varphi,\varepsilon).$$

Therefore

$$P^\lambda_{m,n,f_{1,\infty}}(g_{1,\infty},\varphi,\varepsilon)\leq e^{n\delta}Q^\lambda_{m,n,f_{1,\infty}}(g_{1,\infty},\varphi,\frac{\varepsilon}{2}).$$

It follows that

$$P^\lambda(m,f_{1,\infty},g_{1,\infty},\varphi,\varepsilon)\leq\delta+Q^\lambda(m,f_{1,\infty},g_{1,\infty},\varphi,\frac{\varepsilon}{2}).$$

STEP 2. We prove $P^{\lambda}(m, f_{1,\infty}, g_{1,\infty}, K, \varphi) = \widetilde{P^{\lambda}}(m, f_{1,\infty}, g_{1,\infty}, K, \varphi)$.
By corollary 2.3 (2), we know that

$$P^{\lambda}(m, f_{1,\infty}, g_{1,\infty}, K, \varphi) \leq \widetilde{P^{\lambda}}(m, f_{1,\infty}, g_{1,\infty}, K, \varphi).$$

By step 1, for any $\delta > 0$, we have

$$\widetilde{P^{\lambda}}(m, f_{1,\infty}, g_{1,\infty}, K, \varphi) \leq \delta + P^{\lambda}(m, f_{1,\infty}, g_{1,\infty}, K, \varphi).$$

Hence

$$\widetilde{P^{\lambda}}(m, f_{1,\infty}, g_{1,\infty}, K, \varphi) \leq P^{\lambda}(m, f_{1,\infty}, g_{1,\infty}, K, \varphi).$$

It follows that

$$P^{\lambda}(m, f_{1,\infty}, g_{1,\infty}, K, \varphi) = \widetilde{P^{\lambda}}(m, f_{1,\infty}, g_{1,\infty}, K, \varphi).$$

Taking limit with respect to $m$, we have

$$P^{\lambda}(g_{1,\infty}|_{(f_{1,\infty},K)}, \varphi) = \widetilde{P^{\lambda}}(g_{1,\infty}|_{(f_{1,\infty},K)}, \varphi).$$ □

## 3. Properties of the relative topological pressure

Before giving the properties of the relative topological pressure, we shall present the following lemma.

LEMMA 3.1.

(1) **(Hölder inequality)** *Let* $a_k, b_k \geq 0$, $k = 1, 2, \cdots, n$, $\frac{1}{p} + \frac{1}{q} = 1$, $p > 1$, *then* $\sum_{k=1}^{n} a_k b_k \leq (\sum_{k=1}^{n} a_k^p)^{\frac{1}{p}} \cdot (\sum_{k=1}^{n} b_k^q)^{\frac{1}{q}}$.

(2) *Let* $\{a_j\}$ , $\{b_j\}$ *are collections of positive real numbers, then* $\frac{\sup a_j}{\sup b_j} \leq \sup \frac{a_j}{b_j}$.

(3) *Let* $x, y$ *are two positive real numbers, then* $x^c + y^c \leq (x+y)^c$ *if* $c \geq 1$, *and* $x^c + y^c \geq (x+y)^c$ *if* $c \leq 1$.

REMARK 3.1. The proof of Lemma 3.1 see [15], we omitted it.

Now let us study the properties of the relative topological pressure for non-compact non-autonomous dynamical systems.

THEOREM 3.1. *Let* $(g_{1,\infty}|_{(f_{1,\infty},K)}, X, d)$ *be a relative topological dynamical system. If* $\varphi, \psi \in C(X, R)$ *and* $c \in R$, *then the following hold:*

(1) $\widetilde{P^{\lambda}}(g_{1,\infty}|_{(f_{1,\infty},K)}, 0) = h(g_{1,\infty}|_{(f_{1,\infty},K)})$.

(2) $\lambda_1 \varphi_1 \leq \lambda_2 \varphi_2$ *implies that* $\widetilde{P^{\lambda_1}}(g_{1,\infty}|_{(f_{1,\infty},K)}, \varphi_1) \leq \widetilde{P^{\lambda_2}}(g_{1,\infty}|_{(f_{1,\infty},K)}, \varphi_2)$.

*Particularly, if* $\varphi_1 \leq \varphi_2$ *then* $\widetilde{P^{\lambda}}(g_{1,\infty}|_{(f_{1,\infty},K)}, \varphi_1) \leq \widetilde{P^{\lambda}}(g_{1,\infty}|_{(f_{1,\infty},K)}, \varphi_2)$; *if* $\lambda_1 \leq \lambda_2$ *and* $\varphi$ *is a non-negative continuous function on* $X$ *then*

$$\widetilde{P^{\lambda_1}}(g_{1,\infty}|_{(f_{1,\infty},K)}, \varphi) \leq \widetilde{P^{\lambda_2}}(g_{1,\infty}|_{(f_{1,\infty},K)}, \varphi).$$

(3) *If* $\varphi \in BC(X, R)$, *then* $\widetilde{P^{\lambda}}(g_{1,\infty}|_{(f_{1,\infty},K)}, \varphi)$ *is a real number or it is identically* $+\infty$.

(4) *If* $|\widetilde{P^{\lambda}}(g_{1,\infty}|_{(f_{1,\infty},K)}, \cdot)| < +\infty$, *then* $\widetilde{P^{\lambda}}(g_{1,\infty}|_{(f_{1,\infty},K)}, \cdot)$ *is convex.*

(5) *If* $|\widetilde{P^{\lambda}}(g_{1,\infty}|_{(f_{1,\infty},K)}, \cdot)| < +\infty$ *and* $\varphi, \psi \in BC(X, R)$, *then*

$$|\widetilde{P^{\lambda}}(g_{1,\infty}|_{(f_{1,\infty},K)}, \varphi) - \widetilde{P^{\lambda}}(g_{1,\infty}|_{(f_{1,\infty},K)}, \psi)| \leq \| \varphi - \psi \|_B .$$

(6) $\widetilde{P^{\lambda}}(g_{1,\infty}|_{(f_{1,\infty},K)}, \varphi + \psi) \leq \widetilde{P^{\lambda}}(g_{1,\infty}|_{(f_{1,\infty},K)}, \varphi) + \widetilde{P^{\lambda}}(g_{1,\infty}|_{(f_{1,\infty},K)}, \psi)$.

(7) *If* $c \geq 0$, *then* $\widetilde{P^{\lambda}}(g_{1,\infty}|_{(f_{1,\infty},K)}, \varphi + c) \leq \widetilde{P^{\lambda}}(g_{1,\infty}|_{(f_{1,\infty},K)}, \varphi) + c$.

(8) $\widetilde{P^\lambda}(g_{1,\infty}|_{(f_{1,\infty},K)}, c\varphi) \le c \cdot \widetilde{P^\lambda}(g_{1,\infty}|_{(f_{1,\infty},K)}, \varphi)$ *if* $c \ge 1$,
$\widetilde{P^\lambda}(g_{1,\infty}|_{(f_{1,\infty},K)}, c\varphi) \ge c \cdot \widetilde{P^\lambda}(g_{1,\infty}|_{(f_{1,\infty},K)}, \varphi)$ *if* $c \le 1$.

(9) $|\widetilde{P^\lambda}(g_{1,\infty}|_{(f_{1,\infty},K)}, \varphi)| \le \widetilde{P^\lambda}(g_{1,\infty}|_{(f_{1,\infty},K)}, |\varphi|)$.

PROOF.

(1)

$$\begin{aligned}
\widetilde{P^\lambda}(g_{1,\infty}|_{(f_{1,\infty},K)}, 0) &= \limsup_{m\to 0} \widetilde{P^\lambda}(m, f_{1,\infty}, g_{1,\infty}, K, 0) \\
&= \limsup_{m\to 0}(\lim_{\varepsilon\to 0} P^\lambda(m, f_{1,\infty}, g_{1,\infty}, 0, \varepsilon)) \\
&= \limsup_{m\to 0}(\lim_{\varepsilon\to 0}(\limsup_{n\to\infty} \frac{1}{n} \log P^\lambda_{m,n,f_{1,\infty}}(g_{1,\infty}, 0, \varepsilon))) \\
&= \limsup_{m\to 0}(\lim_{\varepsilon\to 0}(\limsup_{n\to\infty} \frac{1}{n} logs(g_{1,\infty}|_{f_1^m(K)}, n, \varepsilon))) \\
&= \limsup_{m\to \infty}(\lim_{\varepsilon\to 0} s(g_{1,\infty}|_{f_1^m(K)}, \varepsilon)) \\
&= \limsup_{m\to \infty} h(g_{1,\infty}|_{f_1^m(K)}) = h(g_{1,\infty}|_{(f_{1,\infty},K)}).
\end{aligned}$$

(2) Since

$$(S^{\lambda_j}_{m,n}\varphi_j)(x) = \sum_{i=0}^{n-1} \delta^{g_1^i(x)}_{f_1^m(K)} \lambda_j(g_1^i(x))\varphi_j(g_1^i(x)), j = 1, 2 \text{ and } \lambda_1\varphi_1 \le \lambda_2\varphi_2,$$

then

$$(S^{\lambda_1}_{m,n}\varphi_1)(x) \le (S^{\lambda_2}_{m,n}\varphi_2)(x).$$

This implies that

$$P^{\lambda_1}_{m,n,f_{1,\infty}}(g_{1,\infty}, \varphi_1, \varepsilon) \le P^{\lambda_2}_{m,n,f_{1,\infty}}(g_{1,\infty}, \varphi_2, \varepsilon).$$

Hence

$$P^{\lambda_1}(m, f_{1,\infty}, g_{1,\infty}, \varphi_1, \varepsilon) \le P^{\lambda_2}(m, f_{1,\infty}, g_{1,\infty}, \varphi_2, \varepsilon).$$

Thus

$$\widetilde{P^{\lambda_1}}(m, f_{1,\infty}, g_{1,\infty}, K, \varphi_1) \le \widetilde{P^{\lambda_2}}(m, f_{1,\infty}, g_{1,\infty}, K, \varphi_2).$$

Therefore

$$\widetilde{P^{\lambda_1}}(g_{1,\infty}|_{(f_{1,\infty},K)}, \varphi_1) \le \widetilde{P^{\lambda_2}}(g_{1,\infty}|_{(f_{1,\infty},K)}, \varphi_2).$$

Since

$$(S^{\lambda}_{m,n}\varphi_j)(x) = \sum_{i=0}^{n-1} \delta^{g_1^i(x)}_{f_1^m(K)} \lambda(g_1^i(x))\varphi_j(g_1^i(x)), j = 1, 2 \text{ and } \varphi_1 \le \varphi_2,$$

then

$$(S^{\lambda}_{m,n}\varphi_1)(x) \le (S^{\lambda}_{m,n}\varphi_2)(x).$$

This implies that

$$P^{\lambda}_{m,n,f_{1,\infty}}(g_{1,\infty}, \varphi_1, \varepsilon) \le P^{\lambda}_{m,n,f_{1,\infty}}(g_{1,\infty}, \varphi_2, \varepsilon).$$

Hence

$$P^{\lambda}(m, f_{1,\infty}, g_{1,\infty}, \varphi_1, \varepsilon) \le P^{\lambda}(m, f_{1,\infty}, g_{1,\infty}, \varphi_2, \varepsilon).$$

Thus

$$\widetilde{P^{\lambda}}(m, f_{1,\infty}, g_{1,\infty}, K, \varphi_1) \le \widetilde{P^{\lambda}}(m, f_{1,\infty}, g_{1,\infty}, K, \varphi_2).$$

Therefore

$$\widetilde{P^\lambda}(g_{1,\infty}|_{(f_{1,\infty},K)},\varphi_1) \le \widetilde{P^\lambda}(g_{1,\infty}|_{(f_{1,\infty},K)},\varphi_2).$$

Since

$(S^{\lambda_j}_{m,n}\varphi)(x) = \sum_{i=0}^{n-1} \delta^{g_1^i(x)}_{f_1^m(K)}\lambda_j(g_1^i(x))\varphi(g_1^i(x)),\ j=1,2, \lambda_1 \le \lambda_2$ and $\varphi$ is a nonnegative continuous function on X, then

$$(S^{\lambda_1}_{m,n}\varphi)(x) \le (S^{\lambda_2}_{m,n}\varphi)(x).$$

This implies that

$$P^{\lambda_1}_{m,n,f_{1,\infty}}(g_{1,\infty},\varphi,\varepsilon) \le P^{\lambda_2}_{m,n,f_{1,\infty}}(g_{1,\infty},\varphi,\varepsilon).$$

Hence

$$P^{\lambda_1}(m,f_{1,\infty},g_{1,\infty},\varphi,\varepsilon) \le P^{\lambda_2}(m,f_{1,\infty},g_{1,\infty},\varphi,\varepsilon).$$

Thus

$$\widetilde{P^{\lambda_1}}(m,f_{1,\infty},g_{1,\infty},K,\varphi) \le \widetilde{P^{\lambda_2}}(m,f_{1,\infty},g_{1,\infty},K,\varphi).$$

Therefore

$$\widetilde{P^{\lambda_1}}(g_{1,\infty}|_{(f_{1,\infty},K)},\varphi) \le \widetilde{P^{\lambda_2}}(g_{1,\infty}|_{(f_{1,\infty},K)},\varphi).$$

(3) Let

$$\varphi_0 = \varphi\,|_{f_1^m(K)},\ a_m = \inf\{\lambda(x)\varphi_0(x), 0 | x \in f_1^m(K)\},$$
$$b_m = \sup\{\lambda(x)\varphi_0(x), 0 | x \in f_1^m(K)\}.$$

Then

$$na_m \le (S^{\lambda}_{m,n}\varphi)(x) = \sum_{i=0}^{n-1} \delta^{g_1^i(x)}_{f_1^m(K)}\lambda(g_1^i(x))\varphi(g_1^i(x)) \le nb_m.$$

This implies that

$$r(g_{1,\infty}|_{f_1^m(K)},n,\varepsilon)e^{na_m} \le P^{\lambda}_{m,n,f_{1,\infty}}(g_{1,\infty},\varphi,\varepsilon) \le s(g_{1,\infty}|_{f_1^m(K)},n,\varepsilon)e^{nb_m}.$$

Hence,

$$r(g_{1,\infty}|_{f_1^m(K)},\varepsilon) + a_m \le P^{\lambda}(m,f_{1,\infty},g_{1,\infty},\varphi,\varepsilon) \le s(g_{1,\infty}|_{f_1^m(K)},\varepsilon) + b_m.$$

Thus

$$h(g_{1,\infty}|_{f_1^m(K)}) + a_m \le \widetilde{P^{\lambda}}(m,f_{1,\infty},g_{1,\infty},K,\varphi) \le h(g_{1,\infty}|_{f_1^m(K)}) + b_m.$$

Therefore,

$$h(g_{1,\infty}|_{(f_{1,\infty},K)}) + a \le \widetilde{P^{\lambda}}(g_{1,\infty}|_{(f_{1,\infty},K)},\varphi) \le h(g_{1,\infty}|_{(f_{1,\infty},K)}) + b.$$

Where

$$a = \limsup_{m\to\infty} a_m, b = \limsup_{m\to\infty} b_m.$$

If $h(g_{1,\infty}|_{(f_{1,\infty},K)}) = +\infty$ then $\widetilde{P^{\lambda}}(g_{1,\infty}|_{(f_{1,\infty},K)},\varphi) = +\infty$. If $h(g_{1,\infty}|_{(f_{1,\infty},K)})$ is a real number then by $\varphi \in BC(X,R)$ we conclude that $\widetilde{P^{\lambda}}(g_{1,\infty}|_{(f_{1,\infty},K)},\varphi)$ is a real number.

(4) Since $p \in [0,1]$ and $E$ is a finite subset of $X$, then by *Hölder's* inequality,

$$\begin{aligned}
&\sum_{x\in E} e^{(S^{\lambda}_{m,n}(p\varphi+(1-p)\psi))(x)} \\
&\quad = \sum_{x\in E}((e^{p(S^{\lambda}_{m,n}\varphi)(x)})^{\frac{1}{p}})^{p}\cdot((e^{(1-p)(S^{\lambda}_{m,n}\psi)(x)})^{\frac{1}{1-p}})^{1-p} \\
&\quad \le (\sum_{x\in E} e^{(S^{\lambda}_{m,n}\varphi)(x)})^{p}\cdot(\sum_{x\in E} e^{(S^{\lambda}_{m,n}\psi)(x)})^{1-p} \\
&\quad \le (P^{\lambda}_{m,n,f_{1,\infty}}(g_{1,\infty},\varphi,\varepsilon))^{p}\cdot(P^{\lambda}_{m,n,f_{1,\infty}}(g_{1,\infty},\psi,\varepsilon))^{1-p}.
\end{aligned}$$

This implies that

$$\begin{aligned}
&P^{\lambda}_{m,n,f_{1,\infty}}(g_{1,\infty},p\varphi+(1-p)\psi,\varepsilon) \\
&\quad \le (P^{\lambda}_{m,n,f_{1,\infty}}(g_{1,\infty},\varphi,\varepsilon))^{p}\cdot(P^{\lambda}_{m,n,f_{1,\infty}}(g_{1,\infty},\psi,\varepsilon))^{1-p}.
\end{aligned}$$

Thus

$$\begin{aligned}
&P^{\lambda}(m,f_{1,\infty},g_{1,\infty},p\varphi+(1-p)\psi,\varepsilon) \\
&\quad \le p\cdot P^{\lambda}(m,f_{1,\infty},g_{1,\infty},\varphi,\varepsilon)+(1-p)\cdot P^{\lambda}(m,f_{1,\infty},g_{1,\infty},\psi,\varepsilon).
\end{aligned}$$

Since $|\widetilde{P^{\lambda}}(g_{1,\infty}|_{(f_{1,\infty},K)},\cdot)|<+\infty$, then by the above inequality, we have

$$\begin{aligned}
&\widetilde{P^{\lambda}}(m,f_{1,\infty},g_{1,\infty},K,p\varphi+(1-p)\psi) \\
&\quad \le p\cdot\widetilde{P^{\lambda}}(m,f_{1,\infty},g_{1,\infty},K,\varphi)+(1-p)\cdot\widetilde{P^{\lambda}}(m,f_{1,\infty},g_{1,\infty},K,\psi).
\end{aligned}$$

Thus

$$\begin{aligned}
&\widetilde{P^{\lambda}}(g_{1,\infty}|_{(f_{1,\infty},K)},p\varphi+(1-p)\psi) \\
&\quad \le p\cdot\widetilde{P^{\lambda}}(g_{1,\infty}|_{(f_{1,\infty},K)},\varphi)+(1-p)\cdot\widetilde{P^{\lambda}}(g_{1,\infty}|_{(f_{1,\infty},K)},\psi).
\end{aligned}$$

Therefore $\widetilde{P^{\lambda}}(g_{1,\infty}|_{(f_{1,\infty},K)},\cdot)$ is convex.

(5) Since

$$\begin{aligned}
&\frac{P^{\lambda}_{m,n,f_{1,\infty}}(g_{1,\infty},\varphi,\varepsilon)}{P^{\lambda}_{m,n,f_{1,\infty}}(g_{1,\infty},\psi,\varepsilon)} \\
&\quad \le \sup\{\frac{\sum_{x\in E} e^{(S^{\lambda}_{m,n}\varphi)(x)}}{\sum_{x\in E} e^{(S^{\lambda}_{m,n}\psi)(x)}}|\ E \text{ is a } (m,n,g_{1,\infty},\varepsilon)\text{-separated set of } f_1^m(K)\} \\
&\quad \le \sup\{\max_{x\in E}\frac{e^{(S^{\lambda}_{m,n}\varphi)(x)}}{e^{(S^{\lambda}_{m,n}\psi)(x)}}|\ E \text{ is a } (m,n,g_{1,\infty},\varepsilon)\text{-separated set of } f_1^m(K)\} \\
&\quad \le e^{n\|\varphi-\psi\|_B},
\end{aligned}$$

then

$$|P^{\lambda}(m,f_{1,\infty},g_{1,\infty},\varphi,\varepsilon)-P^{\lambda}(m,f_{1,\infty},g_{1,\infty},\psi,\varepsilon)|\le\|\varphi-\psi\|_B.$$

Since $|\widetilde{P^{\lambda}}(g_{1,\infty}|_{(f_{1,\infty},K)},\cdot)|<+\infty$, then by the above inequality, we have

$$|\widetilde{P^{\lambda}}(m,f_{1,\infty},g_{1,\infty},K,\varphi)-\widetilde{P^{\lambda}}(m,f_{1,\infty},g_{1,\infty},K,\psi)|\le\|\varphi-\psi\|_B.$$

Thus

$$|\widetilde{P^{\lambda}}(g_{1,\infty}|_{(f_{1,\infty},K)},\varphi)-\widetilde{P^{\lambda}}(g_{1,\infty}|_{(f_{1,\infty},K)},\psi)|\ \le\|\varphi-\psi\|_B.$$

(6) Since

$$(S^{\lambda}_{m,n}(\varphi+\psi))(x)=\sum_{i=0}^{n-1}\delta^{g_1^i(x)}_{f_1^m(K)}\lambda(g_1^i(x))(\varphi+\psi)(g_1^i(x))=(S^{\lambda}_{m,n}\varphi)(x)+(S^{\lambda}_{m,n}\psi)(x),$$

then

$$\sum_{x\in E}e^{(S^{\lambda}_{m,n}(\varphi+\psi))(x)}=\sum_{x\in E}e^{(S^{\lambda}_{m,n}\varphi)(x)}\cdot e^{(S^{\lambda}_{m,n}\psi)(x)}\leq\sum_{x\in E}e^{(S^{\lambda}_{m,n}\varphi)(x)}\cdot\sum_{x\in E}e^{(S^{\lambda}_{m,n}\psi)(x)}.$$

This implies that

$$P^{\lambda}_{m,n,f_{1,\infty}}(g_{1,\infty},\varphi+\psi,\varepsilon)\leq P^{\lambda}_{m,n,f_{1,\infty}}(g_{1,\infty},\varphi,\varepsilon)\cdot P^{\lambda}_{m,n,f_{1,\infty}}(g_{1,\infty},\psi,\varepsilon).$$

Hence

$$P^{\lambda}(m,f_{1,\infty},g_{1,\infty},\varphi+\psi,\varepsilon)\leq P^{\lambda}(m,f_{1,\infty},g_{1,\infty},\varphi,\varepsilon)+P^{\lambda}(m,f_{1,\infty},g_{1,\infty},\psi,\varepsilon).$$

Thus

$$\widetilde{P^{\lambda}}(m,f_{1,\infty},g_{1,\infty},K,\varphi+\psi)\leq\widetilde{P^{\lambda}}(m,f_{1,\infty},g_{1,\infty},K,\varphi)+\widetilde{P^{\lambda}}(m,f_{1,\infty},g_{1,\infty},K,\psi).$$

Therefore

$$\widetilde{P^{\lambda}}(g_{1,\infty}|_{(f_{1,\infty},K)},\varphi+\psi)\leq\widetilde{P^{\lambda}}(g_{1,\infty}|_{(f_{1,\infty},K)},\varphi)+\widetilde{P^{\lambda}}(g_{1,\infty}|_{(f_{1,\infty},K)},\psi).$$

(7) Since

$$(S^{\lambda}_{m,n}(\varphi+c))(x)=\sum_{i=0}^{n-1}\delta^{g_1^i(x)}_{f_1^m(K)}\lambda(g_1^i(x))(\varphi+c)(g_1^i(x))\leq(S^{\lambda}_{m,n}\varphi)(x)+nc,$$

then

$$\sum_{x\in E}e^{(S^{\lambda}_{m,n}(\varphi+c))(x)}\leq\sum_{x\in E}e^{(S^{\lambda}_{m,n}\varphi)(x)+nc}=e^{nc}\sum_{x\in E}e^{(S^{\lambda}_{m,n}\varphi)(x)}.$$

This implies that

$$P^{\lambda}_{m,n,f_{1,\infty}}(g_{1,\infty},\varphi+c,\varepsilon)\leq e^{nc}\cdot P^{\lambda}_{m,n,f_{1,\infty}}(g_{1,\infty},\varphi,\varepsilon).$$

Hence

$$P^{\lambda}(m,f_{1,\infty},g_{1,\infty},\varphi+c,\varepsilon)\leq P^{\lambda}(m,f_{1,\infty},g_{1,\infty},\varphi,\varepsilon)+c.$$

Thus

$$\widetilde{P^{\lambda}}(m,f_{1,\infty},g_{1,\infty},K,\varphi+c)\leq\widetilde{P^{\lambda}}(m,f_{1,\infty},g_{1,\infty},K,\varphi)+c.$$

Therefore

$$\widetilde{P^{\lambda}}(g_{1,\infty}|_{(f_{1,\infty},K)},\varphi+c)\leq\widetilde{P^{\lambda}}(g_{1,\infty}|_{(f_{1,\infty},K)},\varphi)+c.$$

(8)

(i) Case 1 : $c\geq 1$.
Since any $(m,n,g_{1,\infty},\varepsilon)$-separated set of $f_1^m(K)$ is a finite subset of X , then we have

$$\begin{aligned}\sum_{x\in E}e^{(S^{\lambda}_{m,n}(c\varphi))(x)}&=\sum_{x\in E}(e^{(S^{\lambda}_{m,n}\varphi)(x)})^c\\&\leq(\sum_{x\in E}e^{(S^{\lambda}_{m,n}\varphi)(x)})^c\leq(P^{\lambda}_{m,n,f_{1,\infty}}(g_{1,\infty},\varphi,\varepsilon))^c,\ c\geq 1.\end{aligned}$$

This implies that

$$P^{\lambda}_{m,n,f_{1,\infty}}(g_{1,\infty},c\varphi,\varepsilon)\leq(P^{\lambda}_{m,n,f_{1,\infty}}(g_{1,\infty},\varphi,\varepsilon))^c.$$

Hence

$$P^\lambda(m, f_{1,\infty}, g_{1,\infty}, c\varphi, \varepsilon) \le c \cdot P^\lambda(m, f_{1,\infty}, g_{1,\infty}, \varphi, \varepsilon).$$

Thus

$$\widetilde{P^\lambda}(m, f_{1,\infty}, g_{1,\infty}, K, c\varphi) \le c \cdot \widetilde{P^\lambda}(m, f_{1,\infty}, g_{1,\infty}, K, \varphi).$$

Therefore

$$\widetilde{P^\lambda}(g_{1,\infty}|_{(f_{1,\infty},K)}, c\varphi) \le c \cdot \widetilde{P^\lambda}(g_{1,\infty}|_{(f_{1,\infty},K)}, \varphi).$$

(ii) Case 2 : $c \le 1$.
Since any $(m, n, g_{1,\infty}, \varepsilon)$-separated set of $f_1^m(K)$ is a finite subset of X , then we have

$$\begin{aligned} P^\lambda_{m,n,f_{1,\infty}}(g_{1,\infty}, c\varphi, \varepsilon) &\ge \sum_{x\in E} e^{(S^\lambda_{m,n}(c\varphi))(x)} \\ &= \sum_{x\in E} (e^{(S^\lambda_{m,n}\varphi)(x)})^c \ge (\sum_{x\in E} e^{(S^\lambda_{m,n}\varphi)(x)})^c, \quad c \le 1. \end{aligned}$$

This implies that

$$P^\lambda_{m,n,f_{1,\infty}}(g_{1,\infty}, c\varphi, \varepsilon) \ge (P^\lambda_{m,n,f_{1,\infty}}(g_{1,\infty}, \varphi, \varepsilon))^c.$$

Hence

$$P^\lambda(m, f_{1,\infty}, g_{1,\infty}, c\varphi, \varepsilon) \ge c \cdot P^\lambda(m, f_{1,\infty}, g_{1,\infty}, \varphi, \varepsilon).$$

Thus

$$\widetilde{P^\lambda}(m, f_{1,\infty}, g_{1,\infty}, K, c\varphi) \ge c \cdot \widetilde{P^\lambda}(m, f_{1,\infty}, g_{1,\infty}, K, \varphi).$$

Therefore

$$\widetilde{P^\lambda}(g_{1,\infty}|_{(f_{1,\infty},K)}, c\varphi) \ge c \cdot \widetilde{P^\lambda}(g_{1,\infty}|_{(f_{1,\infty},K)}, \varphi).$$

(9) Since $-|\varphi| \le \varphi \le |\varphi|$ and by (1), we have

$$\widetilde{P^\lambda}(g_{1,\infty}|_{(f_{1,\infty},K)}, -|\varphi|) \le \widetilde{P^\lambda}(g_{1,\infty}|_{(f_{1,\infty},K)}, \varphi) \le \widetilde{P^\lambda}(g_{1,\infty}|_{(f_{1,\infty},K)}, |\varphi|).$$

By (8), we have

$$\widetilde{P^\lambda}(g_{1,\infty}|_{(f_{1,\infty},K)}, -|\varphi|) \ge -\widetilde{P^\lambda}(g_{1,\infty}|_{(f_{1,\infty},K)}, |\varphi|).$$

Hence,

$$-\widetilde{P^\lambda}(g_{1,\infty}|_{(f_{1,\infty},K)}, |\varphi|) \le \widetilde{P^\lambda}(g_{1,\infty}|_{(f_{1,\infty},K)}, \varphi) \le \widetilde{P^\lambda}(g_{1,\infty}|_{(f_{1,\infty},K)}, |\varphi|).$$

That is,

$$|\widetilde{P^\lambda}(g_{1,\infty}|_{(f_{1,\infty},K)}, \varphi)| \le \widetilde{P^\lambda}(g_{1,\infty}|_{(f_{1,\infty},K)}, |\varphi|).$$

□

REMARK 3.2. For the relative topological pressure $P^\lambda(g_{1,\infty}|_{(f_{1,\infty},K)}, \cdot)$, the conclusion (1),(2),(3),(7),(8),(9) in theorem 3.2 are also hold. Since the proof is similar to the case of $\widetilde{P^\lambda}(g_{1,\infty}|_{(f_{1,\infty},K)}, \cdot)$, we omitted it. While the conclusion (4),(5),(6) in theorem 3.2 are merely hold for $P^\lambda(g_{1,\infty}|_{(f_{1,\infty},K)}, \cdot)$. Moreover, (1),(2),(4),(6),(7),(8),(9) of Theorem 3.2 is a natural generalization of the properties of topological pressure in the case of the autonomous dynamical system (see [15]). While by adding additional assumptions, we can obtain the conclusion (3) and (5) of Theorem 3.2.

THEOREM 3.2. *If $\varphi$ is a non-negative continuous function on $X$ then the relative topological pressure $\widetilde{P^{\lambda}}(g_{1,\infty}|_{(f_{1,\infty},K)},\varphi)$ is independent of the choice of base i.e. if $(f_{1,\infty},K)$ and $(\widetilde{f}_{1,\infty},\widetilde{K})$ are two basis of $X$ then $\widetilde{P^{\lambda}}(g_{1,\infty}|_{(f_{1,\infty},K)},\varphi) = \widetilde{P^{\lambda}}(g_{1,\infty}|_{(\widetilde{f}_{1,\infty},\widetilde{K})},\varphi)$.*

PROOF. Since $(f_{1,\infty},K)$ is a base of $X$, then for any $m \geq 1$ , $f_1^m(K)$ is a compact subset of $X$. Since $(\widetilde{f}_{1,\infty},\widetilde{K})$ is a base of $X$, then there exists $m_2 \in N$ such that $f_1^m(K) \subseteq \widetilde{f}_1^{m_3}(\widetilde{K})$ for all $m_3 \geq m_2$. It means any $(m,n,g_{1,\infty},\varepsilon)$-separated set of $f_1^m(K)$ is an $(m_3,n,g_{1,\infty},\varepsilon)$-separated set of $\widetilde{f}_1^{m_3}(\widetilde{K})$ and

$$\sum_{i=0}^{n-1}\delta_{f_1^m(K)}^{g_1^i(x)}\lambda(g_1^i(x))\varphi(g_1^i(x)) \leq \sum_{i=0}^{n-1}\delta_{\widetilde{f}_1^{m_3}(\widetilde{K})}^{g_1^i(x)}\lambda(g_1^i(x))\varphi(g_1^i(x)).$$

According to the above inequality, we have

$$P^{\lambda}_{m,n,f_{1,\infty}}(g_{1,\infty},\varphi,\varepsilon)_K \leq P^{\lambda}_{m_3,n,\widetilde{f}_{1,\infty}}(g_{1,\infty},\varphi,\varepsilon)_{\widetilde{K}}.$$

Hence

$$P^{\lambda}(m,f_{1,\infty},g_{1,\infty},\varphi,\varepsilon)_K \leq P^{\lambda}(m_3,\widetilde{f}_{1,\infty},g_{1,\infty},\varphi,\varepsilon)_{\widetilde{K}}.$$

Therefore

$$\widetilde{P^{\lambda}}(m,f_{1,\infty},g_{1,\infty},K,\varphi) \leq \widetilde{P^{\lambda}}(m_3,\widetilde{f}_{1,\infty},g_{1,\infty},\widetilde{K},\varphi).$$

It follows that

$$\widetilde{P^{\lambda}}(g_{1,\infty}|_{(f_{1,\infty},K)},\varphi) \leq \widetilde{P^{\lambda}}(g_{1,\infty}|_{(\widetilde{f}_{1,\infty},\widetilde{K})},\varphi).$$

Now, replacing $(f_{1,\infty},K)$ with $(\widetilde{f}_{1,\infty},\widetilde{K})$ we have

$$\widetilde{P^{\lambda}}(g_{1,\infty}|_{(\widetilde{f}_{1,\infty},\widetilde{K})},\varphi) \leq \widetilde{P^{\lambda}}(g_{1,\infty}|_{(f_{1,\infty},K)},\varphi).$$

Thus

$$\widetilde{P^{\lambda}}(g_{1,\infty}|_{(f_{1,\infty},K)},\varphi) = \widetilde{P^{\lambda}}(g_{1,\infty}|_{(\widetilde{f}_{1,\infty},\widetilde{K})},\varphi).$$

□

By theorem 3.3, we have the next corollary.

COROLLARY 3.1. *If $\varphi = 0$ then the relative topological entropy $h(g_{1,\infty}|_{(f_{1,\infty},K)})$ is independent of the choice of base i.e. if $(f_{1,\infty},K)$ and $(\widetilde{f}_{1,\infty},\widetilde{K})$ are two basis of $X$ then $h(g_{1,\infty}|_{(f_{1,\infty},K)}) = h(g_{1,\infty}|_{(\widetilde{f}_{1,\infty},\widetilde{K})})$.*

## 4. Uniformly topological conjugate

In this section, we will discuss the relationship between the two dynamical systems which are uniformly topological conjugate.

DEFINITION 4.1. Let $(X,d,g_{1,\infty})$ and $(Y,\rho,\widetilde{g}_{1,\infty})$ be two dynamical systems. If there exists a homeomorphism $\pi : X \to Y$ such that $\widetilde{g}_1^i \circ \pi = \pi \circ g_1^i$, for every $i \geq 1$, then we call $\pi$ is **a topological conjugate** between $(X,d,g_{1,\infty})$ and $(Y,\rho,\widetilde{g}_{1,\infty})$, and the system $(X,d,g_{1,\infty})$ is **topological conjugate with** $(Y,\rho,\widetilde{g}_{1,\infty})$. In particular, if $\pi : X \to Y$ is a uniformly homeomorphism (i.e. the homeomorphism $\pi : X \to Y$ and its inverse homeomorphism $\pi^{-1} : Y \to X$ are uniformly continuous), then we call $\pi$ is **a uniformly topological conjugate** between $(X,d,g_{1,\infty})$ and $(Y,\rho,\widetilde{g}_{1,\infty})$, and the system $(X,d,g_{1,\infty})$ is **uniformly topological conjugate with** $(Y,\rho,\widetilde{g}_{1,\infty})$.

Now we show the relative topological pressure is an invariant with respect to uniformly topological conjugate.

THEOREM 4.1. *Let* $(g_{1,\infty}|_{(f_{1,\infty},K)}, X, d)$ *and* $(Y, \rho, \widetilde{g}_{1,\infty})$ *be two dynamical systems,* $\lambda : X \to [0,1]$, $\widetilde{\lambda} : Y \to [0,1]$, $\varphi \in C(X,R)$, $\widetilde{\varphi} \in C(Y,R)$. *If the system* $(X, d, g_{1,\infty})$ *is uniformly topological conjugate with* $(Y, \rho, \widetilde{g}_{1,\infty})$ *(denote the uniformly conjugate by* $\pi$*), and* $\lambda = \widetilde{\lambda} \circ \pi, \varphi = \widetilde{\varphi} \circ \pi$, *then*

(1) $(\pi(K), h_{1,\infty})$ *is a base of* $Y$, *where* $h_{1,\infty} = \{h_i\}_{i=1}^{\infty}$ *and* $h_i = \pi \circ f_i \circ \pi^{-1} : Y \to Y$, *that is,* $(\widetilde{g}_{1,\infty}|_{(h_{1,\infty},\pi(K))}, Y, \rho)$ *is a relative topological dynamical system;*

(2) $\widetilde{P^{\lambda}}(g_{1,\infty}|_{(f_{1,\infty},K)}, \varphi) = \widetilde{P^{\widetilde{\lambda}}}(\widetilde{g}_{1,\infty}|_{(h_{1,\infty},\pi(K))}, \widetilde{\varphi})$.

PROOF.

(1) For all compact set $C \subseteq Y$ there exists $m \in N$ such that $f_1^n(K) \supseteq \pi^{-1}(C)$ for all $n \geq m$. Hence

$$h_1^n(\pi(K)) = (\pi \circ f_1^n \circ \pi^{-1})(\pi(K)) = \pi(f_1^n(K)) \supseteq \pi(\pi^{-1}(C)) = C.$$

This shows $(\pi(K), h_{1,\infty})$ is a base of $Y$.

(2) Since $\pi : X \to Y$ is a uniformly topological conjugate between $(X, d, g_{1,\infty})$ and $(Y, \rho, \widetilde{g}_{1,\infty})$, then $\pi : X \to Y$ is a uniformly homeomorphism such that $\widetilde{g}_1^i \circ \pi = \pi \circ g_1^i$, for every $i \geq 1$. Then by the hypothesis of $\lambda = \widetilde{\lambda} \circ \pi$ , $\varphi = \widetilde{\varphi} \circ \pi$, we have

$$\begin{aligned} g_1^i(x) \in f_1^m(K) \Leftrightarrow \widetilde{g}_1^i(\pi(x)) &= \pi(g_1^i(x)) \\ &\in \pi(f_1^m(K)) = (\pi \circ f_1^m \circ \pi^{-1})(\pi(K)) = h_1^m(\pi(K)), \end{aligned}$$

and

$$\begin{aligned} \lambda(g_1^i(x)) &= (\widetilde{\lambda} \circ \pi)(g_1^i(x)) = \widetilde{\lambda}(\pi(g_1^i(x))) = \widetilde{\lambda}(\widetilde{g}_1^i(\pi(x))), \\ \varphi(g_1^i(x)) &= (\widetilde{\varphi} \circ \pi)(g_1^i(x)) = \widetilde{\varphi}(\pi(g_1^i(x))) = \widetilde{\varphi}(\widetilde{g}_1^i(\pi(x))). \end{aligned}$$

Thus

$$\sum_{i=0}^{n-1} \delta_{f_1^m(K)}^{g_1^i(x)} \lambda(g_1^i(x))\varphi(g_1^i(x)) = \sum_{i=0}^{n-1} \delta_{h_1^m(\pi(K))}^{\widetilde{g}_1^i(\pi(x))} \widetilde{\lambda}(\widetilde{g}_1^i(\pi(x)))\widetilde{\varphi}(\widetilde{g}_1^i(\pi(x))),$$

i.e. $(S_{m,n}^{\lambda}\varphi)(x) = (S_{m,n}^{\widetilde{\lambda}}\widetilde{\varphi})(\pi(x))$. Since the map $\pi : X \to Y$ is uniformly continuous, then for any $\varepsilon > 0$ there exists $\delta > 0$ such that for all $x, y \in X$, if $\rho(\pi(x), \pi(y)) > \varepsilon$, then $d(x,y) > \delta$. Here, we take $\delta \leq \varepsilon$. Let $\widetilde{E}$ be an $(m, n, \widetilde{g}_{1,\infty}, \varepsilon, \rho)$-separated set of $h_1^m(\pi(K))$, then $E = \pi^{-1}(\widetilde{E})$ is an $(m, n, g_{1,\infty}, \delta, d)$-separated set of $f_1^m(K)$. Thus

$$\sum_{y \in \widetilde{E}} e^{(S_{m,n}^{\widetilde{\lambda}}\widetilde{\varphi})(y)} = \sum_{\pi(x) \in \widetilde{E}} e^{(S_{m,n}^{\widetilde{\lambda}}\widetilde{\varphi})(\pi(x))} = \sum_{x \in \pi^{-1}(\widetilde{E})} e^{(S_{m,n}^{\widetilde{\lambda}}\widetilde{\varphi})(\pi(x))} = \sum_{x \in E} e^{(S_{m,n}^{\lambda}\varphi)(x)}.$$

In view of the above inequality, we have

$$P_{m,n,h_{1,\infty}}^{\widetilde{\lambda}}(\widetilde{g}_{1,\infty}, \widetilde{\varphi}, \varepsilon) \leq P_{m,n,f_{1,\infty}}^{\lambda}(g_{1,\infty}, \varphi, \delta).$$

Hence

$$P^{\widetilde{\lambda}}(m, h_{1,\infty}, \widetilde{g}_{1,\infty}, \widetilde{\varphi}, \varepsilon) \leq P^{\lambda}(m, f_{1,\infty}, g_{1,\infty}, \varphi, \delta).$$

Therefore

$$\lim_{\varepsilon \to 0} P^{\widetilde{\lambda}}(m, h_{1,\infty}, \widetilde{g}_{1,\infty}, \widetilde{\varphi}, \varepsilon) \leq \lim_{\varepsilon \to 0} P^{\lambda}(m, f_{1,\infty}, g_{1,\infty}, \varphi, \delta).$$

Since $\delta \leq \varepsilon$, then

$$\lim_{\varepsilon\to 0} P^{\lambda}(m, f_{1,\infty}, g_{1,\infty}, \varphi, \delta) = \lim_{\delta\to 0} P^{\lambda}(m, f_{1,\infty}, g_{1,\infty}, \varphi, \delta).$$

Thus

$$\lim_{\varepsilon\to 0} P^{\widetilde{\lambda}}(m, h_{1,\infty}, \widetilde{g}_{1,\infty}, \widetilde{\varphi}, \varepsilon) \leq \lim_{\delta\to 0} P^{\lambda}(m, f_{1,\infty}, g_{1,\infty}, \varphi, \delta),$$

that is,

$$\widetilde{P^{\widetilde{\lambda}}}(m, h_{1,\infty}, \widetilde{g}_{1,\infty}, \pi(K), \widetilde{\varphi}) \leq \widetilde{P^{\lambda}}(m, f_{1,\infty}, g_{1,\infty}, K, \varphi).$$

It follows that

$$\widetilde{P^{\widetilde{\lambda}}}(\widetilde{g}_{1,\infty}|_{(h_{1,\infty},\pi(K))}, \widetilde{\varphi}) \leq \widetilde{P^{\lambda}}(g_{1,\infty}|_{(f_{1,\infty},K)}, \varphi).$$

Since $\pi^{-1} : Y \to X$ is a uniformly topological conjugate between $(Y, \rho, \widetilde{g}_{1,\infty})$ and $(X, d, g_{1,\infty})$, then

$$\widetilde{P^{\lambda}}(g_{1,\infty}|_{(f_{1,\infty},K)}, \varphi) \leq \widetilde{P^{\widetilde{\lambda}}}(\widetilde{g}_{1,\infty}|_{(h_{1,\infty},\pi(K))}, \widetilde{\varphi}).$$

Therefore

$$\widetilde{P^{\lambda}}(g_{1,\infty}|_{(f_{1,\infty},K)}, \varphi) = \widetilde{P^{\widetilde{\lambda}}}(\widetilde{g}_{1,\infty}|_{(h_{1,\infty},\pi(K))}, \widetilde{\varphi}).$$

□

COROLLARY 4.1. *Let $(g_{1,\infty}|_{(f_{1,\infty},K)}, X, d)$ and $(\widetilde{g}_{1,\infty}|_{(\widetilde{f}_{1,\infty},\widetilde{K})}, Y, \rho)$ be two relative topological dynamical systems, $\lambda : X \to [0,1]$, $\widetilde{\lambda} : Y \to [0,1]$, $\varphi \in C(X, R)$, $\widetilde{\varphi} \in C(Y, R)$, $\widetilde{\varphi} \geq 0$. If the system $(X, d, g_{1,\infty})$ is uniformly topological conjugate with $(Y, \rho, \widetilde{g}_{1,\infty})$ (denote the uniformly conjugate by $\pi$), and $\lambda = \widetilde{\lambda} \circ \pi, \varphi = \widetilde{\varphi} \circ \pi$, then*

$$\widetilde{P^{\lambda}}(g_{1,\infty}|_{(f_{1,\infty},K)}, \varphi) = \widetilde{P^{\widetilde{\lambda}}}(\widetilde{g}_{1,\infty}|_{(\widetilde{f}_{1,\infty},\widetilde{K})}, \widetilde{\varphi}).$$

PROOF. By theorem 3.3, we have

$$\widetilde{P^{\widetilde{\lambda}}}(\widetilde{g}_{1,\infty}|_{(\widetilde{f}_{1,\infty},\widetilde{K})}, \widetilde{\varphi}) = \widetilde{P^{\widetilde{\lambda}}}(\widetilde{g}_{1,\infty}|_{(h_{1,\infty},\pi(K))}, \widetilde{\varphi}).$$

By theorem 4.1, we have

$$\widetilde{P^{\lambda}}(g_{1,\infty}|_{(f_{1,\infty},K)}, \varphi) = \widetilde{P^{\widetilde{\lambda}}}(\widetilde{g}_{1,\infty}|_{(h_{1,\infty},\pi(K))}, \widetilde{\varphi}).$$

Therefore

$$\widetilde{P^{\lambda}}(g_{1,\infty}|_{(f_{1,\infty},K)}, \varphi) = \widetilde{P^{\widetilde{\lambda}}}(\widetilde{g}_{1,\infty}|_{(\widetilde{f}_{1,\infty},\widetilde{K})}, \widetilde{\varphi}).$$

□

## References


[1] R. L. Adler, A. G. Konheim, and M. H. McAndrew, *Topological entropy*, Trans. Amer. Math. Soc. **114** (1965), 309–319, DOI 10.2307/1994177. MR175106

[2] Rufus Bowen, *Entropy for group endomorphisms and homogeneous spaces*, Trans. Amer. Math. Soc. **153** (1971), 401–414, DOI 10.2307/1995565. MR274707

[3] Rufus Bowen, *Hausdorff dimension of quasicircles*, Inst. Hautes Études Sci. Publ. Math. **50** (1979), 11–25. MR556580

[4] Rufus Bowen, *Equilibrium states and the ergodic theory of Anosov diffeomorphisms*, Second revised edition, Lecture Notes in Mathematics, vol. 470, Springer-Verlag, Berlin, 2008. With a preface by David Ruelle; Edited by Jean-René Chazottes. MR2423393

[5] E. I. Dinaburg, *A connection between various entropy characterizations of dynamical systems* (Russian), Izv. Akad. Nauk SSSR Ser. Mat. **35** (1971), 324–366. MR286091

[6] Gerhard Keller, *Equilibrium states in ergodic theory*, London Mathematical Society Student Texts, vol. 42, Cambridge University Press, Cambridge, 1998, DOI 10.1017/CBO9781107359987. MR1618769

[7] Sergiĭ Kolyada and ubomír Snoha, *Topological entropy of nonautonomous dynamical systems*, Random Comput. Dynam. **4** (1996), no. 2-3, 205–233. MR1402417

[8] M. Malziri and M. R. Molaei, *An extension of the notion of topological entropy*, Chaos Solitons Fractals **36** (2008), no. 2, 370–373, DOI 10.1016/j.chaos.2006.06.061. MR2382567

[9] M. R. Molaei and H. Molaei, *Relative topological pressure*, Math. Rep. (Bucur.) **12(62)** (2010), no. 1, 31–36. MR2666011

[10] David Ruelle, *Statistical mechanics on a compact set with $Z^v$ action satisfying expansiveness and specification*, Trans. Amer. Math. Soc. **187** (1973), 237–251, DOI 10.2307/1996437. MR417391

[11] David Ruelle, *Thermodynamic formalism: The mathematical structures of classical equilibrium statistical mechanics*, Encyclopedia of Mathematics and its Applications, vol. 5, Addison-Wesley Publishing Co., Reading, MA, 1978. With a foreword by Giovanni Gallavotti and Gian-Carlo Rota. MR511655

[12] David Ruelle, *Repellers for real analytic maps*, Ergodic Theory Dynam. Systems **2** (1982), no. 1, 99–107, DOI 10.1017/s0143385700009603. MR684247

[13] Chuanjun Tian and Guanrong Chen, *Chaos of a sequence of maps in a metric space*, Chaos Solitons Fractals **28** (2006), no. 4, 1067–1075, DOI 10.1016/j.chaos.2005.08.127. MR2212790

[14] Peter Walters, *An introduction to ergodic theory*, Graduate Texts in Mathematics, vol. 79, Springer-Verlag, New York-Berlin, 1982. MR648108

[15] Peter Walters, *A variational principle for the pressure of continuous transformations*, Amer. J. Math. **97** (1975), no. 4, 937–971, DOI 10.2307/2373682. MR390180

[16] Yujun Zhu, Zhaofeng Liu, Xueli Xu, and Wenda Zhang, *Entropy of nonautonomous dynamical systems*, J. Korean Math. Soc. **49** (2012), no. 1, 165–185, DOI 10.4134/JKMS.2012.49.1.165. MR2907548



College of Chemical Engineering, Xi'an University, Xi'an, Shaanxi, 710065, People's Republic of China

*Email address*: silbert2013@163.com

Analysis, Probability and Fractals Laboratory LR18ES17, Department of Mathematics, Faculty of Sciences of Monastir, University of Monastir, 5000-Monastir, Tunisia

*Email address*: bilel.selmi@fsm.rnu.tn

School of Mathematics, Northwest University Xi'an, Shaanxi, 710127, People's Republic of China

*Email address*: china-lizhiming@163.com

Contemporary Mathematics
Volume **839**, 2026
https://doi.org/10.1090/conm/839/16723

# Self-similar GIFS tilings of Euclidean space

Andrew Vince

ABSTRACT. An approach to tiling Euclidean space using a graph iterated function system (GIFS) was introduced by Barnsely and Vince [Geometriae Dedicata, 212 (2021), 299–324]. The main concept in this paper is called a *self-similar GIFS* in which all the similarity mappings in the GIFS have the same scaling ratio. Despite the simplification, the tilings associated with a self-similar GIFS have properties that are expected of a self-similar tiling. Any GIFS based tiling by congruent copies of a finite set of prototiles that may reasonably be considered self-similar must be one of our self-similar GIFS-tiling. Moreover, the space of tilings in this paper is the closure, in an appropriate metric space, of the self-similar tilings of W. Thurston [AMS Colloq. Lect., 1989], generalized from tiling by translated copies of a set of prototiles to tilings by congruent copies of a set of prototiles.

## 1. Introduction

The tilings in this paper are tilings of Euclidean space by congruent copies of a finite set of prototiles, in particular those that have self-replicating properties. The mathematical investigation of such tilings has a long history. Consider, for example, the prescient 1619 "monster" tiling of Johannes Kepler in Figure 1. Research has been particularly robust since the discovery of the Penrose tilings [**29**] in 1974 and quasicrystals in 1984.

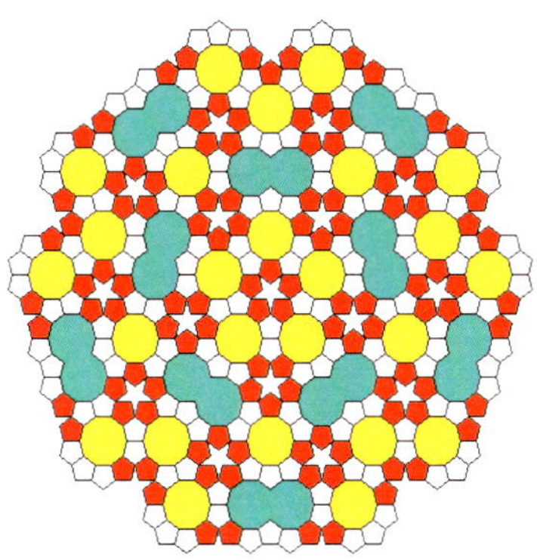

FIGURE 1. Kelpler's monster

2020 *Mathematics Subject Classification.* Primary 52C22, 52C23, 28A80, 05B45.

Tilings with self-referential properties have a vast literature, the subject approached from geometric, topological, analytic, and dynamical points of view. Such tilings, for example, appear early on in work of Mandelbrot [**25**], Bedford [**6**], Thurston [**39**], Kenyon [**20**] and from a dynamical systems point of view in, for example [**30, 33, 37**]. Rauzy type tilings appear in [**32**] and surveys by Berthè and Siegel [**8**] and by Siegel and Thuswaldner [**36**], both containing well over 100 references.

The goal of this paper is to provide a unifying overview of this fascinating theory from a graph iterated function system (GIFS) point of view. A GIFS is a generalization of an iterated function system (IFS). The concept of a GIFS originated in the construction of deterministic fractals [**26**] but is implicit in the concept of a Bratteli diagram [**10**] introduced in 1972 in the context of operator algebras. A GIFS is related to the earlier notion of a rep-tile, the term coined by S. Golomb [**14**] and popularized by Martin Gardner in 1963 in Scientific American [**12**]. A rep-tile was originally defined as a single polygon that can be tiled by smaller similar and congruent copies of itself. If we generalize from a single polygon to a finite prototile set $Q$ such that each tile in $Q$ is, in turn, tiled by smaller similar copies of prototiles in $Q$, then we are close to the concept of a GIFS (see Remark 3.5). Systematic use of an IFS in tiling theory gained traction in the early 1990s with the construction of lattice and crystallographic tilings by copies of a single, usually fractal, tile; see for example [**4, 13, 15, 21, 22, 38, 40**]. The use of a GIFS expands the possibilities to tilings by various shapes and to non-periodic tilings. It makes precise the notion of a geometric substitution tiling; see the survey by Frank [**11**] and the many references therein.

A GIFS $\mathcal{F} = (G, F)$ consists of a digraph $G$ and a set $F$ of functions assigned to the edges of $G$. Every GIFS for which the functions are contractions has a unique attractor consisting of a set $(A_1, A_2, \dots, A_n)$ of compact subsets of $\mathbb{R}^d$, one for each vertex of $G$. The functions in $F$ explicitly indicate how each attractor component is the union of smaller copies of components of $\mathbf{A}$. Formal definitions are provided in Section 3. Figure 2 shows, in the top row, the three attractor components of the last GIFS in Example 9.7. The second row of the figure illustrates that each attractor component is the union of smaller similar copies of the attractor components. In this example, the boundary of each attractor component is a fractal.

The main object in this paper is a *self-similar GIFS* $\mathcal{F}$ and the associated *self-similar GIFS-tilings*. Associated with any self-similar GIFS $\mathcal{F}$ with attractor $(A_1, A_2, \dots, A_n)$ is

- an uncountable *parameter space* $\mathcal{P} := \mathcal{P}(\mathcal{F})$,
- a *tiling space* $\mathbb{T} := \mathbb{T}(\mathcal{F})$ consisting of a set of tilings of $\mathbb{R}^d$, each with prototile set $\{A_1, A_2, \dots, A_n\}$, and
- a *tiling map* $\mathcal{T} : \mathcal{P} \to \mathbb{T}$ from the parameter space to the tiling space.

Given a self-similar GIFS $\mathcal{F}$, each tiling in $\mathbb{T}(\mathcal{F})$ will be referred to as an $\mathcal{F}$*-tiling*. Both the parameter space and the tiling space are metric spaces, and the tiling map is continuous with respect to these metrics. Via the tiling map there are potentially infinitely many tilings that can be obtained from a single given self-similar GIFS.

The tilings in Barnsley and Vince [**5**] use a GIFS, but are very general, including tilings that are somewhat arbitrary. The self-similar GIFS-tilings in this paper are at once a specialization and a simplification - a simplification in that every similarity in the GIFS has the same scaling ratio. Nevertheless, it is shown in Section 6 that

FIGURE 2. Prototile set for the self-similar GIFS of Example 9.7.

any GIFS based tiling by congruent copies of a finite set of prototiles that may reasonably be considered self-similar must be a self-similar GIFS-tiling. Moreover, the self-similar tilings of Thurston and Kenyon [**20**, **39**], generalized from tilings by translation to tilings by congruent copies, are self-similar GIFS-tilings.

A tiling $T$ can be either periodic or non-periodic, periodic if there is a translational symmetry of $T$, otherwise non-periodic. Figure 3 shows two well-known tilings, the twin dragon tiling and the original Rauzy tiling [**32**]. The twin dragon tiling is periodic and the Rauzy tiling is non-periodic. Both tilings can be obtained from our systematic method for the construction of self-similar GIFS-tilings. The two additional fractal self-similar GIFS-tilings that appear in Figure 4 are also non-periodic and will be discussed in Section 9.1.

FIGURE 3. Twin dragon (periodic) and Rauzy (non-periodic) self-similar GIFS-tilings.

Since the paper is intended to be self-contained and we touch on topics that have appeared previously in some form, not necessarily GIFS related, this paper is, to some extent, expository. In Section 2 we give the organization of the paper and indicate results that are new.

FIGURE 4. Self-similar GIFS-tilings from Examples 9.7.

## 2. Organization and results

Basic notions about tiling Euclidean space, about rep-tiles and rep-sets, and about graph directed iterated function systems are covered in Section 3.

*Self-similar GIFS* and the associated *self-similar GIFS-tilings* are defined in Section 4. In particular, the parameter space, tiling space, and the tiling map are defined.

Basic properties of self-similar GIFS-tilings are covered in Theorem 5.5 of Section 5. These involve familiar notions in tiling theory like tile frequency, quasiperiodicity, hierarchy, and periodicity. Proofs that are new in a GIFS setting are included.

Requiring the scaling ratios of all similarity functions in a self-similar GIFS to be equal may seem excessively restrictive. Theorem 6.1 in Section 6, however, essentially states that any GIFS-tiling whose prototile set consists of components of the attractor of the GIFS must be a self-similar GIFS-tiling.

The notion of a self-similar tiling by translates of a set of prototiles was formulated, without reference to a GIFS, by Thurston and Kenyon [**20**, **39**]. Theirs is a global self-similarity. Basically, a tiling $T$ of $\mathbb{R}^d$ is self-similar in their terminology if there exists a similarity transformation $\phi : \mathbb{R}^d \to \mathbb{R}^d$ with scaling ratio greater than 1 such that, for all $t \in T$, the large tile $\phi(t)$ is tiled, in turn, by tiles in $T$. In Section 7 Thurston and Kenyon's concept is extended from tilings by translation to tilings by isometric (congruent) copies of a set of prototiles. Calling such a tiling *globally self-similar* we show that:

- Every globally self-similar tiling is a self-similar GIFS-tiling (Theorem 7.4).
- If the parameter of a self-similar GIFS-tiling is eventually periodic, then the associated self-similar GIFS-tiling is globally self-similar (Theorem 7.7).
- For a given self-similar GIFS $\mathcal{F}$, the set of globally self-similar tilings in the tiling space $\mathbb{T}(\mathcal{F})$ is dense in $\mathbb{T}(\mathcal{F})$ (Theorem 7.8).

Thus, for a given self-similar GIFS $\mathcal{F}$, the subset of globally self-similar tilings in the set of all self-similar GIFS-tilings play a role analogous to the subset of rational numbers in the reals.

The subject of Section 8 is a tiling dynamical system $(\mathbb{T}, H)$ of a self-similar GIFS. The function $H$ acts on the tiling space $\mathbb{T}$ by taking a tiling $T \in \mathbb{T}$ one level up, after scaling, in its hierarchy. Theorem 8.1 states that, for a self-similar GIFS

$\mathcal{F}$ satisfying natural assumptions, $(\mathbb{T}, H)$ is topologically conjugate to the discrete dynamical system $(\mathcal{P}, S)$, where $S$ is the shift map acting on the parameter space $\mathcal{P}$. The section ends with Theorems 8.2 and 8.5, applying the notions of topological entropy and the Artin-Masur zeta function.

If a GIFS is non-overlapping and has attractor components with nonempty interior, then there are potentially infinitely many associated tilings. In dimensions $d \geq 2$, however, GIFSs with these properties are hard to come by. Section 9 concerns the existence of such self-similar GIFSs. We conclude with a short exposition of an elegant construction, due to Rao, Wen and Yang [**31**], of Rauzy type tilings using a dual GIFS.

## 3. Graph iterated function system

**3.1. Tilings, prototile sets, and rep-sets.** In this paper, a *tile* is a compact subset of $\mathbb{R}^d$, and a *tiling of a set* $X \subseteq \mathbb{R}^d$ is a set of pairwise non-overlapping tiles whose union is $X$. *Non-overlapping* means that the intersection of any two distinct tiles has measure zero. Two tilings $T$ and $T'$ are *isometric* or *congruent* if there is an isometry of $\mathbb{R}^d$ taking one onto the other, and this is denoted $T \cong T'$. Two tilings $T$ and $T'$ are *equal* denoted $T = T'$ if they are identical.

For a finite set $Q$ of tiles, a *$Q$-tiling* $T$ is a tiling of $\mathbb{R}^d$ in which each tile is congruent to a tile in $Q$. The set $Q$ is called a *prototile set* for $T$. A tile $t$ that is congruent to $q \in Q$ will be referred to as a *type* $q$ tile.

A *rep-set* is a finite multiset $Q$ of tiles in $\mathbb{R}^d$, each tiled by smaller similar copies of tiles in $Q$. In the recreational literature, these are referred to as irreptiles. We allow multiple copies of tiles to allow for the same shape to be tiled by similar copies of tiles in $Q$ in different ways. For example, a $2 \times 2$ square can be tiled by four $1 \times 1$ squares or the $2 \times 2$ square can be tiled by two $1 \times 1$ squares and a $1 \times 2$ rectangle. How a GIFS captures the notion of a rep-set is explained in Remark 3.5.

**3.2. Directed graphs and adjacency matrices.** Let $G = (V, E)$ be a finite, strongly connected, directed graph (digraph) with vertex set $V = \{1, 2, \dots, n\}$ and edge set $E$. A digraph $G$ is *strongly connected* if, for any two vertices $i$ and $j$, there is a directed path from $i$ to $j$. In this paper, path always means a directed path, and a path can have repeated vertices and/or edges. The digraph $G$ may have loops and/or multiple edges. A strongly connected digraph $G$ will be called *primitive* if the greatest common divisor of the lengths of all closed paths of $G$ is 1. Equivalently, a strongly connected digraph is primitive if and only if, for $k$ sufficiently large, there is a path of length $k$ joining any two vertices.

For an edge $e = (i, j)$, directed from vertex $i$ to vertex $j$, the vertex $i$ is denoted $e^-$ and the the vertex $j$ is denoted $e^+$. Let $E_i$ denote the set of all edges $e$ such that $e^- = i$, i.e., the set of vertices directed out of vertex $i$.

Associated to a digraph $G$ is its *adjacency matrix* $M := M(G) = (m_{i,j})$, where $m_{i,j}$ is the number of edges from vertex $i$ to vertex $j$. It is well known that $G$ is strongly connected if and only if the matrix $M$ is irreducible, and $G$ is primitive if and only if $M$ is primitive. A square non-negative matrix $M$ is *primitive* if there is an integer $k \geq 0$ such that all entries of $M^k$ are positive, and $M$ is *irreducible* if for all $i, j$ there is a $k = k(i, j)$ such that $M^k_{i,j} > 0$. Clearly a primitive matrix is irreducible.

The spectral radius of $M(G)$, denoted $\rho(M)$, is an eigenvalue of $M$ called the Perron-Frobenius eigenvalue. The corresponding left and right eigenspaces are 1-dimensional, and $\rho(M)$ has left and right eigenvectors all of whose coordinates are positive.

### 3.3. Graph directed iterated function system.

DEFINITION 3.1 (GIFS). A *graph directed iterated function system (GIFS)* on $\mathbb{R}^d$ is a pair $\mathcal{F} = (G, F)$, where $G = (V, E)$ is a strongly connected digraph and

$$F = \{f_e : e \in E\},$$

where each function $f_e : \mathbb{R}^d \to \mathbb{R}^d$ is a similarity transformation with similarity ratio denoted $\lambda(f_e)$. It is assumed that $\lambda(f_e) < 1$ for all $e \in E$, i.e., the functions are contractions. The function $f_e$ can be considered as a label on the edge $e$.

Let $\mathbb{H}$ denote the set of nonempty compact subsets of $\mathbb{R}^d$, and define $\mathbf{F} : \mathbb{H}^n \to \mathbb{H}^n$ as follows. If $\mathbf{X} = (X_1, X_2, \dots, X_n) \in \mathbb{H}^n$, then

$$\mathbf{F}(\mathbf{X}) = \big(F_1(\mathbf{X}), F_2(\mathbf{X}), \dots, F_n(\mathbf{X})\big),$$

where, for $i = 1, 2, \dots, n$,

$$F_i(\mathbf{X}) = \bigcup_{e \in E_i} f_e(X_{e^+}).$$

The following is a well-known result in the theory of graph iterated function systems, which is a generalization of a fundamental result of Hutchinson [**17**]. In the theorem $\mathbf{F}^k$ denotes the $k$-fold iteration of $\mathbf{F}$.

THEOREM 3.2 ([**26**]). *If $(G, F)$ is a GIFS such that each function in $F$ is a contraction, then there exists a unique $\mathbf{A} = (A_1, A_2, \dots, A_n) \in \mathbb{H}^n$ such that*

$$\mathbf{A} = \mathbf{F}(\mathbf{A}) \qquad \text{and} \qquad \mathbf{A} = \lim_{k\to\infty} \mathbf{F}^k(\mathbf{B}) \tag{3.1}$$

*independent of $\mathbf{B} \in \mathbb{H}^n$, where convergence is with respect to the Hausdorff metric on $\mathbb{H}^n$.*

DEFINITION 3.3 (Attractor). The set $\mathbf{A}$ is called the *attractor* of the GIFS, and $\{A_1, A_2, \dots, A_n\}$ is its set of *attractor components*, each of which is compact. The first condition in Equation (3.1) can be restated as

$$A_i = \bigcup_{e \in E_i} f_e(A_{e^+}) \tag{3.2}$$

for $i \in \{1, 2, \dots, n\}$. If, in Equation (3.2), each distinct pair $f_e(A_{e^+}), f_{e'}(A_{e'^+})$ is non-overlapping, then $A_i$ is called *non-overlapping*. In this case, $\{f_e(A_{e^+}) : e \in E_i\}$ is a tiling of $A_i$. If every attractor component is non-overlapping, then the GIFS is called *non-overlapping*.

Although generally not required of a GIFS $(G, F)$, *from here on in this paper all function in $F$ are assumed to be similarity transformations.* The scaling ratio of a similarity $f : \mathbb{R}^d \to \mathbb{R}^d$ is denoted $\lambda(f)$.

REMARK 3.4. An ordinary iterated function system (IFS) is the special case of a GIFS whose graph consists of a single vertex with loops and whose attractor consists of a single component.

REMARK 3.5 (Every rep-set is the attractor of a GIFS). For a rep-set $Q$, let $(G, F)$ be the GIFS where the vertices of $G$ are the elements of $Q$. Each $p \in Q$ is tiled by smaller similar copies of tiles in $Q$. If one of those tiles is a smaller similar copy of $p' \in Q$, let $f_{(p,p')} \in F$ be a similarity transformation taking $p'$ onto $p$. Then

$$p = \bigcup_{e \in E_p} f_e(p_{e^+}),$$

in accordance with Equation (3.2). Conversely, if the attractor components of a GIFS $(G, F)$ have nonempty interior, the GIFS is non-overlapping, and the functions in $F$ are similarities, then the attractor $\mathbf{A}$ of $(G, F)$ is a rep-set.

**3.4. Admissible patch.** It is assumed in this section that all GIFSs are non-overlapping and that all components of the attractor have nonempty interior. Naturally associated to any GIFS are finite patches of tiles that we call admissible patches. These are natural subdivisions of the attractor components of the GIFS. For a GIFS $\mathcal{F} = (G, F)$, let $r$ be any vertex of $G$, referred to as a *root*. Referring to Definition 3.3,

$$W_r(\mathcal{F}) := \{f_e(A_{e^+}) : e \in E_r\}$$

is a tiling of the attractor component $A_r$.

DEFINITION 3.6 (Admissible patch). A set $X_r$ of finite directed paths in $G$ starting at $r$ is called *admissible* if

(1) no proper subpath of a path in $X_r$ lies in $X_r$, and
(2) if $\sigma \in X_r$ and $\sigma'$ is any proper subpath of $\sigma$ starting at vertex $r$, then $\sigma' e$ is a subpath (not necessarily proper) of a path in $X_r$ for all edges $e$ such that $\sigma'^+ = e^-$.

Applying Equation (3.2) recursively we arrive at the fact that

$$W_r(\mathcal{F}, X_r) := \left\{ f_\sigma(A_{\sigma^+}) : \sigma \in X_r \right\}$$

is a tiling of $A_r$. An *admissible patch* is any patch that is, a possibly scaled, copy of $W_r(\mathcal{F}, X_r)$ for some GIFS $\mathcal{F} = (G, F)$, some vertex $r$ of $G$, and some admissible set $X_r$ of paths.

## 4. Self-similar GIFS-tiling

**4.1. Notation.** Denote by $\Sigma^* := \Sigma^*(G)$ the set of paths of finite length in a digraph $G = (V, E)$ and $\Sigma := \Sigma(G)$ the set of all infinite paths. An infinite path has a starting vertex but no terminal vertex. A path $\sigma = e_1 e_2 \cdots$ will be written as its ordered string of edges $e_i \in E$, $i = 1, 2 \ldots$. The starting vertex of a path $\sigma$ will be denoted $\sigma^-$, and the terminal vertex of a finite path by $\sigma^+$. The length of a finite path $\sigma$, i.e., the number of edges, will be denoted $|\sigma|$. A path consisting of a single vertex has length zero.

For $\sigma = e_1 e_2 \cdots \in \Sigma$ let

$$\sigma|k = e_1 e_2 \cdots e_k \in \Sigma^*,$$

and $\sigma|0$ the path that is just the vertex $\sigma^-$. For any edge $e$ in $G$, let $\overleftarrow{e}$ be the oppositely directed edge.

Let $\mathcal{F} = (G, F)$ be a GIFS. For any function $f_e \in F$ define

$$f_{\overleftarrow{e}} := (f_e)^{-1}.$$

Denote by $\overleftarrow{G}$ the digraph obtained from $G$ by reversing the direction on all edges. Define $\overleftarrow{\Sigma}^* := \overleftarrow{\Sigma}^*(G)$ and $\overleftarrow{\Sigma} := \overleftarrow{\Sigma}(G)$ as the set of all finite and infinite paths, respectively, in $\overleftarrow{G}$. For $\sigma = e_1 e_2 e_3 \cdots e_k \in \Sigma^*$, define

$$f_\sigma := f_{e_1} \circ f_{e_2} \circ f_{e_3} \circ \cdots \circ f_{e_k}.$$

For $\overleftarrow{\sigma} = \overleftarrow{e_1}\,\overleftarrow{e_2}\,\overleftarrow{e_3} \cdots \overleftarrow{e_k} \in \overleftarrow{\Sigma}^*$, let

$$f_{\overleftarrow{\sigma}} := f_{\overleftarrow{e_1}} \circ f_{\overleftarrow{e_2}} \circ f_{\overleftarrow{e_3}} \circ \cdots \circ f_{\overleftarrow{e_k}} = f_{e_1}^{-1} \circ f_{e_2}^{-1} \circ f_{e_3}^{-1} \circ \cdots \circ f_{e_k}^{-1}.$$

**4.2. The parameter space.** Let $\mathcal{F} = (G, F)$ be a GIFS. Any path $\overleftarrow{\theta} \in \overleftarrow{\Sigma}$ will be referred to as a *parameter* of $\mathcal{F}$. To simplify notation, denote the set of parameters by

$$\mathcal{P} = \mathcal{P}(\mathcal{F}) := \overleftarrow{\Sigma}.$$

Define a metric $d$ on $\mathcal{P}$ by

$$d(\overleftarrow{\sigma}, \overleftarrow{\omega}) = \begin{cases} 0 & \text{if } \overleftarrow{\sigma} = \overleftarrow{\omega}, \\ 2^{-k} & \text{otherwise, where } k \text{ is the first integer such that } \overleftarrow{\sigma}_k \neq \overleftarrow{\omega}_k. \end{cases}$$

This makes $(\mathcal{P}, d)$ a compact metric space, which we call the *parameter space* of the self-similar GIFS. A parameter $\overleftarrow{\theta} \in \mathcal{P}$ is *eventually periodic* if there exist $\overleftarrow{\theta}_0, \overleftarrow{\theta}_1 \in \overleftarrow{\Sigma}^*$ such that $\overleftarrow{\theta} = \overleftarrow{\theta}_0\,\overleftarrow{\theta}_1\,\overleftarrow{\theta}_1\,\overleftarrow{\theta}_1 \cdots$.

**4.3. GIFS-tilings.** According to Theorem 3.2, any GIFS whose functions are contractions has an attractor. An issue, however, is that an attractor component that has empty interior cannot serve as a tile. Even if the attractor components have nonempty interior, it may occur that the GIFS is overlapping (Definition 3.3), resulting in overlap in a tiling obtained from the GIFS. Therefore, it is assumed in the following definition that all GIFSs are non-overlapping and that the attractor components have nonempty interior.

DEFINITION 4.1 (GIFS-tiling). Given a GIFS $\mathcal{F} = (G, F)$ and a parameter $\overleftarrow{\theta} = \overleftarrow{\theta}_0\,\overleftarrow{\theta}_1\,\overleftarrow{\theta}_2 \cdots \in \mathcal{P}$, let $X_k$ be an admissible set of paths rooted at $\theta_k^-$ such that

$$\{\theta_k\,\sigma : \sigma \in X_{k-1}\} \subset X_k \quad \text{for } k = 1, 2, \dots. \tag{4.1}$$

Let $\mathcal{X} = \mathcal{X}_{\overleftarrow{\theta}} = \{X_k, k = 1, 2, \dots\}$. Define a tiling $T(\mathcal{F}, \overleftarrow{\theta}, \mathcal{X})$ as follows.

$$\begin{aligned} &\text{(patch of tiles)} \qquad T(\overleftarrow{\theta}, k) = \{f_{\overleftarrow{\theta}|k}(t) : t \in W(\mathcal{F}, X_k)\} \\ &\text{(tiling)} \qquad T(\mathcal{F}, \overleftarrow{\theta}, \mathcal{X}) = \bigcup_{k=0}^{\infty} T(\overleftarrow{\theta}, k). \end{aligned} \tag{4.2}$$

The tiling $T(\mathcal{F}, \overleftarrow{\theta}, \mathcal{X})$ will be referred to as a *GIFS-tiling* or, more generally, an $\mathcal{F}$-tiling.

Notation aside, the idea is simple. We are just blowing up admissible patches $W(\mathcal{F}, X_k)$ along the vertices of the path $\overleftarrow{\theta}$ that is the parameter. That these patches are nested, i.e, $T(\overleftarrow{\theta}, k) \subset T(\overleftarrow{\theta}, k+1)$ follows from Equation (4.1). Their union gives the tiling $T(\mathcal{F}, \overleftarrow{\theta})$.

**4.4. Self-similar GIFS-tilings.** GIFS-tilings are very general and can be quite varied and irregular. Although a GIFS-tiling can have at most finitely many tile shapes up to similarity, it can have infinitely many tile shapes up to congruence. For example, there are GIFS-tilings of the plane into squares of arbitrarily large size or arbitrarily small size. And there are GIFS-tilings of the plane into squares of two sizes, but the small squares are at arbitrary positions in the tiling. Definition 4.3 below places a strong restriction of the kinds of tilings in Definition 4.1.

DEFINITION 4.2 (Self-similar GIFS). Call a GIFS $\mathcal{F} = (G, F)$ a *self-similar GIFS* if

(1) every function $f \in F$ is a similarity transformation with $\lambda(f)$ independent of $f \in F$, the common value denoted $\lambda(\mathcal{F}) < 1$;
(2) $\mathcal{F}$ is non-overlapping and each component of the attractor of $\mathcal{F}$ has non-empty interior.

The set $\{A_1, \dots, A_n\}$ of attractor components of $\mathcal{F}$ will be called the *prototile set* of $\mathcal{F}$, denoted $Q(\mathcal{F})$.

DEFINITION 4.3 (Self-similar GIFS-tiling). Given a self-similar GIFS $\mathcal{F}$, for each parameter $\overleftarrow{\theta} \in \mathcal{P}$, the collection $\mathcal{Y} = \{Y_k : k \geq 1\}$ of admissible sets of paths

$$Y_k = \{\sigma \in \Sigma^* : \sigma^- = \theta_k^-, \, |\sigma| = k\},$$

satisfies inclusion property (4.1). In words, the path $\sigma$ starts where the path $\overleftarrow{\theta}|k$ ends. Referring to Equations (4.2), let $T(\mathcal{F}, \overleftarrow{\theta}, \mathcal{Y})$ be the tiling for the above particular choice $\mathcal{Y}$ and the corresponding admissible sets $W(\mathcal{F}, Y_k)$, $k = 1, 2, \dots$, of patches. Since $\mathcal{Y}$ is completely determined by the parameter $\overleftarrow{\theta}$, we may write $T(\mathcal{F}, \overleftarrow{\theta})$ instead of $T(\mathcal{F}, \overleftarrow{\theta}, \mathcal{Y})$. The tiling $T(\mathcal{F}, \overleftarrow{\theta})$ will be called a *self-similar GIFS-tiling* and will be referred to as an $\mathcal{F}$-tiling. If the self-similar GIFS $\mathcal{F}$ is understood, we may write $T(\overleftarrow{\theta})$ instead of $T(\mathcal{F}, \overleftarrow{\theta})$.

DEFINITION 4.4 (Tiling space). The *tiling space* $\mathbb{T} = \mathbb{T}(\mathcal{F})$ of a self-similar GIFS $\mathcal{F}$ is the set of all $\mathcal{F}$-tilings of $\mathbb{R}^d$ endowed with the following metric $d$:

$$d(T, T') = \inf \big\{\epsilon : T \text{ and } T' \text{ coincide on a patch covering } B(1/\epsilon)\big\}.$$

The *tiling map*

$$\begin{aligned} \mathcal{T} : \mathcal{P}(\mathcal{F}) &\to \mathbb{T}(\mathcal{F}) \\ \overleftarrow{\theta} &\mapsto T(\mathcal{F}, \overleftarrow{\theta}) \end{aligned}$$

is a continuous map from the parameter space $\mathcal{P}(\mathcal{F})$ onto the tiling space $\mathbb{T}$. For each self-similar GIFS there are potentially uncountably many self-similar GIFS-tilings, one for each parameter, although some, and possible all, may coincide.

REMARK 4.5. Almost all self-similar GIFS-tilings fill the whole space $\mathbb{R}^d$. More specifically, for all $\theta$ in a dense subset of the parameter space $\mathcal{P}$, the tiling $T(\mathcal{F}, \overleftarrow{\theta})$ covers $\mathbb{R}^d$ [**5**]. There are cases, however, where $T(\mathcal{F}, \overleftarrow{\theta})$ tiles a subset of $\mathbb{R}^d$. For example, consider the self-similar GIFS $(G, F)$ on $\mathbb{R}$, where $G$ consists of a single vertex and two loops $e_1, e_2$ and $F = \{f_{e_1}, f_{e_2}\}$, where $f_{e_1}(x) = (1/2)\,x$, $f_{e_2}(x) = (1/2)\,x + 1/2$, and $\overleftarrow{\theta} = \overleftarrow{e}_1 \overleftarrow{e}_1 \cdots$. Then $T(\mathcal{F}, \overleftarrow{\theta})$ tiles the half line $\{x \in \mathbb{R} : x \geq 0\}$. *We assume hereafter, unless stated otherwise, that self-similar GIFS-tiling means a tiling of the whole Euclidean space.* See also [**19**] for Kellendonk's related notion of "forcing the border of a supertile".

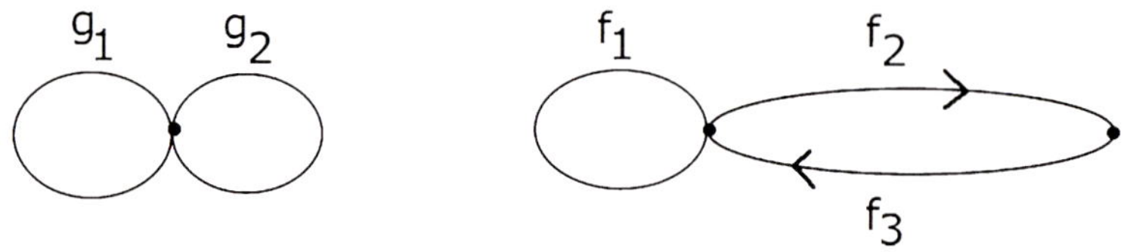


FIGURE 5. A digraph of a GIFS and the digraph of the companion self-similar GIFS.

FIGURE 6. Ammann chair tiling: prototiles, second level patch, self-similar GIFS-tiling.

EXAMPLE 4.6 (Ammann chair tiling). We illustrate Definition 4.3 with the Ammann chair tiling. The shape, called the A2 tile or sometimes the "golden bee", was discovered by R. Ammann in 1977 and is shown on the left in Figure 6. Figure 5 shows two digraphs. The digraph on the left will be relevant in Section 6. Consider now only the digraph $G$ on the right. With $s = 1/\sqrt{\tau}$, where $\tau = (1+\sqrt{5})/2$ is the golden ratio, the functions are:

$$f_1\begin{pmatrix} x \\ y \end{pmatrix} = \begin{pmatrix} 0 & -s \\ s & 0 \end{pmatrix}\begin{pmatrix} x \\ y \end{pmatrix} + \begin{pmatrix} s \\ 0 \end{pmatrix},$$

$$f_2\begin{pmatrix} x \\ y \end{pmatrix} = \begin{pmatrix} s & 0 \\ 0 & -s \end{pmatrix}\begin{pmatrix} x \\ y \end{pmatrix} + \begin{pmatrix} 0 \\ 1 \end{pmatrix},$$

$$f_3\begin{pmatrix} x \\ y \end{pmatrix} = \begin{pmatrix} s & 0 \\ 0 & s \end{pmatrix}\begin{pmatrix} x \\ y \end{pmatrix}.$$

The respective scaling ratios are $\lambda(f_1) = \lambda(f_2) = \lambda(f_3) = s$. The two attractor components, shown in orange at the left in Figure 6 have nonempty interior. The GIFS $(G, F)$ is non-overlapping. Therefore $\mathcal{F} = (G, \{f_1, f_2, f_3\})$ is a self-similar GIFS. A patch $T(\overleftarrow{\theta}, 2)$ for some $\overleftarrow{\theta} \in \mathcal{P}$ and part of larger patch of a self-similar GIFS-tiling $T(\mathcal{F}, \overleftarrow{\theta})$ are shown at the right in Figure 6.

## 5. Properties of self-similar GIFS tilings

Theorem 5.5 below lists nine properties of self-similar GIFS-tilings. Some of the proofs are omitted because the methods are standard. Statement 3 follows immediately from Lemma 5.4. The proofs of statements 4, 5, 6, and 7 are more involved and are included. We start with the definitions of the concepts involved.

DEFINITION 5.1 (Shift map). The *shift map* $S : \mathcal{P} \to \mathcal{P}$ on the parameter space of a GIFS is defined by $(S\overleftarrow{\theta})_i = \theta_{i+1}$, i.e., $S(e_1e_2e_3\cdots) = e_2e_3\cdots$, and $S^k$ denotes its $k^{th}$ iterate.

That certain tilings have a hierarchical structure has been known at least since Berger's 1966 proof that his set of prototiles is aperiodic [**7**].

DEFINITION 5.2 (Hierarchy). A *hierarchy* for a self-similar GIFS-tiling $T$ with prototile set $Q$ is a sequence $T_0, T_1, T_2, \dots$ of tilings such that $T_0 = T$ and, for all integers $k \geq 0$, the following properties hold:

(1) $T_k$ is a tiling with prototile set $\{(1/\lambda)^k p : p \in Q\}$, up to isometry.
(2) Every tile in $T_k$ is contained in a tile of $T_{k+1}$.

Call the tiling $T_k$ the $k^{th}$ *level* in the hierarchy of $T$. A self-similar GIFS $\mathcal{F}$ for which every $\mathcal{F}$-tiling has exactly one hierarchy is called *uniquely hierarchical.* The standard tiling of the plane by squares is an example of a self-similar GIFS-tiling that has many hierarchies.

DEFINITION 5.3 (Tile frequencies). It has been long known that in any Penrose tiling there are $\tau \approx 1.618$ times as many kites as darts; see [**16**]. In general, for a prototile $p \in Q(\mathcal{F})$ define $N_{k,\theta}$ and $N_{k,\theta}(p)$ as the number of tiles in the patch $T(\overleftarrow{\theta}, k)$ and the number of tiles of type $p$ in $T(\overleftarrow{\theta}, k)$, respectively. Letting $Q = Q(\mathcal{F}) = \{p_1, p_2, \dots, p_n\}$, define the *frequency* $\beta_\theta(p)$ of prototile $p \in Q$ in tiling $T(\mathcal{F}, \overleftarrow{\theta})$ and the *frequence vector* $\boldsymbol{\beta}_\theta(\mathcal{F})$ as

$$\beta_\theta(p) := \lim_{k\to\infty} \frac{N_{k,\theta}(p)}{N_{k,\theta}} \qquad \text{and} \qquad \boldsymbol{\beta}_\theta(\mathcal{F}) := \big(\beta_\theta(p_1), \beta_\theta(p_2), \dots, \beta_\theta(p_n)\big),$$

respectively.

A tiling $T$ is *repetitive*, also called *quasiperiodic*, if, for every patch $T_0$ of $T$, there is a real number $R$ such that every ball of radius $R$ contains a patch congruent to $T_0$.

Call a self-similar GIFS $\mathcal{F}$ *redundant* if $W_i(\mathcal{F}) = W_j(\mathcal{F})$ for some vertices $i \neq j$. Call a self-similar GIFS $\mathcal{F}$ *asymmetric* if, for all vertices $i$, the only symmetry of $A_i$ that preserves the tiling of the admissible patch $W_i(\mathcal{F})$ is the identity.

LEMMA 5.4. *Let $\mathcal{F}$ be a GIFS on $\mathbb{R}^d$ such that all of its functions have common scaling ratio and all of its attractor components have nonempty interior. Then $\mathcal{F}$ is non-overlapping if and only if $\lambda(\mathcal{F}) = 1/\sqrt[d]{\rho}$, where $\rho$ is the Perron-Frobenius eigenvalue of $M(G)$.*

PROOF. Assume that $\lambda(f) = 1/\sqrt[d]{\rho}$ for all $f \in F$. Denote by $x_i$ the Lebesgue measure of the attractor component $A_i$, $i = 1, 2, \dots, n$, of $\mathcal{F}$, and let $x = (x_1, \dots, x_n)^t$, where $t$ denotes the transpose. The Lebesgue measure of $f_e(A_{e^+})$ is then $(1/\rho)x_i$. If $A_i$ is overlapping for some $i$, then $x < (1/\rho)Mx = x$, a contradiction. Here the vector inequality $x < y$ means that $x_i \leq y_i$ for all $i$ and $x_i < y_i$ for at least one $i$.

Conversely, assume that $\mathcal{F} = (G, F)$ is non-overlapping. Then $M(\lambda^d x) = x$, where $\lambda := \lambda(\mathcal{F})$ and $M = M(G)$ is the adjacency matrix of $G$. Equivalently the eigen-equation $Mx = (1/\lambda^d)x$ holds. Since $x$ is positive, it must be an eigenvector corresponding to $\rho$. Therefore $\lambda(\mathcal{F}) = 1/\rho$. □

THEOREM 5.5. *For a given self-similar GIFS $\mathcal{F} = (G, F)$ on $\mathbb{R}^d$ with prototile set $Q(\mathcal{F})$ and for every pair $\overleftarrow{\theta}, \overleftarrow{\theta}' \in \mathcal{P}(\mathcal{F})$ the following properties hold for the self-similar GIFS-tiling $T(\mathcal{F}, \overleftarrow{\theta})$.*

1. *Each tile in $T(\mathcal{F}, \overleftarrow{\theta})$ is congruent to a tile in $Q(\mathcal{F})$.*
2. *If $\mathcal{F}$ is primitive, then a congruent copy of each tile in $Q(\mathcal{F})$ occurs in $T(\mathcal{F}, \overleftarrow{\theta})$.*
3. *$\lambda(\mathcal{F}) = 1/\sqrt[d]{\rho(\mathcal{F})}$, where $\rho(\mathcal{F})$ is the Perron-Frobenius eigenvalue of the adjacency matrix of $G$.*
4. *$\boldsymbol{\beta}_\theta(\mathcal{F})$ is equal to the normalized (unit vector with respect to the 1-norm) positive left eigenvector corresponding to the Perron-Frobenius eigenvalue of $\mathcal{F}$, independent of the parameter $\overleftarrow{\theta} \in \mathcal{P}$.*
5. *Every $\mathcal{F}$-tiling $T$ has a hierarchy $T = T_0, T_1, T_2, \dots$, such that each $T_k$, $k = 0, 1, 2, \dots$, is a self-similar GIFS-tiling. Moreover, $T_k = f_{\overleftarrow{\theta}|k} T(\mathcal{F}, S^k \overleftarrow{\theta})$ provides such a hierarchy.*
6. *If $S^k(\overleftarrow{\theta}) = S^k(\overleftarrow{\theta}')$ for some $k$, then $T(\overleftarrow{\theta}) \cong T(\overleftarrow{\theta}')$. Moreover, if $\mathcal{F}$ is uniquely hierarchical, asymmetric and not redundant, then $T(\overleftarrow{\theta}) \cong T(\overleftarrow{\theta}')$ if and only if $S^k(\overleftarrow{\theta}) = S^k(\overleftarrow{\theta}')$ for some $k$.*
7. *$T(\mathcal{F}, \overleftarrow{\theta})$ is repetitive for all $\overleftarrow{\theta} \in \mathcal{P}$.*
8. *If $\mathcal{F}$ is uniquely hierarchical, then every $\mathcal{F}$-tiling is non-periodic.*
9. *If $\mathcal{F}$ is uniquely hierarchical, asymmetric, and not redundant, then there are uncountable many $\mathcal{F}$-tilings up to congruence.*

REMARK 5.6 (Primitivity is needed in statement 2 of Theorem 5.5). The following is an example of a self-similar GIFS that is not primitive for which a prototile does not appear in a self-similar GIFS-tiling. Let $\mathcal{F} = (G, F)$ be the 1-dimensional self-similar GIFS with digraph given by its adjacency matrix $M$ and $F = \{f_{1,2}, f_{1,3}, f_{2,1}, f_{3,1}\}$ where

$$M = \begin{pmatrix} 0 & 1 & 1 \\ 1 & 0 & 0 \\ 1 & 0 & 0 \end{pmatrix} \qquad \begin{array}{ll} f_{1,2}(x) = \dfrac{x}{\sqrt{2}} & f_{1,3}(x) = \dfrac{x+1}{\sqrt{2}} \\ f_{2,1}(x) = f_{3,1}(x) = \dfrac{x}{\sqrt{2}}, & \end{array}$$

and $f_{i,j}$ denotes the function on the edge $(i, j)$. The attractor components are intervals
$\{[0, \sqrt{2}], [0, 1], [0, 1]\}$, but $T(\mathcal{F}, \overleftarrow{\theta})$ is a tiling of the real line by intervals of just length $\sqrt{2}$ if $\overleftarrow{\theta} = 1212\cdots$.

EXAMPLE 5.7 (Tile frequencies for the Ammann chair tiling of Example 4.6). The adjacency matrix of the self-similar GIFS digraph is $\begin{pmatrix} 1 & 1 \\ 1 & 0 \end{pmatrix}$. The Perron-Frobenius eigenvalue is $\tau = (1 + \sqrt{5})/2$, the golden ratio. The normalized corresponding left eigenvector is $(1/\tau, 1/\tau^2)$. Therefore, asymptotically about 61.80% of the tiles in an Ammann chair tiling are the large prototile, and about 38.20% are the small prototile.

PROOF OF PROPERTY 4. Denote the $(i, j)$ entry of $M^k$ by $m_{i,j}^{(k)}$, which is the number of paths in the digraph of $\mathcal{F}$ of length $k$ from vertex $i$ t vertex $j$. For a fixed parameter $\overleftarrow{\theta}$, let $i(k) = \theta_k^-$, i.e., the last vertex in the path $\overleftarrow{\theta}|k$. For ease of notation, let $N_k = N_{k,\theta}$ and $N_k(p) = N_{k,\theta}(p)$. Referring to the Definition 4.3 of an $\mathcal{F}$-tiling we have

$$N_k(p_j) = m^{(k)}_{i(k),j} \qquad \text{and} \qquad N_k = \sum_{j=1}^{n} m^{(k)}_{i(k),j}.$$

Noting that $N_k = \|(N_k(p_1), N_k(p_2), \ldots, N_k(p_n))\|_1$ we have

$$\Big(\frac{N_k(p_1)}{N_k}, \frac{N_k(p_2)}{N_k}, \ldots, \frac{N_k(p_n)}{N_k}\Big) = \frac{(N_k(p_1), N_k(p_2), \ldots, N_k(p_n))}{\|(N_k(p_1), N_k(p_2), \ldots, N_k(p_n))\|_1} = \frac{e_{i(k)} \cdot M^k}{\|e_{i(k)} \cdot M^k\|_1}.$$

A known consequence [**27**] of the Perron-Frobenius theorem, with Perron-Frobenius eigenvalue $\rho$, is

$$\lim_{k\to\infty} \frac{M^k}{\rho^k} = v^t \cdot w,$$

where $v$ is the right and $w$ is the left positive eigenvector of $M$ corresponding to eigenvalue $\rho$ and normalized so that $w \cdot v^t = 1$. In fact, we may assume that $\|w\|_1 = 1$. For any matrix $B$, let $B_i$ denote the $i^{th}$ row of $B$. Then $e_{i(k)} \cdot (M/\rho)^k = [(M/\rho)^k]_i$. As $k \to \infty$, the sequence $\big(e_{i(k)} \cdot (M/\rho)^k\big)$ of vectors gets close to the sequence of vectors whose terms are

$$e_{i(k)} \cdot (v^t \cdot w) = (e_{i(k)} \cdot v^t) \cdot w = c_k w,$$

where $c_k$ is a constant depending on $k$ that can take on at most $n$ values. Upon normalization we have

$$\boldsymbol{\beta}_\theta(\mathcal{F}) = \lim_{k\to\infty} \frac{e_{i(k)} \cdot M^k}{\|e_{i(k)} \cdot M^k\|_1} = \lim_{k\to\infty} \frac{e_{i(k)} \cdot (\frac{M}{\rho})^k}{\|e_{i(k)} \cdot (\frac{M}{\rho})^k\|_1} = w,$$

independent of the particular standard basis vector independent of the parameter $\overleftarrow{\theta}$. □

Proof of Property 5. The set of attractor components $Q = \{A_1, A_2, \ldots, A_n\}$ is a prototile set for $T$. Every tile $t \in T_k$, as defined explicitly in statement 5 of Theorem 5.5 has the form

$$t := f_{\overleftarrow{\theta}|k} \circ f^{-1}_{\theta_{k+1}} \circ \cdots \circ f^{-1}_{\theta_{k+j}} \circ f_\sigma(A_{\sigma^+}), \qquad \text{where} \qquad \sigma_1^- = \theta_j^-, \ |\sigma| = j-k.$$

Note that $t$ is isometric to $(1/\lambda)^k(A\sigma^+)$. Therefore condition (1) in Definition 5.2 is satisfied.

It follows from the formula above for $t$ and from Equation (3.2) that

$$\begin{aligned} t &= f_{\overleftarrow{\theta}|k-1} \circ f^{-1}_{\theta_k} \circ \cdots \circ f^{-1}_{\theta_j} \circ f_\sigma\Big(\bigcup_{e\in E_{\sigma^+}} f_e(A_{e^+})\Big) \\ &= \bigcup\Big\{f_{\overleftarrow{\theta}|k-1} \circ f^{-1}_{\theta_k} \circ \cdots \circ f^{-1}_{\theta_j} \circ f_\sigma \circ f_e(A_{e^+}) : e \in E_{\sigma^+}\Big\} \\ &\subset \bigcup\Big\{f_{\overleftarrow{\theta}|k-1} \circ f^{-1}_{\theta_k} \circ \cdots \circ f^{-1}_{\theta_j} \circ f_\omega(A_{\omega^+}) : \omega^- = {\theta_{k+j}}^-, \ |\omega| = j-(k-1)\Big\}. \end{aligned}$$

Because each $f_{\overleftarrow{\theta}|k-1} \circ f^{-1}_{\theta_k} \circ \cdots \circ f^{-1}_{\theta_j} \circ f_\omega(A_{\omega^+})$ in the line above is a tile in $T_{k-1}$, each tile in $T_k$ is, in turn, tiled by a patch in $T_{k-1}$. This proves condition (2) in Definition 5.2. Therefore $T = T_0, T_1, T_2, \ldots$ is indeed a hierarchy. □

Proof of Property 6. If $S^k(\overleftarrow{\theta}) = S^k(\overleftarrow{\theta}')$ for some $k$, then an isometry taking $T(\overleftarrow{\theta})$ onto $T(\overleftarrow{\theta}')$ is $f_{\overleftarrow{\theta}'|k} \circ (f_{\overleftarrow{\theta}|k})^{-1}$.

In the other direction, we will show, under the conditions assumed in the statement, that if two tilings $T(\overleftarrow{\theta})$ and $T(\overleftarrow{\theta}')$ are congruent, then there is an integer $j$ such that $S^j(\overleftarrow{\theta}) = S^j(\overleftarrow{\theta}')$. Assume that $T(\overleftarrow{\theta}) \cong T(\overleftarrow{\theta}')$. Let $t_0$ be an arbitrary tile in $T_0 = T(\overleftarrow{\theta})$ and $t_0'$ the corresponding tile in $T_0' = T(\overleftarrow{\theta}')$ under the isometry; call it $h : \mathbb{R}^d \to \mathbb{R}^d$. In the hierarchy, let $t_k \in T_k(\overleftarrow{\theta})$ be such that $t_k \subset t_{k+1}$ for all $k \geq 0$; define $t_k'$ similarly.

Let $j \geq k$ be sufficiently large that $t_0 \in T(\theta, j)$ and $t_0' \in T(\theta', j)$. Because it is assumed that $\mathcal{F}$ is uniquely hierarchical, the hierarchy must be the one provided in statement (5) of Theorem 5.5. Therefore, for all $j \geq k \geq 0$, if $t \in T_k$ and $t' \in T_k$ we have

$$t_k = f_{\overleftarrow{\theta}|j} f_\sigma(A_{\sigma^+}) \qquad t_k' = f_{\overleftarrow{\theta}'|j} f_\omega(A_{\omega^+}),$$

where $|\sigma| = j-k$, $\sigma^- = (\theta_j)^-$ and $|\omega| = j-k$, $\omega^- = (\theta_j')^-$. Let

$$\sigma = \sigma_{j-k}\,\sigma_{j-k-1}\cdots\sigma_2\,\sigma_1 \qquad \text{and} \qquad \omega = \omega_{j-k}\,\omega_{j-k-1}\cdots\omega_2\,\omega_1.$$

Note that, for $i = 1, 2, \dots, j-k$, the function $f_{\sigma_i}$ gives the embedding of $t_{i-1}$ into $t_i$. The same is true for $t_k'$. If $h$ is an isometry that takes $T(\overleftarrow{\theta})$ onto $T(\overleftarrow{\theta}')$, then $h$ takes $t_{k+1}$ onto $t_{k+1}'$ and the tiling of $t_{k+1}$ by a patch in $T_k$ onto the tiling of $t_{k+1}'$ by a patch in $T_k'$. By the assumption of non-redundancy and asymmetry, it is the case that $\sigma_i = \omega_i$, $i = 1, 2, \dots, j-k$.

Now take any integer $J > j$, so that

$$t_k = f_{\overleftarrow{\theta}|J} f_\sigma(A_{\sigma^+}) \qquad \text{where} \qquad \sigma = \theta_J\,\theta_{J-1}\cdots\theta_{j+1}\,\sigma_{j-k}\,\sigma_{j-k-1}\cdots\sigma_2\,\sigma_1$$

$$t_k' = f_{\overleftarrow{\theta}'|J} f_\omega(A_{\omega^+}) \qquad \text{where} \qquad \omega = \theta_J'\,\theta_{J-1}'\cdots\theta_{j+1}'\,\omega_{j-k}\,\omega_{j-k-1}\cdots\omega_2\,\omega_1.$$

As above, we have $\theta_i' = \theta_i$ for all $i > j$. □

Proof of Property 7. We use Property 5 of Theorem 5.5 in this proof. Any patch $X$ of an $\mathcal{F}$-tiling $T$ is contained in patch $T(\overleftarrow{\theta}, k)$ for some $k$. Given $n \geq 1$, there exists a real number $R$ such that every ball of radius $R$ contains a tile of $T_n(\overleftarrow{\theta}) = T_n(\mathcal{F}, \overleftarrow{\theta})$, the $n^{th}$ level of the hierarchy of $T(\overleftarrow{\theta}) = T(\mathcal{F}, \overleftarrow{\theta})$. Therefore, it suffices to show that there is an $n$ such that every tile of $T_n(\overleftarrow{\theta})$ contains an isometric copy of $T(\overleftarrow{\theta}, k)$.

Let $m$ be the greatest common divisor of all closed paths in $G$. We claim that there exists an $M$ such that if $n \geq M$ and $n \equiv 0 \pmod m$ the following holds: for any vertex $v$ of $G$ there is a circuit through $v$ of length $n$. To prove the claim, let $C$ be a circuit that contains every vertex of $G$. There are circuits $C_1, C_2, \dots, C_k$ of lengths $mq_1, mq_2, \dots, mq_k$ such that that $\gcd(q_1, q_2, \dots, q_k) = 1$. An elementary result in number theory states there exists an $N$ such that if integer $s \geq N$, then there exists positive integers $a_1, a_2, \dots, a_k$ such that $\sum_{i=1}^k a_i q_i = s$. Therefore, by traversing $C$ with detours around the circuits $C_1, C_2, \dots, C_k$ sufficiently many times, for any vertex $v$, there is a circuit containing $v$ of length $ms + |C| = m(s + |C|/m)$, an integer multiple of $m$. Taking $M = mN + |C|$, the claim is proved.

Given $k$, we now show that there is an integer $n$ such that every tile of $T_n(\overleftarrow{\theta})$ contains an isometric copy of $T(\overleftarrow{\theta}, k)$. Let $D$ denote the diameter of $G$, i.e., the greatest directed distance between any two vertices. Let $n = M + k + D$, where $M$

is as in the paragraph above. Let $u = \theta_k^-$, i.e., the last vertex of $\overleftarrow{\theta}|k$ in $\overleftarrow{\Sigma}$. Let $K \geq n$ and let $w = \theta_K^-$. A tile $t$ of $T_n(\overleftarrow{\theta})$ by definition has the form

$$t = f_{\overleftarrow{\theta}|K} \circ f_\sigma(A_{\sigma^+}) \qquad \text{where } |\sigma| = K - n,\, \sigma^- = \theta_K^+.$$

Let $v = \sigma^+$. We claim that there exists a path $\gamma$ from $v$ to $u$ in $G$ of length $n - k$. Assume that the claim is true, and consider the patch of tiles

$$T = \{f_{\overleftarrow{\theta}|K} \circ f_\sigma \circ f_\gamma \circ f_\omega(A_{\omega^+}) \,:\, |\omega| = k,\, \omega^- = \gamma^+\}.$$

Since $|\overleftarrow{\theta}|K| = |K| = (K - n) + (n - k) + k = |\sigma| + |\gamma| + |\omega|$, each tile in $T$ is a tile in $T(\overleftarrow{\theta})$. Moreover, since $\gamma^+ = u$, the tiling $T$ is an isometric copy of $T(\overleftarrow{\theta}, k)$. It now only remains to prove the claim.

Because $G$ is strongly connected, there is a simple (no crossing) path $\widehat{\gamma}$ from $v$ to $u$ of length at most $D$, and also a simple path $\delta$ from $u$ to $w$. The concatenation of $\delta$ and the path $\theta_K\, \theta_{k-1} \cdots \theta_{k+1}$ form a cycle which implies that $(K - k) + |\delta| \equiv 0 \pmod m$. Similarly, the concatenation of the paths $\widehat{\gamma}, \delta$ and $\sigma$ form a cycle which implies that $(K - n) + |\widehat{\gamma}| + |\delta| \equiv 0 \pmod m$. The two congruences yields $n - k - |\widehat{\gamma}| \equiv 0 \pmod m$. Therefore, by the definition of $M$, there is a circuit $\beta$ containing vertex $v$ such that $|\beta| = n - k - |\widehat{\gamma}|$ if $n - k - |\widehat{\gamma}| \geq M = n - k - D$, which is equivalent to $D \geq |\widehat{\gamma}|$, which is clearly true. Taking $\gamma = \beta\,\widehat{\gamma}$ we have

$$|\gamma| = |\beta| + |\widehat{\gamma}| = (n - k - |\widehat{\gamma}|) + |\widehat{\gamma}| = n - k. \qquad \square$$

## 6. Every GIFS-based tiling is a self-similar GIFS-tiling

The main result of this section is as follows.

THEOREM 6.1. *Let $\mathcal{F}$ be a primitive GIFS with no two attractor components that are similar but not congruent. Any $\mathcal{F}$-tiling $T$ (Definition* 4.1*) whose prototile set is the set of components of the attractor of $\mathcal{F}$ must be a self-similar $\mathcal{F}$-tiling (Definition* 4.3*).*

REMARK 6.2. The necessity of assuming $\mathcal{F}$ primitive is explained in Remark 6.7 below. The reason for the assumption on the components of the attractor of $\mathcal{F}$ is as follows. Assume that there exist attractor components $p, p'$ of self-similar GIFS $\mathcal{F}$ that are similar but not congruent with $\lambda(p) > \lambda(p')$. Assume further that $T$ is a self-similar $\mathcal{F}$-tiling as in Definition 4.3. In $T$ it is possible that there exists a (possibly infinite) set $T'$ of tiles congruent to $p$ in the first level hierarchy of $T$. Take any *arbitrary* subset $T''$ of tiles in $T'$ and remove from each tile in $T''$ the tiles in $T$ that it contains. What results is still an $\mathcal{F}$-tiling, but a type of tiling that we eliminate from consideration due to its arbitrary nature.

The proof of Theorem 6.1 hinges on the concept of commensurability and on several lemmas. The notion of commensurability in the context of tiling arose as early as 1976 in a paper of Kakutani [**18**] and in Sadun's 1998 generalization of the pinwheel tiling [**34**].

DEFINITION 6.3 (Commensurable GIFS). Let $F$ be a set of similarity transformations from $\mathbb{R}^d$ onto $\mathbb{R}^d$. For $f, g \in F$, call $f$ and $g$ *commensurable* if

$$\frac{\log\big(\lambda(f)\big)}{\log\big(\lambda(g)\big)} \in \mathbb{Q}.$$

Call $F$ *commensurable* if every pair of functions in $F$ is commensurable, and call a GIFS $(G, F)$ *commensurable* if $F$ is commensurable.

PROPOSITION 6.4. *A set $F$ of similarities is commensurable if and only if there is a real number $s > 0$ and a set $\{b_f \in \mathbb{N} : f \in F\}$ of positive integers such that $\lambda(f) = s^{b_f}$ for $f \in F$.*

PROOF. The existence of a real $s > 0$ and a set $\{b_f \in \mathbb{N} : f \in F\}$ of positive integers such that $\lambda(f) = s^{b_f}$ for all $f \in F$ clearly implies that $F$ is commensurable.

In the other direction, let

$$\alpha_f = \log_s \big(\lambda(f)\big) \qquad \text{so that} \qquad \lambda(f) = s^{\alpha_f}.$$

Let $f_0 \in F$. By the assumption that $F$ is commensurable, there is a $d \in \mathbb{N}$ and $b_f \in \mathbb{N}$ for all $f \in F$ such that $\alpha_f / \alpha_{f_0} = b_f / d$. Let $s' = s^{\frac{\alpha_{f_0}}{d}}$. Then, for all $f \in F$,

$$\lambda(f) = s^{\alpha_f} = \left(s^{\frac{\alpha_{f_0}}{d}}\right)^{b_f} = (s')^{b_f}. \qquad \square$$

Let $G$ be a strongly connected digraph whose edges are colored using $q$ colors, $q \geq 2$. For an admissible set $X_r$ of paths rooted at a vertex $r$ of $G$, call two paths *equivalent* if they contain the same number of edges of each color, and let $|X_r|_\equiv$ denote the number of equivalence classes.

LEMMA 6.5. *Let $G$ be a strongly connected, primitive digraph whose edges are colored using $q$ colors, $q \geq 2$. For every integer $N$ there exists an $M$ such that, if $X_r$ is an admissible set of paths rooted at vertex $r$ with $|X_r| \geq M$, then $|X_r|_\equiv \geq N$.*

PROOF. If the lemma holds for every 2-coloring, then it holds for every $q$-coloring. To see this, let the colors be $\{1, 2, \dots, q\}$. For all edges colored $3, 4, \dots q$, change the colors to color 2. The number $|X_r|$ does not change, and $|X_r|_\equiv$ cannot increase.

We now prove the result for every 2-coloring (say red and blue) of its edge set. By way of contradiction assume that there is a 2-coloring, a vertex $r$, a natural number $N$, and a sequence $(X_k)_{k \geq 1}$ of admissible paths rooted at $r$ such that

(1) $|X_k| \to \infty$ as $k \to \infty$, and
(2) $|X_k|_\equiv \leq N$.

In particular, from (2) it follows that there must be a bound $m$, independent of $k$, such that the set of lengths satisfies $|\{|\sigma| : \sigma \in X_k\}| \leq m$. We claim that there is a sequence $(Y_k)$ of admissible sets of paths satisfying properties (1) and (2) above and such that, for each $k$, all paths in $Y_k$ have the same length. To prove the claim, denote the lengths of paths in $X_k$ by $l_1(k) > l_2(k) > \cdots > l_m(k)$. We now prove the claim by induction on $m$, the number of distinct lengths of paths in the $X_k$. The claim is trivially true for $m = 1$. Assume it true for $m-1$ and let the sequence $(X_k)_{k \geq 1}$ have paths of $m$ different lengths.

Let $Y_k$ be the set of paths obtained from $X_k$ by replacing (pruning) each path $\sigma$ of length $l_1(k)$ by its subpath $\sigma'$ rooted at $r$ and having length $l_2$. Note that $Y_k$ remains a set of admissible paths. We will show that the sequence $(Y_k)$ of sets of admissible paths satisfies conditions (1) and (2). This will complete the induction argument because $Y_k$ has one less path length than $X_k$.

Concerning condition (1), it follows from the definition of an admissible set of paths that, if $v$ is the second to last vertex of $\sigma$ and $v'$ is the last vertex of $\sigma'$, then the outdegree in $G$ of all vertices on the path from $v'$ to $v$, including $v'$ but not including $v$, have outdegree 1. This implies that $|Y_k| \leq |X_k|/\Delta$, where $\Delta$ is the maximum outdegree of vertices in $G$. Therefore $|Y_k| \to \infty$ as $k \to \infty$.

Concerning condition (2), first note that there cannot exist a cycle $C$ in $G$ such that the outdegree of every vertex on $C$ is 1. Otherwise there would exist an attractor component with empty interior, which we are assuming in this section is not the case. Therefore, the longest path $\gamma$ such that every vertex, except the last, has outdegree 1 is less than $n$, the order of $G$. Assume that, in going from $X_k$ to $Y_k$ we prune just one prune just one length at a time, obtaining a sequence $X+k = X_k^0, X_k^1, \dots, X_k^q = Y_k$, where $q = l_1(k) - l_1(k) < n$. Since adjoining a single edge to a path can change the number of red edges (or blue edges) by at most 1, we have $|X_k^{i+1}|_\equiv \leq 2|X_k^i|_\equiv$. Therefore $|Y_k|_\equiv \leq 2^n |Y_k|_\equiv \leq 2^n N$.

It remains to prove a contradiction in the case that, for each $k$, all paths in $Y_k$ have the same length $n(k)$. If all paths in $Y_k$ have the same length, then by the definition of admissible path, $Y_k$ is the set of all paths of length $n(k)$ rooted at vertex $r$ in $G$. There exists a closed path $c_1$ in $G$ containing $r$ that is not monochromatic (recall that both red and blue are used in the coloring). Let $L_1$ denote the length of $c_1$. By primitivity of $G$, there exists another closed path $c_2$ in $G$ containing $r$ whose length $L_2$ is relatively prime to $L_1$. Any non-negative integer solution $x, y$ to the equation

$$xL_1 + yL_2 = n(k)$$

provides a path $c_k(x, y)$ in $Y_k$ obtained by winding $x$ times around $c_1$ followed by winding $y$ times around $c_2$. Call such a path a $(c_1, c_2)$-path. For clarity we omit the index $k$ in what follows, i.e., $k$ fixed and, for example, $n = n(k)$. If $n$ is sufficiently large, then from elementary number theory there are positive integers $x_0, y_0$ such that $x_0 L_1 + y_0 L_2 = n$. It follows that $x = x_0 - jL_2$, $y = y_0 + jL_1$ is also a solution for any $j \in \mathbb{Z}$. Since we seek non-negative solutions, the condition

$$\frac{x_0}{L_2} \geq j \geq -\frac{y_0}{L_1}$$

must be satisfied, which implies that there are

$$\left\lfloor \frac{x_0}{L_2} + \frac{y_0}{L_1} \right\rfloor = \left\lfloor \frac{n}{L_1 L_2} \right\rfloor \xrightarrow[n \to \infty]{} \infty$$

solutions. In other words, the number of $(c_1, c_2)$-paths in $Y_k$ goes to infinity with $k$.

Denote by $a_1, a_2$ the number of red edges on $L_1$ and $L_2$, respectively. Two of the $(c_1, c_2)$-paths in $Y_k$ are in the same color equivalence class if and only if they contain the same number of red edges. Counting the number of red edges on the path corresponding to solution $x, y$, i.e., to each valid $j$, we obtain $xa_1 + ya_2 = (x_0 - ja_2)a_1 + (y_0 + ja_1)a_2$ red edges. Therefore, two $(c_1, c_2)$-paths, which we denote by $c(i)$ and $d(j)$, are in the same equivalence class if and only if

$$(x_0 - ja_2)a_1 + (y_0 + ja_1)a_2 = (x_0 - ia_2)a_1 + (y_0 + ia_1)a_2$$

which simplifies to $(i-j)(L_1 a_2 - L_2 a_1) = 0$. If $i = j$, then $c(i) = c(j)$. That $L_1 a_2 = L_2 a_1$ is impossible since $L_1$ and $L_2$ are relatively prime and $0 < a_1 < L_1$. We have shown that $(c_1, c_2)$-path in $Y_k$ is in its own equivalence class. Since the number of $(c_1, c_2)$-path in $Y_k$ goes to infinity with $k$, we have the desired contradiction to condition (2). □

LEMMA 6.6. *Let $\mathcal{F}$ be a primitive GIFS. If there exists a tiling $T$ of $\mathbb{R}^d$ (not necessarily a GIFS-tiling) having a finite prototile set and containing admissible $\mathcal{F}$-patches of arbitrary large cardinality, then $\mathcal{F}$ must be commensurable.*

PROOF. For a set $W$ of tiles, let $|W|$ denote the cardinality of $W$, and let $|W|_{\equiv}$ denote the number of tiles up to congruence. Because it is assumed that $T$ has a finite prototile set and contains admissible (scaled) $\mathcal{F}$-patches of arbitrary large cardinality, there must be a sequence $\{W_r(X_k)\}$ of (unscaled) admissible patches such that $|W_r(X_k)|_{\equiv}$ is bounded but $\lim_{k\to\infty} |W_r(X_k)| = \infty$.

By way of contradiction, assume that $\mathcal{F}$ is not commensurable. We will show that, for every $N$ there exists a $k_0$ such that $|W_r(X_{k_0})|_{\equiv} \geq N$, a contradiction.

Let $\lambda_1, \lambda_2, \dots, \lambda_m$ be the scaling ratios of the functions in $F$. Let $s = \lambda_1$, and define $\alpha_i$, $i = 1, 2, \dots, m$, by $\lambda_i = s^{\alpha_i}$. Note that $\alpha_1 = 1$. The commensurable relation is an equivalence relation. Partition the set $\{\alpha_1, \dots, \alpha_m\}$ into equivalence classes, and call two edges of $G$ equivalent if the corresponding $\alpha' s$ are equivalent. Let the number of equivalence classes be $q$, which is at least 2 by Proposition 6.4. Color the edges of $G$ in $q$ colors according to their equivalence class. For an edge $e$, denote the color by $\alpha(e)$. For a path $\sigma \in \Sigma^*(G)$, define $\alpha(\sigma) := \sum_{e\in\sigma} \alpha(e)$.

Let $\sigma, \omega \in X_k$. Because a set of pairwise incommensurable numbers are linearly independent over $\mathbb{Q}$, we have that $\alpha(\sigma) = \alpha(\omega)$ if and only if, for every color, the number of occurrences of that color in $\sigma$ equals the the number of occurrences of that color in $\omega$. Now $\lambda(f_\sigma) = \lambda(f_\omega)$ if and only if $\alpha(\sigma) = \alpha(\omega)$ if and only if $\sigma$ and $\omega$ are in the same color equivalence class. Because $|X_k| = |W_r(X_k)|$ we have $\lim_{k\to\infty} |X_k| = \infty$. Call $\lambda(f_\sigma)$ the scaling ratio of the path $\sigma$. By Lemma 6.5, for every $N$ there exists a $k_0$ such that if $k \geq k_0$, then $|X_k|_{\equiv} \geq N$. Therefore, for every $N$ there exists a $k_0$ such that if $k \geq k_0$, then there exists at least $N$ paths in $X_k$ with pairwise different scaling ratios $\lambda$. For $\sigma \in X_k$, there are at most $n$ (order of digraph $G$) possibilities for $\sigma^+$, which implies that, for $k \geq k_0$, there are at least $N/n$ distinct tiles $f_\sigma(A_{\sigma^+})$ in $W_k(X_k)$, i.e., $|W_r(X_{k_0})|_{\equiv} \geq N/n$. □

REMARK 6.7 (Primitivity in Theorem 6.6 is necessary). The following is a counterexample to Lemma 6.6 if the assumption of primitivity is removed. Consider the GIFS $\mathcal{F} = (G, F)$ on $\mathbb{R}$, where $G$ is the graph consisting of two vertices $r, r'$, with edges $e_1, e_2$ from $r$ to $r'$ and edges $e_3, e_4$ from $r'$ to $r$. Note that $G$ is not primitive; the lengths of all closed paths are divisible by 2. Let $F = \{f_{e_1}, f_{e_2}, f_{e_3}, f_{e_4}\}$ where $f_{e_1}(x) = 3/4\,x$, $f_{e_2}(x) = 3/4\,x + 1/2$, $f_{e_3}(x) = 1/3\,x$, $f_{e_4}(x) = 1/3\,x + 1/3$. The attractor components of $\mathcal{F}$ are the intervals $A_r = [0, 1]$, $A_{r'} = [0, 2/3]$. Note that the scaling ratios $3/4$ and $1/3$ are not commensurable; thus $\mathcal{F}$ is not commensurable. Let $X_r(k)$ be the set of all paths in $G$ rooted at $r$ of length $2k$; this set of paths is admissible. It is routine to check that the admissible patch $W(\mathcal{F}, X_r(k))$ consists of the interval $[0, 4^k]$ subdivided into $4^k$ unit intervals. Let $T$ be the tiling of the line by unit intervals. Thus $T$ contains admissible patches of arbitrary large cardinality.

DEFINITION 6.8 (Companion GIFS). Let $\mathcal{F} = (G, F)$ be a commensurable GIFS. By Proposition 6.4 there is an $s > 0$ and a set $\{a_e \in \mathbb{N} : e \in E\}$ of positive integers associated with each edge $e \in E$ of $G$ such that $\lambda(f_e) = s^{a_e}$ for all $e \in E$. Attach the label $a_e$ to each edge of $G$. Construct a new GIFS $\mathcal{F}' = (G', F')$, called the *companion* of $\mathcal{F}$, as follows. To obtain the graph $G'$, consider each edge $e = (u, v)$ of $G$ with $a_e > 1$. Replace $e$ by a path $\sigma(e) := e_1 e_2 \cdots e_{a_e}$ from $u$ to $v$. Note that no vertex of $G$ has been removed. Also note that $G'$ is strongly connected if and only if $G$ is strongly connected. It is not hard to see that there exist functions $f_{e_1}, f_{e_2}, \dots, f_{e_{a_e}}$ on the respective new edges $e_1, e_2, \dots, e_{a_e}$ such that $\lambda(f_{e_i}) = s$ for $i = 1, 2, \dots, a_e$ and therefore $f_{\sigma(e)} = f_e$. The graph $G'$ and function set $F'$ is the result of the above alterations for all edges $e$ with $a_e > 1$.

EXAMPLE 6.9 (Companion GIFS). On the left in Figure 5 is the digraph of a GIFS $\mathcal{F} = (G, F)$ where $G$ has one vertex and two loops and $F = \{g_1, g_2\}$, where

$$g_1 \begin{pmatrix} x \\ y \end{pmatrix} = \begin{pmatrix} 0 & -s \\ s & 0 \end{pmatrix} \begin{pmatrix} x \\ y \end{pmatrix} + \begin{pmatrix} s \\ 0 \end{pmatrix}, \qquad g_2 \begin{pmatrix} x \\ y \end{pmatrix} = \begin{pmatrix} s^2 & 0 \\ 0 & -s^2 \end{pmatrix} \begin{pmatrix} x \\ y \end{pmatrix} + \begin{pmatrix} 0 \\ 1 \end{pmatrix},$$

with $s = 1/\sqrt{\tau}$, where $\tau$ is the golden ratio. By Proposition 6.4, $\mathcal{F}$ is commensurable, the scaling ratios being $\lambda(g_1) = s$, $\lambda(g_2) = s^2$. The attractor is the Ammann chair tile shown at the left in Figure 6 (either orange polygon). Let $\mathcal{F}' = (G', F)$, where $G'$ is the graph on the right in Figure 5 and $F' = \{f_1, f_2, f_3\}$ is as given in Example 4.6. This is the companion of $\mathcal{F}$. The scaling ratios are equal: $\lambda(f_1) = \lambda(f_2) = \lambda(f_3) = s$; $\mathcal{F}$ is a self-similar GIFS. The two attractor components of $\mathcal{F}'$ are the polygons in orange at the left in Figure 6. The self-similar GIFS $\mathcal{F}'$ is the companion of the GIFS $\mathcal{F}$. Every $\mathcal{F}$-tiling $T(\overleftarrow{\theta})$ with prototile set consisting of the two orange polygons on the left in Figure 6 is also a self-similar $\mathcal{F}'$-tiling $T(\overleftarrow{\theta}')$ for an appropriate parameter $\overleftarrow{\theta}'$ as constructed in the proof of Lemma 6.10.

LEMMA 6.10. *The companion GIFS $\mathcal{F}' = (G', F')$ of a commensurable GIFS $\mathcal{F} = (G, F)$ satisfies the following properties:*

(1) *$\mathcal{F}'$ is a self-similar GIFS, and*
(2) *every $\mathcal{F}$-tiling is a $\mathcal{F}'$-tiling.*

PROOF. Concerning statement (1) and referring to Definition 6.8, $\lambda(f_e) = s$ for all edges in $G'$. Therefore $\lambda(\mathcal{F}') = s$.

Concerning statement (2), denote the set of vertices of $G$ by $\{1, 2, \dots, n\}$ and the set of corresponding attractor components by $\{A_1, A_2, \dots, A_n\}$. Each edge $e = (i, j)$ of $G$ with $a_e > 1$ is replaced in $G'$ by a path whose successive vertices we denote by $u = u_0, u_1, u_2, \dots, u_k = v$. Note that the outdegree of $u_i$ is 1 for $i = 0, 1, 2, \dots, k-1$. The successive edges are $e_1 = (u, u_1) = (u_0, u_1), e_2 = (u_1, u_2), \dots, e_k = (u_{k-1}, u_k) = (u_{k-1}, v)$. It is routine to check that in $\mathcal{F}'$, the attractor component of all vertices from $G$ remain the same, namely $A_i' = A_i$ for $i = 1, 2, \dots, n$. The attractor component $A'_{u_i}$ of each new vertex is defined recursively by $A'_{u_k} = A_{u_k}$ and $A'_{u_i} = f_{e_i}(A'_{u_{i+1}})$ for $i = k-1, k-2, \dots, 1$. Note that the attractor components $A'_{u_1}, A'_{u_2}, \dots, A'_{u_k} = A_v$ are all similar, each scaled down from its successor by a factor $s$.

Let $\overleftarrow{\theta} \in \mathcal{P}(\mathcal{F})$ and $\mathcal{X} = \{X_1, X_2, \dots\}$ a corresponding collection of admissible sest of paths for $k = 1, 2, \dots$. Let $T(\mathcal{F}, \overleftarrow{\theta}, \mathcal{X})$ be the associated GIFS-tiling. Define $\overleftarrow{\theta}' \in \mathcal{P}(\mathcal{F}')$ by replacing edges $e$ in $\overleftarrow{\theta}$ with $a_e > 1$ by paths as in the definition of the companion GIFS. On vertices $u$ of $\overleftarrow{\theta}'$ that correspond to vertices of $\overleftarrow{\theta}$ retain the admissible set $X_u$ of paths. On each new vertex $u_i$, $1 \le i < k$, define the admissible set $X_{u_i}$ of paths by $e_i\, e_{i-1}, \dots e_1$ followed by the admissible set $X_u$ of paths at $u$. It is now routine that statement (2) of Lemma 6.10 holds. □

PROOF OF THEOREM 6.1. Theorem 6.1 follows from Lemmas 6.6 and 6.10 and the following. For a GIFS-tiling to be a self-similar GIFS-tiling it is required in

Definition 4.3 that if $\sigma \in Y(k)$, then $|\sigma| = k$, i.e., the length of the admissible path $\sigma$ must equal the length $k$ of the initial path of the parameter $\overleftarrow{\theta}$. This is insured by the assumption that no two attractor components of the original GIFS are similar but not congruent. Those attractor components introduced in a companion GIFS are not relevant since the introduced vertices of the graph $G$ of the GIFS have outdegree 1. □

## 7. Every globally self-similar tiling is a self-similar GIFS-tiling

This section concerns the self-similar tilings introduced by Thurston [**39**] in the context of tilings by translation of a set of prototiles. Basically, a self-similar tiling $T$ is a tiling of $\mathbb{R}^d$ for which there exists a similarity transformation $\phi : \mathbb{R}^d \to \mathbb{R}^d$ with scaling ratio $\lambda(\phi) > 1$ such that, for all $t \in T$, the large tile $\phi(t)$ is tiled, in turn, by tiles in $T$. The following example shows that an additional requirement is needed in a formal definition of self-similar.

EXAMPLE 7.1. Let $Q$ consists of two intervals on the real line, $I_1$ of length 1 and $I_2$ of length 2. Let $T$ be *any* tiling of $\mathbb{R}$ with prototile set $Q$ such that

(1) the origin is located at an endpoint of a tile, and
(2) the endpoints of each tile of length 2 are located at even coordinates on $\mathbb{R}$.

Let $\phi(x) = 2x$ for all $x \in \mathbb{R}$. Then for every tile $t \in T$, its image $\phi(t)$ is tiled by tiles in $T$ for all $t \in T$. The issue is that $T$ can be quite random. For example, construct a tiling as follows. Moving along the positive real line starting at the origin, place a random number of tiles of length 2. Follow that by a random even number of tiles of length 1, and repeat in this fashion moving in the positive direction; similarly on the negative real line.

To avoid the randomness in Example 7.1 and similar examples and to extend from tilings by translation to tilings by isometric images of the prototiles, we will require that, for any tile $t \in T$, the tiling of $\phi(t)$ by tiles in $T$ depends uniquely on the prototile type of $t$. To impose this requirement rigorously requires additional notation.

DEFINITION 7.2 (Induced tiling). Let $T$ be a tiling of $\mathbb{R}^d$ with prototile multiset $Q$, and let $\phi : \mathbb{R}^d \to \mathbb{R}^d$ be a similarity with scaling ratio $\lambda(\phi) > 1$. We allow $Q$ to be a multiset to allow for multiple ways for $\phi(p)$, $p \in Q$, to be tiled. For each $p \in Q$, let $T_p$ be a tiling of $\phi(p)$ by tiles in $Q$. Call the set $\{T_p : p \in Q\}$ of patches a *$\phi$-tiling rule*. Let $H := \{h_t : t \in T\}$ be a set of isometries, called *tiling isometries*, such that $t = h_t(p)$, where $p \in Q$ is the type of tile $t \in T$. If every $p \in Q$ has trivial symmetry group, then $H$ is uniquely determined. This is the case, for example, in the definition of self-similar in [**20, 39**] where the tilings are by copies of the prototiles by translation rather than by isometry as is the case here.

For each $t \in T$, the isometry $\widehat{h}_t := \phi\, h_t\, \phi^{-1}$ maps $\phi(p)$ onto $\phi(t)$. Now

$$\phi(t) = \phi\big(h_t(p)\big) = \widehat{h}_t\, \phi(p) = \bigcup_{q \in T_p} \widehat{h}_t(q).$$

Note that $p := p_t$ in the equation above depends on $t$. Given the tiling $T$ and tile $t \in T$, we now obtain the *induced tiling* of $\phi(t)$ and the *induced tiling* of $T$ defined by

$$T_{\phi(t)} := \{\widehat{h}_t(q) : q \in T_{p_t}\} \qquad \text{and} \qquad T_\phi := \bigcup_{t \in T} T_t.$$

The following definition is basically that of Thurston and Kenyon [**20**], removing the restriction that the isometries in $H$ be translations.

DEFINITION 7.3 (Global self-similarity). Given a prototile set $Q$, a $Q$-tiling $T$ of $\mathbb{R}^d$ is *globally self-similar* if there is a similarity $\phi$, a $\phi$-tiling rule, and a set $H$ of tiling isometries such that $T_\phi = T$. In particular, the image $\phi(t)$ is tiled by tiles in $T$ for all $t \in T$.

Call a similarity transformation $\phi$ *proper with respect to a tiling* $T$ if the fixed point of $\phi$ lies in the interior of a tile. As shown in Example 7.5 below, Theorem 7.4 may fail without this condition.

THEOREM 7.4. *Every globally self-similar tiling with proper self-similarity $\phi$ is a self-similar GIFS-tiling.*

PROOF. Given a self-similar tiling $T$, define a self-similar GIFS $\mathcal{F}(T) = (G, F)$ as follows. The vertex set of $G$ is $Q := Q(\mathcal{F})$. The edge set $E$ of $G$ is defined by the $\phi$-rule as follows. The $\phi$-tiling rule can be expressed as

$$\phi(p) = \bigcup_{q \in T_p} h_{p,q}(q) \tag{7.1}$$

for some isometries $h_{p,q}$. In $G$ add an edge $e$ directed from vertex $p$ to vertex $q$ and let $f_e = \phi^{-1} h_{p,q}$. Note that $\lambda(f_e) = 1/\lambda(\phi) < 1$ for all $e \in E$. Equation (7.1) then becomes

$$p = \bigcup_{e \in E_p} f_e(p_{e^+}),$$

corresponding to Equation (3.2) in the definition of a GIFS attractor. Take $F = \{f_e : e \in E\}$. Now $\mathcal{F}(T) = (G, F)$ is a self-similar GIFS with scaling ratio $1/\lambda(\phi)$. Note that the prototiles of $Q$ are isometric copies of the attractor components of $\mathcal{F}(T)$. It remains to shown that there exists a $\overleftarrow{\theta} \in \overleftarrow{\Sigma}$ such that $T = T(\overleftarrow{\theta})$.

Let $t_0$ be the tile in the self-similar tiling $T$ that contains the fixed point $O$ of $\phi$ in its interior. If $t_0$ is of type $p \in Q$, then there is no loss of generality in assuming that $t_0 = p$, and consequently there is an edge (loop) $e$ in $G$ from $p$ to $p$ labeled $f_e = \phi^{-1}$. Consider the parameter $\overleftarrow{\theta} := e\,e\,e \cdots \in \overleftarrow{\Sigma}$ that winds infinitely many times around the loop $e$. Then $f_{\overleftarrow{\theta}|k} := f_e \circ f_e \circ \cdots \circ f_e = \phi^{-k}$ for all non-negative integers $k$. We will show that $T = T(\overleftarrow{\theta})$.

Let $H$ be a set of tiling isometries for $T$ such that $T_\theta = T$. We claim that there a set $H' = \{h'_t : t \in T\}$ of tiling isometries such that each $h'_t(p_t) = h_t(p_t)$ for all $t \in T$, and each $h'_t \in H'$ has the form

$$h'_t = \phi^k \circ f_\sigma, \tag{7.2}$$

where $\sigma \in \overleftarrow{\Sigma}^*$ and $|\sigma| = k$. If this is the case, then

$$t = (\phi^k \circ f_\sigma)(p) \in T(\overleftarrow{\theta}, k),$$

where $p = p_{\sigma^+} \in Q$. This is exactly as in Equation (4.2) in Definition 4.1, showing that $T$ is an $\mathcal{F}$-tiling and completing the proof.

It only remains to prove the claim. Since the fixed point $O$ lies interior to $t_0$, any tile $t \in T$ is contained in $X_k := \phi^k(t_0)$ for some integer $k$. Note that the sets $X_k$ are nested. The existence of such a set $H'$ of tiling isometries of the form in Equation (7.2) is proved by induction on $k$. If $t \in X_1 = \phi(t_0) = \phi(p)$, then by Equation (7.1) we have $t = h_{p,p_1}(p_1) = \phi\, f_e(p_1)$ for an edge $e$ directed from $p$ to $p_1$, where $p_1$ is the type of tile $t$. Take $h'_t = h_{p,p_1} = \phi\, f_{e_1}$, which is of the form in Equation (7.2). Assume that $h'_t$ has been defined in the form of Equation (7.2) for all $t \in X_k$, and let $t' \in X_{k+1}$. Then $t' \in \phi(t)$ for some $t \in X_k$, where, by the induction hypothesis, $h'_t = \phi^k \circ f_\sigma$ holds, where $|\sigma| = k$ and $\sigma$ is a path from $p$ to $p_t$. By the requirement that $T_\phi = T$ we have $t' = \widehat{h}_t(q)$, where $q \in T_{\phi(p_t)}$ is the type of tile $t'$, i.e., $q = p_{t'}$. By the tiling rule $q = h_{p_t,p_{t'}}(p_{t'}) = (\phi^{-1} \circ f_e)(p_{e^+})$, where $e$ is directed from $p_t$ to $p_{t'}$. Therefore

$$
\begin{aligned}
t' = \widehat{h}_t(q) = t' &= \big(\phi\, h_t\, (\phi)^{-1}\big)\big(\phi\, f_e(p_{e^+})\big) = \big(\phi\, h'_t\, (\phi)^{-1}\big)\big(\phi\, f_e(p_{e^+})\big) \\
&= \phi\big(\phi^k \circ f_\sigma\big) f_e(p_{e^+}) = \phi^{k+1} \circ f_{\sigma'}(p_{t'}) = \phi^{k+1} \circ f_{\sigma'}(A_{\sigma'^+}),
\end{aligned}
$$

where $\sigma' = \sigma\, e$. □

REMARK 7.5 (Theorem 7.4 may not hold when the self-similarity is not proper). For the Ammann chair self-similar GIFS $(G, F)$ in Example 4.6, $T := T(\overleftarrow{e}_1\, \overleftarrow{e}_1, \cdots)$ tiles the first quandrant of the plane. The union of $T$ and copies of $T$ obtained by reflected in the $x$ and $y$-axes and by rotation by $\pi$ about the origin tiles the plane and is self-similar but is not a self-similar GIFS-tiling.

REMARK 7.6. In [**20**], it is part of the definition of a self-similar tiling $T$ that $T$ be repetitive. That every self-similar tiling with a proper self-similarity must be repetitive follows from Theorem 7.4 and statement (7) of Theorem 5.5.

THEOREM 7.7. *For a self-simlar GIFS-tiling $T(\mathcal{F}, \overleftarrow{\theta})$, if the parameter $\overleftarrow{\theta}$ is eventually periodic, then $T(\mathcal{F}, \overleftarrow{\theta})$ is globally self-similar.*

PROOF. We must produce a similarity $\phi$, a $\phi$-tiling rule, and a set $H$ of tiling isometries such that $T_\phi = T$. Since $\overleftarrow{\theta}$ is eventually periodic, there exist $\overleftarrow{\theta_0}, \overleftarrow{\theta_1} \in \Sigma^*$ such that $\overleftarrow{\theta} = \overleftarrow{\theta}_0\, \overleftarrow{\theta}_1\, \overleftarrow{\theta}_1\, \overleftarrow{\theta}_1 \cdots$. Let $\phi = f_{\overleftarrow{\theta}_0} \circ f_{\overleftarrow{\theta}_1} \circ (f_{\overleftarrow{\theta}_0})^{-1}$, which is a similarity. For $t \in T(\mathcal{F}, \overleftarrow{\theta})$ there is a least integer $k$ such that $t \in T(\overleftarrow{\theta}, |\overleftarrow{\theta}_0| + k|\overleftarrow{\theta}_1|)$, in which case $t = f_{\overleftarrow{\theta}_0} \circ (f_{\overleftarrow{\theta}_1})^k \circ f_\sigma(A_\sigma^+)$ for an appropriate $\sigma \in \Sigma^*$ with $|\sigma| = k+1$. Let $h_t = f_{\overleftarrow{\theta}_0} \circ (f_{\overleftarrow{\theta}_1})^k \circ f_\sigma$ and $H = \{h_t : t \in T\}$ the set of tiling isometries. Since each prototile (attractor component) $p$ is tiled by $\{f_e(p_{e^+}) : e \in E_p\}$, the $\phi$-tiling rule is taken to be $T_{\phi(p)} = \{\phi \circ f_e(p_{e^+}) : e \in E_p : e \in E_p\}$. Now an arbitrary tile in the induced tiling $T_\phi$ is of the form

$$
\begin{aligned}
t = \widehat{h}_t\, \phi\, f_e(p_{e^+}) &= \phi\, h_t\, \phi^{-1}\, \phi\, f_e(p_{e^+}) = \phi\, f_{\overleftarrow{\theta}_0} (f_{\overleftarrow{\theta}_1})^k\, \sigma\, \phi\, f_e(p_{e^+}) \\
&= f_{\overleftarrow{\theta}_0} (f_{\overleftarrow{\theta}_1})^{k+1}\, \sigma\, \phi\, f_e(p_{e^+}) = f_{\overleftarrow{\theta}_0} (f_{\overleftarrow{\theta}_1})^{k+1}\, \sigma'(p_{\sigma'^+}),
\end{aligned}
$$

where $\sigma' = \sigma\varepsilon$. The last expression in the equality is again a tile in $T(\mathcal{F}, \overleftarrow{\theta})$. □

THEOREM 7.8. *For a given self-similar GIFS $\mathcal{F}$, the set of globally self-similar $\mathcal{F}$-tilings is dense in the tiling space $\mathbb{T}(\mathcal{F})$.*

PROOF. Let $T = T(\mathcal{F}, \overleftarrow{\theta}) \in \mathbb{T}(\mathcal{F})$ be given. Let $\overleftarrow{\theta}(k) \in \mathcal{P}$ be chosen to have the property that $\theta(k)$ is eventually periodic and such that the first $k$ edges of $\overleftarrow{\theta}(k)$ and $\overleftarrow{\theta}$ are the same. By the continuity of the tiling map $\mathcal{T}$, it follows from $\lim_{k\to\infty} \overleftarrow{\theta}(k) = \overleftarrow{\theta}$ that $\lim_{k\to\infty} T\big(\mathcal{F}, \overleftarrow{\theta}(k)\big) = T(\mathcal{F}, \overleftarrow{\theta}) = T$. Because $T\big(\mathcal{F}, \overleftarrow{\theta}(k)\big)$ is self-similar by Theorem 7.7, the tiling $T$ is in the closure of the set of self-similar $\mathcal{F}$-tilings. □

## 8. A GIFS tiling dynamical system

For a self-similar GIFS $\mathcal{F}$, two dynamical systems are considered in this section. One is $(\mathbb{T}, H)$, the action $H : \mathbb{T}(\mathcal{F}) \to \mathbb{T}(\mathcal{F})$ on the tiling space $\mathcal{P}$ of $\mathcal{F}$ defined as follows. For a self-similar $\mathcal{F}$-tiling $T = T(\mathcal{F}, \overleftarrow{\theta}) \in \mathbb{T}$ with parameter $\overleftarrow{\theta} = \overleftarrow{\theta}_1 \overleftarrow{\theta}_2 \cdots \in \mathcal{P}(\mathcal{F})$, let $T = T_0, T_1, T_2, \dots$ denote its hierarchy and let

$$H(T) = f_{\theta_1}(T_1)$$

for all $T \in \mathbb{T}$. In other words, $H$ is a map that takes a self-similar GIFS-tiling to a scaled down tiling at the next level in its hierarchy.

Throughout this section it will be assumed that all self-similar GIFSs are non-redundant, asymmetric and uniquely hierarchical. It will also be assumed, without loss of generality, that the origin is contained in the interior of each attractor component and, if two distinct attractor components are congruent, then they are placed so as not to coincide. In this case it can be shown, similarly to Property 6 in Theorem 5.5, that if $\mathcal{F}$ is a uniquely hierarchical, asymmetric and not redundant, self-similar GIFS and $\overleftarrow{\theta}$ and $\overleftarrow{\theta}'$ are two parameters, then $T(\mathcal{F}, \overleftarrow{\theta}) = T(\mathcal{F}, \overleftarrow{\theta}')$ if and only if $\overleftarrow{\theta} = \overleftarrow{\theta}'$.

THEOREM 8.1. *For a self-similar GIFS the two dynamical systems $(\mathcal{P}, S)$ and $(\mathbb{T}, H)$ are topologically conjugate discrete dynamical systems, the topological conjugation being the tiling map $\mathcal{T}$. As a commuting diagram we have*

$$\begin{array}{ccccc} & & S & & \\ & \mathcal{P} & \rightarrow & \mathcal{P} & \\ \mathcal{T} & \downarrow & & \downarrow & \mathcal{T} \\ & \mathbb{T} & \rightarrow & \mathbb{T} & \\ & & H & & \end{array}$$

PROOF. By the comments prior to the statement of the theorem, the tiling map $\mathbb{T}$ is bijective. A bijective continuous map on a metric space is a homeomorphism. Property 5 of Theorem 5.5 implies that the diagram commutes:

$$H\big(T(\mathcal{F}, \overleftarrow{\theta})\big) = f_{\theta_1} T_1(\mathcal{F}, \overleftarrow{\theta}) = f_{\theta_1} f_{\theta_1}^{-1} T(\mathcal{F}, S\overleftarrow{\theta}) = T(\mathcal{F}, S\overleftarrow{\theta}). \qquad \square$$

**8.1. Application 1.** Let $\mathcal{F}$ be a self-similar GIFS. Throughout this section, it is assumed that $\mathcal{F}$ is non-redundant, asymmetric and uniquely hierarchical, or at least that the tiling map $\mathcal{T}$ induces a topological conjugacy as in the diagram above.

Let $p_k := p_k(\mathcal{F})$ denote the number of $\mathcal{F}$-tilings $T$ such that $H^k(T) = T$. In other words, $p_k$ is the number of $\mathcal{F}$-tilings such that its $n^{th}$-level hierarchical tiling $T_k$, scaled down by a factor $(f_{\overleftarrow{\theta}|k})^{-1}$, is equal to the original tiling $T$. In this sense, $p_k$ counts the number of $\mathcal{F}$-tilings whose hierarchy $T = T_0, T_1, T_2 \dots$ cycles with period $k$.

THEOREM 8.2. *For a self-similar GIFS $\mathcal{F}$ as above, a generating function for the sequence $\{p_k\}_{n=1}^{\infty}$ is given by*

$$\sum_{k=1}^{\infty} \frac{p_k}{k}\, x^k = \log\Big(\frac{1}{\det(I - xM)}\Big),$$

*where $M$ is the adjacency matrix of the digraph of $\mathcal{F}$.*

PROOF. An element in the parameter space $\mathcal{P}$ of $\mathcal{F}$ can be viewed as a one-sided word in the alphabet $V$, where $V$ is the set of vertices of the digraph $\overleftarrow{G}$. The parameter space $\mathcal{P}$ is clearly shift invariant. In the terminology of symbolic dynamics, the dynamical system $(\mathcal{P}, S)$ is a 1-step shift of finite type. This means that there is a finite set $W$ of ordered pairs of elements of $V$, i.e. a set of edges in the complete digraph on $V$, such that $\mathcal{P}$ consists of all words (paths in $\overleftarrow{G}$) that do not contain an ordered pair (edge) in $W$.

The Artin-Mazur zeta function of a dynamical system $(X, g)$ is defined by

$$\zeta(x) = \exp\Big(\sum_{k=1}^{\infty} \frac{q_k}{k}\, x^k\Big),$$

where $q_k$ is the number of points of period $n$ of $X$ under the action of $g$. (Note that a point of period $n$ is also a point of period any multiple of $n$.) For our shift of finite type $(\mathcal{P}, S)$, the number $q_k$ is thus the number of parameters $\overleftarrow{\theta}$ such that $S^k \overleftarrow{\theta} = \overleftarrow{\theta}$, equivalently the number of closed paths in $\overleftarrow{G}$ of length $k$. The zeta function is a well-known invariant of topological conjugacy and, for a shift of finite type, can be computed by the Bowen-Lanford formula:

$$\zeta(x) = \frac{1}{\det(I - xM^t)} = \frac{1}{\det(I - xM)},$$

where $M$ is the adjacency matrix of the digraph $G$ and its transpose $M^t$ is the adjacency matrix of the digraph $\overleftarrow{G}$. For our shift of finite type $(\mathbb{T}, H)$, the number $q_k$ is thus the number of tilings $T \in \mathbb{T}$ such that $H^k(T) = T$, equivalently $q_k = p_k$.

According to Theorem 8.1, the two dynamical systems $(\mathbb{T}, H)$ and $(\mathcal{P}, S)$ are topologically conjugate and therefore have the same zeta function. Hence

$$\exp\Big(\sum_{k=1}^{\infty} \frac{p_k}{k}\, x^k\Big) = \frac{1}{\det(I - xM)}$$

and Theorem 8.2 follows. □

We note that the Artin-Mazur zeta invariant used in the prove above is also investigated by Anderson and Putnam [**2**] for substitution tiling spaces.

EXAMPLE 8.3. A simple computer calculation gives the following series for the Ammann chair tilings of Example 4.6:

$$\sum_{k=1}^{\infty} \frac{p_k}{n}\, x^k = x + \frac{3x^2}{2} + \frac{4x^3}{3} + \frac{7x^4}{4} + \frac{11x^5}{5} + \frac{18x^6}{6} + \frac{29x^7}{7} + \frac{47x^8}{8} + \frac{76x^9}{9} + \cdots .$$

For example, there are 4 Ammann chair tilings for which the hierarchy cycles with period 3. Referring to the graph in the right panel of Figure 5, these correspond in $(\mathcal{P}, S)$, via topological conjugacy, to the 4 parameters $\overline{111}, \overline{132}, \overline{213}, \overline{321}$, where the individual digits are shorthand for the edges in $\overleftarrow{G}$ with those function subscripts, and the bar over the numbers means that the three numbers repeat.

### 8.2. Application 2.

LEMMA 8.4. *Let $\mathcal{F}$ be a self-similar GIFS, and let $\overleftarrow{\theta}, \overleftarrow{\theta}' \in \mathcal{P}$. Then $T(\overleftarrow{\theta}) = T(\overleftarrow{\theta}')$ if and only if $T(\theta, k) = T(\theta', k)$ for $k = 0, 1, 2, \dots$.*

PROOF. Since $T(\overleftarrow{\theta})$ is the nested union of the $T(\theta, k)$, clearly $T(\theta, k) = T(\theta', k)$ for $k = 0, 1, 2, \dots$ implies that $T(\overleftarrow{\theta}) = T(\overleftarrow{\theta}')$.

In the other direction, it follows from the proof of of Property 5 of Theorem 5.5 that the union of the tiles in $T(\theta, k)$ is itself a tile $t_k \in T_k$; similarly the union of the tiles in $T(\theta', k)$ is a tile $t'_k \in T'_k$. Since $0 \in t_k \cap t'_k$ and $T := T(\overleftarrow{\theta}) = T(\overleftarrow{\theta}')$ has a unique hierarchy, it follows that $T(\theta, k) = T(\theta', k)$. □

For a self-similar GIFS $\mathcal{F}$, let $N_k := N_k(\mathcal{F})$ denote the number of distinct (pairwise unequal) patches $T(\overleftarrow{\theta}, k)$ over all $\overleftarrow{\theta} \in \mathcal{P}$. In view of Lemma 8.4, if two $\mathcal{F}$-tilings $T(\overleftarrow{\theta})$ and $T(\overleftarrow{\theta}')$ are not equal, then, for some $k$, the patches $T(\overleftarrow{\theta}, k)$ and $T(\overleftarrow{\theta}', k)$ will not be equal. Therefore, the growth of the sequence $\{N_k\}_{k=1}^{\infty}$ is a measure of how fast $\mathcal{F}$-tilings can be distinguished by looking at these increasingly large finite patches.

THEOREM 8.5. *For a self-similar GIFS $\mathcal{F}$, let $\rho := \rho(\mathcal{F})$ denote the Perron-Frobenius eigenvalue of the adjacency matrix of the digraph of $\mathcal{F}$. Then asymptotically*

$$N_k \simeq e^{k \log \rho}, \qquad i.e.,$$

$$\lim_{k \to \infty} \sqrt[k]{N_k} = \rho.$$

PROOF. Given a self-similar GIFS $\mathcal{F} = (G, F)$, the topological entropy of the shift of finite type $(\mathcal{P}, S)$ is defined by

$$h(\mathcal{P}) = \lim_{k \to \infty} \frac{1}{k} \log \widehat{N}_k,$$

where $\widehat{N}_k$ is the number of paths of length $k$ in the digraph $\overleftarrow{G}$, which equals the number of paths of length $k$ in the digraph $G$. In the same way that $T(\overleftarrow{\theta}) = T(\overleftarrow{\theta}')$ if and only if $\overleftarrow{\theta} = \overleftarrow{\theta}'$, we have $T(\overleftarrow{\theta}, k) = T(\overleftarrow{\theta}', k)$ if and only if $\overleftarrow{\theta}|k = \overleftarrow{\theta}'|k$. Therefore $N_k = \widehat{N}_k$ for all $k$.

For $(\mathcal{P}, S)$ (and more generally for any shift of finite type) it is well known that $h(\mathcal{P}) = \rho(\mathcal{F})$, the Perron-Frobenius eigenvalue of $M$. Therefore

$$\lim_{k \to \infty} \frac{1}{k} \log N_k = \lim_{k \to \infty} \frac{1}{k} \log \widehat{N}_k = \rho,$$

which is equivalent to $\lim_{k \to \infty} \sqrt[k]{N_k} = \rho$. □

EXAMPLE 8.6. Applying Theorem 8.5 to the Ammann chair self-similar GIFS $\mathcal{F}$ of Example 4.6 yields

$$N_k \simeq \tau^k \approx e^{.4812\, k},$$

where $\tau$ is the golden ratio.

## 9. The existence of self-similar GIFSs

According to Theorem 9.2 below, self-similar GIFSs are easy to construct in dimension 1. Polygonal rep-sets in dimension 2 may lead to self-similar GIFSs. These have appeared in both recreational websites and mathematical journals, mostly discovered in an ad hoc manner, some very clever. They include the Ammann chair tile in Example 4.6, Robinson's triangle variant of the Penrose tiles, and the pinwheel tile (the tile due to J. Conway, the tiling due to C. Radin). According to Theorem 9.3 below, for every dimension $d$ and for every Perron number $\rho$, there exists a self-similar GIFS on $\mathbb{R}^d$ whose scaling ratio is $1/\rho$; the attractor components are boxes. All of the above examples give self-similar GIFS tilings by the method in Definitions 4.1 and 4.3. The digit IFSs referred to in the introduction [**41**] are self-similar IFSs and produce lattice tilings by translates of a single prototile.

Non-polyhedral, aperiodic tilings in dimension $d \geq 2$ with self-replicating properties are hard to come by. The difficulty in producing such a self-similar GIFS is that the attractor is required to be non-overlapping and the attractor components to have non-empty interior. Rauzy type fractals, often called central tiles, do admit non-period tilings. They are usually obtained from symbolic dynamics [**9**] or from numeration systems ($\beta$-expansions) [**1**] and are also related to model sets obtained by the cut-and-project method [**28**]. Although a general discussion of Rauzy tilings does not fall within the scope of this paper, we give a shot exposition in Section 9.1 of the elegant GIFS approach due to Rao, Wen and Yang [**31**].

A real number $\rho > 1$ is a *Perron number* if it is a real algebraic integer such that the moduli of all other Galois conjugates (roots of the minimal polynomial of $\rho$) are less than $|\rho|$. Call a real algebraic integer $\rho > 1$ a *weak-Perron number* if the moduli of all of its Galois conjugates are less than or equal to $|\rho|$. A real number $\rho > 1$ is a *Pisot number* if it is a real algebraic integer such that the moduli of all of its other Galois conjugates are less than 1. It is a *unit pisot number* if, in addition, the constant term in its minimal polynomial is $\pm 1$.

PROPOSITION 9.1. *If $\mathcal{F}$ is a self-similar GIFS on $\mathbb{R}^d$ with scaling ratio $\lambda$, then $1/\lambda$ is a weak-Perron number.*

PROOF. Let $\rho = 1/\lambda$. By condition (2) in Statement 3 of Theorem 5.5 we know that $\rho^d$ is the Perron-Frobenius eigenvalue of $M(G)$. If the characteristic polynomial of $M(G)$ is $p(x)$, then $\rho$ is a root of $\widehat{p}(z) = p(z^d)$, all of whose roots are less than or equal to $\rho$. □

Note that there exist no self-similar GIFS whose digraph is a directed cycle.

THEOREM 9.2. (1) *For any $0 < \lambda < 1$ such that $1/\lambda$ is a Perron number, there exists a primitive self-similar GIFS on $\mathbb{R}$ whose scaling ratio is $\lambda$.*
(2) *For every strongly connected digraph $G = (V, E)$, not a directed cycle, there exists a primitive self-similar GIFS $\mathcal{F} = (G, F)$ on $\mathbb{R}$ such that $f_e(x) = (1/\rho)(x + d_e)$ for all $e \in E$, where $\rho$ is the Perron-Frobenius eigenvalue of the adjacency matrix $M(G)$ of $G$ and $d_e \in \mathbb{Q}(\rho)$.*

PROOF. Concerning statement (1), let $0 < \lambda < 1$ and let $\rho = 1/\lambda$. The result [**23**, Theorem 1] states that if $\rho$ is a Perron number, then there is a primitive non-negative integral matrix $M = (m_{i,j})$, $i, j \in \{1, 2, \dots, n\}$ whose spectral radius is $\rho$. A self-similar GIFS on $\mathbb{R}$, whose adjacency matrix is $M$, can be obtained as follows. Let the vertex set of $G$ be $V = \{1, 2, \dots, n\}$. For each pair $(i, j)$ of vertices, add $m_{i,j}$ edges from vertex $i$ to vertex $j$, and label edge $e$ with a function of the form $f_e(x) = (1/\rho)(x + d_e)$, where $d_e$ is defined as follows. Let $(x_1, x_2, \ldots, x_n)$ be a positive right eigenvector corresponding to the eigenvalue $\rho$, which exists by the Perron-Frobenius theorem. The set $\{x_1, x_2, \ldots, x_n\}$ of coordinates is a solution to the linear system

$$x_j = \sum_{i=1}^{n} \lambda\, m_{j,i}\, x_i, \qquad j = 1, 2, \ldots, n,$$

and hence can be chosen to lie in $\mathbb{Q}(\rho)$. Consider the intervals $I_j = [0, x_j]$, $j = 1, 2, \ldots, n$, on the real line. From the system of linear equations above, the interval $I_j$ can be tiled (often in many different ways) by translated copies of the intervals $[0, \lambda\, x_i]$, $i = 1, 2, \ldots, n$. The $d_e$ are the translation distances multiplied by $\rho$, which are sums of the $x_i$s. Therefore $d_e \in \mathbb{Q}(\rho)$ for all $e \in E$.

The adjacency matrix $M(G)$ determines the digraph $G$, so, as long as the Perron-Frobenius eigenvalue of $M(G)$ does not equal 1, the proof of statement (2) is as in the paragraph above. Since $M(G)$ is a non-negative, integral matrix, the Perron-Frobenius eigenvalue equals 1 if and only if $M$ is orthogonal which can occur if and only if $M(G)$ is a permutation matrix if and only if $G$ is a directed cycle. $\square$

THEOREM 9.3. *For any $0 < \lambda < 1$ such that $1/\lambda$ is a Perron number, there exists a primitive self-similar GIFS on $\mathbb{R}^d$ whose scaling ratio is $\lambda$. The components of the attractor are boxes.*

PROOF. The case $d = 1$ is Theorem 9.2. So let $\mathcal{F} = (G, F)$ be such a 1-dimensional self-similar GIFS with scaling ratio $\lambda = 1/\rho$ and with attractor components $I_j = [0, x_j]$, $j = 1, 2, \ldots, n$, intervals on the real line. Denote the vertex set of $G$ by $V = \{1, 2, \ldots n\}$ and the edge set by $E$.

The result is extended to arbitrary dimension by a product construction. In dimension $d$, let $\mathcal{F}_d = (G_d, F_d)$ be the GIFS, where $G_d = (V_d, E_d)$ is the digraph with vertex set $V_d = V^d$ and edge set $E_d = E^d$. The edge $\mathbf{e} := (e_1, e_2, \ldots, e_d) \in E_d$ joins vertex $\mathbf{i} := (i_1, i_2, \ldots, i_d)$ to vertex $\mathbf{j} := (j_1, j_2, \ldots, j_d)$ where $e_k = (i_k, j_k)$ for $k = 1, 2, \ldots d$. Label edge $\mathbf{e}$ with the function

$$f_{\mathbf{e}}(y_1, y_2, \ldots y_d) = \big(f_{e_1}(y_1), f_{e_2}(y_2), \ldots, f_{e_d}(y_d)\big)$$

to obtain $F_d$.

It is routine to check that $\mathcal{F}$ and $\mathcal{F}_d$ have the same scaling ratio. The GIFS $\mathcal{F}_d$ is primitive for the following reason. Since $G$ is primitive, there exists a positive integer $m$ such that, for any two vertices of $G$ there is a directed path of length $m$ from one to the other. Let $\mathbf{i} := (i_1, i_2, \ldots, i_d)$ and $\mathbf{j} := (j_1, j_2, \ldots, j_d)$ be two arbitrary vertices of $G_d$. For $k = 1, 2, \ldots, d$, let $e_{k,1}\, e_{k,2} \cdots e_{k,m}$ be a path of length $m$ in $G$ from vertex $i_k$ to vertex $j_k$. For $q = 1, 2, \ldots, m$, let $\mathbf{e_q} = (e_{1,q}, e_{2,q}, \ldots, e_{d,q})$. Then $\mathbf{e_1}\, \mathbf{e_2} \cdots \mathbf{e_m}$ is a path of length $m$ from vertex $\mathbf{i}$ to vertex $\mathbf{j}$ in $G_d$.

For each $\mathbf{i} := (i_1, i_2, \ldots, i_d) \in V_d$, let $B_{\mathbf{i}} = I_{i_1} \times I_{i_2} \times \cdots \times I_{i_d}$, which is a box in $\mathbb{R}^d$. We leave it as an exercise to verify that

$$B_{\mathbf{i}} = \bigcup_{\mathbf{e} \in (E_d)_{\mathbf{i}}} f_{\mathbf{e}}(B_{\mathbf{e}^+})$$

for all $\mathbf{i} \in V_d$ and that the union is non-overlapping. This shows that $\{B_{\mathbf{i}} : \mathbf{i} \in V_d\}$ is the set of attractor components of $\mathcal{F}_d$. $\square$

Theorem 9.2 suggests the following questions.

QUESTION 9.4. For which strongly connected digraphs $G$ and integers $d \geq 2$, does there exist a self-similar GIFS on $\mathbb{R}^d$ whose digraph is $G$?

### 9.1. The dual of a self-similar GIFS on $\mathbb{R}$.

DEFINITION 9.5 (Dual GIFS). A slightly more general definition of the dual appears in [**31**]. Let $G = (V, E)$ be a strongly connected digraph of order $d+1$, $d \geq 2$, not a cycle, and let $\rho$ be the Perron-Frobenius eigenvalue of the adjacency matrix of $G$. According to Theorem 9.2, there is a self-similar GIFS $\mathcal{F} = (G, F)$ on $\mathbb{R}$ (often many) such that

$$f_e(x) = \frac{1}{\rho}\,(x + d_e),$$

where $d_e \in \mathbb{Q}(\rho)$ for all $e \in E$. Such 1-dimensional self-similar GIFSs can be obtained explicitly and easily using the method provided in the proof of Theorem 9.2. Call any such self-similar GIFS a 1-*dimensional self-similar GIFS* or simply a 1-*dim self-similar GIFS*.

Let $\mathcal{F} = (G, F)$ be a 1-dim self-similar GIFS. If $\rho$ is a Pisot unit, then the *dual GIFS* $\mathcal{F}^* = (\overleftarrow{G}, F^*)$ of $\mathcal{F}$ is defined as follows. The reverse digraph $\overleftarrow{G}$ is as defined in Section 4. Let $\rho = \rho_0, \rho_1, \dots, \rho_d$ be the Galois conjugates of $\rho$ ordered as follows:

$$\rho > 1 > |\rho_1| \geq |\rho_2| \geq \cdots \geq |\rho_d|,$$

where complex conjugates appear consecutively. Let

$$B' = diag(\rho_1, \rho_2, \dots, \rho_d),$$

be the diagonal matrix and replace each pair $z, \overline{z}$ of complex conjugates in $B'$ by the $2 \times 2$ real block $\begin{pmatrix} Re\, z & -Im\, z \\ Im\, z & Re\, z \end{pmatrix}$ to obtain the matrix $B$. Since $1, \rho, \rho^2, \dots, \rho^d$ is a basis for $\mathbb{Q}(\rho)$, we have $x = \sum_{i=0}^d x_i \rho^i$ for any $x \in \mathbb{Q}(\rho)$, where the $x_i \in \mathbb{Q}$. The *dual* of $x$ is defined as

$$x^* = \sum_{i=0}^{d} x_i (\rho_1^i, \dots, \rho_d^i)^t \in \mathbb{C}^d,$$

where pair $z, \overline{z}$ of complex conjugates is replaced with $Re\, z, Im\, z$. Equivalently, the star operator is the unique linear map $* : \mathbb{Q}(\rho) \to \mathbb{R}^d$ such that $(\rho\, x)^* = Bx^*$ for all $x \in \mathbb{Q}(\rho)$. Now set $F^* = \{f_{\overleftarrow{e}}(x) = Bx + d_e^* : e \in E\}$. The *dual* of $\mathcal{F}$ is $\mathcal{F}^* := (\overleftarrow{G}, F^*)$, which is a GIFS on $\mathbb{R}^d$.

The following theorem is essentially [**31**, Theorem 1.2].

THEOREM 9.6. *The dual of a 1-dim self-similar GIFS $\mathcal{F}$ for which $1/\lambda(\mathcal{F})$ is a Pisot unit is also a self-similar GIFS.*

PROOF SKETCH. A GIFS satisfies the OSC if there exists open sets $U_1, \dots, U_n$ such that

$$\bigcup_{e \in E_i} f_e(U_{e+}) \subset U_i$$

for $i = 1, 2, \dots, n$, where $n$ is the order of the digraph of the GIFS. Three results lead to the proof of Theorem 9.6.

First, necessary and sufficient conditions for a GIFS to satisfy the OSC, in terms of certain "digits" is given in [**24**].

Second, a result in [**31**], using the above necessary and sufficient conditions, immediately implies that if 1-dim self-similar GIFS $\mathcal{F}$ satisfies (1) the open set condition (OSC) and (2) that $1/\lambda(\mathcal{F})$ is a Pisot number, then $\mathcal{F}^*$ also satisfies the OSC. Concerning condition (1), the attractor components of a 1-dim self-similar GIFS are closed intervals; hence the OSC is satisfied by taking the $U_i$ as the set of corresponding open intervals. Condition (2) is, by assumption, satisfied. Therefore $\mathcal{F}^*$ satisfies the OSC.

Third, a result in [**22**] immediately implies that if $\mathcal{F}^* = (\overleftarrow{G}, F^*)$ is such that (1) the functions in $F^*$ are affine of the form $f_e(x) = B(x) + d_e$ for all $e \in E$, (2) $\mathcal{F}^*$ satisfies the OSC, and (3) $1/\rho^* = |\det(B)|$, where $\rho^*$ is the Perron-Frobenius eigenvalue of $M(\overleftarrow{G})$, then the attractor components of $\mathcal{F}^*$ have nonempty interior. Condition (1) is clearly true, and condition (2) follows immediately from the previous result from [**24**]. Concerning condition (3), we have

$$|\det(B)| = |\rho_1 \cdot \rho_2 \cdots \rho_d| = \frac{1}{\rho} = \frac{1}{\rho^*}.$$

The second equality follows from the fact that $\rho$ is a unit and the last equality from the fact that $M(\overleftarrow{G})$ is the transpose of $M(G)$.

Therefore the attractor components of $\mathcal{F}^*$ have nonempty interior. Moreover, because $\lambda(\mathcal{F}^*) = \sqrt{|\det(B)|} = 1/\sqrt{\rho^*}$, Lemma 5.4. implies that $\mathcal{F}^*$ is a self-similar GIFS. □

EXAMPLE 9.7 (Rauzy tilings from a dual self-similar GIFS). Duality is used here to obtain fractal, non-periodic self-similar GIFS-tilings of $\mathbb{R}^2$. We start with an initial 1-dim self-similar GIFS $\mathcal{F} = (G, F)$ on $\mathbb{R}$, where $G$ is an order 3 strongly connected digraph and the Perron-Frobenius eigenvalue $\rho$ of the adjacency matrix $M(G)$ is a Pisot unit. It is not hard to find many such 1-dim self-similar GIFSs as follows. Regarding the 9 integer entries of $M(G)$ as variables, it is easy to solve for values of these variables so that (1) $\det(M(G)) = \pm 1$, (2) the degree 3 characteristic polynomial of $M(G)$ has a pair of complex roots, and (3) $G$ is strongly connected. That $\rho > 1$ together with condition (2) immediately implies that $\rho$ is a Pisot number. Condition (1) further implies that $\rho$ is a unit. Because $M(G)$ is a $3 \times 3$ matrix, the dual $\mathcal{F}^*$ is a self-similar GIFS on $\mathbb{R}^2$. With the digraph $G$ in hand, we use Theorem 9.2 (and its proof) to obtain possibilities for $F$ and hence possibilities for the 1-dim self-similar GIFS $\mathcal{F}$. We then use Theorem 9.6 to obtain $F^*$ and hence $\mathcal{F}^* = (\overleftarrow{G}, F^*)$. The tiling method of this paper is then used to produce the $\mathcal{F}^*$-tiling.

For each of the following examples we provide an adjacency matrix $M = M(G)$, which determines the digraph $G$, the characteristic polynomial $p(x)$ of $M$, whose Perron-Frobenius eigenvalue $\rho$ is a Pisot unit and equals $1/\lambda(\mathcal{F})$, and the functions in $F$.

(1) The original Rauzy fractal is the non-overlapping union of three smaller similar copies of itself. A tiling based on the Rauzy fractal appears on the

right in Figure 3. This self-similar GIFS-tiling can be constructed from the dual of this 1-dim self-similar GIFS:

$$M = \begin{pmatrix} 1 & 1 & 1 \\ 1 & 0 & 0 \\ 0 & 1 & 0 \end{pmatrix} \qquad p(x) = x^3 - x^2 - x - 1 \qquad \begin{array}{l} f_{2,1}(x) = f_{31}(x) = f_{1,1}(x) = (1/\rho)x \\ f_{1,2}(x) = f_{2,3}(x) = (x-1)/\rho, \end{array}$$

where $f_{i,j}$ is the function in $F$ that is the label of edge $(i,j)$.

(2) A tiling from the the dual of the data below appears on the left in Figure 7.

$$M = \begin{pmatrix} 1 & 0 & 1 \\ 1 & 0 & 0 \\ 0 & 1 & 0 \end{pmatrix} \qquad p(x) = x^3 - x^2 - 1 \qquad \begin{array}{l} f_{2,1}(x) = f_{3,2}(x) = f_{1,1}(x) = (1/\rho)x \\ f_{1,3}(x) = (x-1)/\rho \end{array}$$

FIGURE 7. Tilings from duals given in Example 9.7.

(3) The tilings on the left and right, respectively, in Figure 4 are from the duals of these 1-dim self-similar GIFSs:

$$M = \begin{pmatrix} 1 & 1 & 1 \\ 1 & 0 & 1 \\ 1 & 0 & 0 \end{pmatrix} \qquad p(x) = x^3 - x^2 - 2x - 1 \qquad \begin{array}{l} f_{3,1}(x) = f_{2,1}(x) = f_{1,1}(x) = (1/\rho)x \\ f_{1,3}(x) = f_{2,3}(x) = (x+\rho)/\rho \\ f_{1,2}(x) = (x+\rho+1)/\rho \end{array}$$

(4)

$$M = \begin{pmatrix} 1 & 1 & 1 \\ 1 & 1 & 0 \\ 0 & 1 & 0 \end{pmatrix} \qquad p(x) = x^3 - 2x^2 - x - 1 \qquad \begin{array}{l} f_{2,3}(x) = f_{3,1}(x) = f_{1,1}(x) = (1/\rho)x \\ f_{2,1}(x) = (x-1)/\rho \\ f_{1,2}(x) = (x+\rho)/\rho \\ f_{2,2}(x) = (x+\rho+1)/\rho \end{array}$$

(5) The tiling on the right in Figure 7 is from the dual of the 1-dim self-similar GIFS below. The three prototiles also appear in Figure 2.

$$M = \begin{pmatrix} 1 & 1 & 1 \\ 1 & 1 & 0 \\ 0 & 1 & 1 \end{pmatrix} \qquad p(x) = x^3 - 3x^2 + 2x - 1 \qquad \begin{array}{l} f_{2,3}(x) = f_{3,3}(x) = f_{1,1}(x) = (1/\rho)x \\ f_{3,1}(x) = f_{2,1}(x) = (x-1)\rho \\ f_{2,2}(x) = (x+\rho)/\rho \\ f_{1,2}(x) = (x+\rho-1)/\rho \end{array}$$

## 10. Conclusion

Tilings of Euclidean space exhibiting a self-similar property appear in the work of Kepler and recur in a myriad guises throughout the second half of the $20^{th}$ century and up to the present. There is a vast literature on the subject, both academic and recreational. The notion of a rep-tiling, a term coined by Golomb, appeared in the early 1960s; the Penrose tilings in the 1970's; the Thurston-Kenyon tilings and Rauzy tilings in the 1980s; digit and crystallographic tilings in the 1990s. Since then connections have been established to many other fields, for example Markov partitions, numeration systems, $C^*$-algebras, and dynamical systems.

The concept of an iterated function system (IFS), a basic method for the construction of fractal sets, was developed by Hutchinson in 1981. To allow for tiles of multiple shapes, the IFS concept was extended to a graph directed IFS (GIFS) by Maulden and Williams in 1988.

This paper provides a unifying overview of the fascinating theory of self-similar tilings from the GIFS point of view. The functions in the GIFS are similarity transformations of Euclidean space; the prototile set of the tilings are the attractor components of the GIFS; and, for a given GIFS $\mathcal{F}$, the set of self-similar GIFS tilings themselves form a metric space $\mathbb{T}(\mathcal{F})$, called the tiling space, that is the image under a tiling map $\mathcal{T}(\mathcal{F})$ of a parameter space $\mathcal{P}(\mathcal{F})$ consisting of infinite paths in the graph; thus $\mathcal{T} : \mathcal{P} \to \mathbb{T}$. In our approach we make only two simple assumptions on the GIFS: (1) that the attractor components have nonempty interior and do not overlap - an assumption necesary for the resulting tilings to fill Euclidean space without overlap, and (2) that the scaling ratio of the similarity transformations in the GIFS are all equal. Despite its simplicity, we prove that essentially any tiling by a finite set of prototiles that may reasonably be considered self-similar must be a self-similar GIFS tiling of the type considered in this paper. In particular, this includes rep-tilings, digit tilings, Penrose tilings, Thurston-Kenyon tilings, and Rauzy tilings.

The basic self-similar GIFS tiling theory includes results (with GIFS-based proofs) on familiar notions from tiling theory like the scaling ratio, tile frequencies, hierarchical structure, periodicity and quasiperiodicity. Moreover, for a given GIFS $\mathcal{F}$, the space $\mathbb{T}(\mathcal{F})$ of GIFS tilings is proved to be the closure, with respect to the tiling space metric, of the Thurston-Kenyon tilings, suitable generalized. Results are obtained on a tiling space dynamical system that is topologically conjugate to the action of the the shift map on the parameter space. The discovery of GIFSs satisfying property (1) in the definition (above) of a self-similar GIFS - allowing for tilings that fill Euclidean space without overlap - has been somewhat sporadic. Several recently discovered families are discussed in the last section of the paper. Finding a systematic approach to this issue would definitely be a worthy avenue of research.

## Acknowledgment

I thank Professor Chun-Kit Lai for valuable conversations about topics covered in this paper during my visit to San Francisco State University.

## References


[1] S. Akiyama, *Self affine tiling and Pisot numeration system*, Number theory and its applications (Kyoto, 1997), Dev. Math., vol. 2, Kluwer Acad. Publ., Dordrecht, 1999, pp. 7–17. MR1738803

[2] J. E. Anderson and I. F. Putnam, *Topological invariants for substitution tilings and their associated $C^*$-algebras*, Ergodic Theory Dynam. Systems **18** (1998), no. 3, 509–537, DOI 10.1017/S0143385798100457. MR1631708

[3] M. Baake and B. Sing, *Kolakoski-$(3,1)$ is a (deformed) model set*, Canad. Math. Bull. **47** (2004), no. 2, 168–190, DOI 10.4153/CMB-2004-018-6. MR2059413

[4] C. Bandt, *Self-similar sets. V. Integer matrices and fractal tilings of* $\mathbf{R}^n$, Proc. Amer. Math. Soc. **112** (1991), no. 2, 549–562, DOI 10.2307/2048752. MR1036982

[5] M. Barnsley and A. Vince, *Tilings from graph directed iterated function systems*, Geom. Dedicata **212** (2021), 299–324, DOI 10.1007/s10711-020-00560-4. MR4251675

[6] T. Bedford, *Generating special Markov partitions for hyperbolic toral automorphisms using fractals*, Ergodic Theory Dynam. Systems **6** (1986), no. 3, 325–333, DOI 10.1017/S0143385700003527. MR863197

[7] R. Berger, *The undecidability of the domino problem*, Mem. Amer. Math. Soc. **66** (1966), 72. MR216954

[8] V. Berthé and A. Siegel, *Tilings associated with beta-numeration and substitutions*, Integers **5** (2005), no. 3, A2, 46. MR2191748

[9] V. Berthé, A. Siegel, and J. Thuswaldner, *Substitutions, Rauzy fractals and tilings*, Combinatorics, automata and number theory, Encyclopedia Math. Appl., vol. 135, Cambridge Univ. Press, Cambridge, 2010, pp. 248–323. MR2759108

[10] O. Bratteli, *Inductive limits of finite dimensional $C^*$-algebras*, Trans. Amer. Math. Soc. **171** (1972), 195–234, DOI 10.2307/1996380. MR312282

[11] N. P. Frank, *A primer of substitution tilings of the Euclidean plane*, Expo. Math. **26** (2008), no. 4, 295–326, DOI 10.1016/j.exmath.2008.02.001. MR2462439

[12] M. Gardner, On reptiles, polygons that can make larger and smaller copies of themselves, *Scientific American, Mathematical Games*, 208 1963.

[13] G. Gelbrich, *Crystallographic reptiles*, Geom. Dedicata **51** (1994), no. 3, 235–256, DOI 10.1007/BF01263995. MR1293800

[14] S. W. Golomb, Replicating figures in the plane, *Math.Gaz.*, 48 (1964), 403–412.

[15] K. Gröchenig and A. Haas, *Self-similar lattice tilings*, J. Fourier Anal. Appl. **1** (1994), no. 2, 131–170, DOI 10.1007/s00041-001-4007-6. MR1348740

[16] B. Grünbaum and G. C. Shephard, *Tilings and patterns*, W. H. Freeman and Company, New York, 1987. MR857454

[17] J. E. Hutchinson, *Fractals and self-similarity*, Indiana Univ. Math. J. **30** (1981), no. 5, 713–747, DOI 10.1512/iumj.1981.30.30055. MR625600

[18] S. Kakutani, *A problem of equidistribution on the unit interval* $[0,1]$, Measure theory (Proc. Conf., Oberwolfach, 1975), Lecture Notes in Math., Vol. 541, Springer, Berlin-New York, 1976, pp. 369–375. MR457678

[19] J. Kellendonk, *Noncommutative geometry of tilings and gap labelling*, Rev. Math. Phys. **7** (1995), no. 7, 1133–1180, DOI 10.1142/S0129055X95000426. MR1359991

[20] R. Kenyon, *The construction of self-similar tilings*, Geom. Funct. Anal. **6** (1996), no. 3, 471–488, DOI 10.1007/BF02249260. MR1392326

[21] J. C. Lagarias and Y. Wang, *Self-affine tiles in* $\mathbf{R}^n$, Adv. Math. **121** (1996), no. 1, 21–49, DOI 10.1006/aima.1996.0045. MR1399601

[22] J. C. Lagarias and Y. Wang, *Integral self-affine tiles in* $\mathbf{R}^n$. *I. Standard and nonstandard digit sets*, J. London Math. Soc. (2) **54** (1996), no. 1, 161–179, DOI 10.1112/jlms/54.1.161. MR1395075

[23] D. A. Lind, *The entropies of topological Markov shifts and a related class of algebraic integers*, Ergodic Theory Dynam. Systems **4** (1984), no. 2, 283–300, DOI 10.1017/S0143385700002443. MR766106

[24] L. Jun and Y.-M. Yang, *On single-matrix graph-directed iterated function systems*, J. Math. Anal. Appl. **372** (2010), no. 1, 8–18, DOI 10.1016/j.jmaa.2010.07.001. MR2672506

[25] B. B. Mandelbrot, *The fractal geometry of nature*, Schriftenreihe für den Referenten. [Series for the Referee], W. H. Freeman and Co., San Francisco, CA, 1982. MR665254

[26] R. D. Mauldin and S. C. Williams, *Hausdorff dimension in graph directed constructions*, Trans. Amer. Math. Soc. **309** (1988), no. 2, 811–829, DOI 10.2307/2000940. MR961615
[27] C. D. Meyer, *Matrix analysis and applied linear algebra*, Society for Industrial and Applied Mathematics (SIAM), Philadelphia, PA, [2023] ©2023. Second edition [of 1777382]. MR4625860
[28] L. V. Moody, Model sets: a survey, in *From Quasicrystals to More Complex Systems*, Centre de Physique des Houches, (2000) 145-166.
[29] R. Penrose, *Pentaplexity: a class of nonperiodic tilings of the plane*, Math. Intelligencer **2** (1979/80), no. 1, 32–37, DOI 10.1007/BF03024384. MR558670
[30] C. Radin, *Miles of tiles*, Student Mathematical Library, vol. 1, American Mathematical Society, Providence, RI, 1999, DOI 10.1090/stml/001. MR1707270
[31] H. Rao, Z.-Y. Wen, and Y.-M. Yang, *Dual systems of algebraic iterated function systems*, Adv. Math. **253** (2014), 63–85, DOI 10.1016/j.aim.2013.11.010. MR3148546
[32] G. Rauzy, *Nombres algébriques et substitutions* (French, with English summary), Bull. Soc. Math. France **110** (1982), no. 2, 147–178. MR667748
[33] L. Sadun, *Topology of tiling spaces*, University Lecture Series, vol. 46, American Mathematical Society, Providence, RI, 2008, DOI 10.1090/ulect/046. MR2446623
[34] L. Sadun, *Some generalizations of the pinwheel tiling*, Discrete Comput. Geom. **20** (1998), no. 1, 79–110, DOI 10.1007/PL00009379. MR1626703
[35] J. H. Schmerl, *Dividing a polygon into two similar polygons*, Discrete Math. **311** (2011), no. 4, 220–231, DOI 10.1016/j.disc.2010.10.021. MR2739907
[36] A. Siegel and J. M. Thuswaldner, *Topological properties of Rauzy fractals* (English, with English and French summaries), Mém. Soc. Math. Fr. (N.S.) **118** (2009), 140. MR2721985
[37] B. Solomyak, *Dynamics of self-similar tilings*, Ergodic Theory Dynam. Systems **17** (1997), no. 3, 695–738, DOI 10.1017/S0143385797084988. MR1452190
[38] R. S. Strichartz, *Wavelets and self-affine tilings*, Constr. Approx. **9** (1993), no. 2-3, 327–346, DOI 10.1007/BF01198010. MR1215776
[39] W. P. Thurston, *Groups, tilings, and finite state automata*, AMS Colloquium Lectures, Boulder, CO (1989).
[40] A. Vince, *Replicating tessellations*, SIAM J. Discrete Math. **6** (1993), no. 3, 501–521, DOI 10.1137/0406040. MR1229702
[41] A. Vince, *Digit tiling of Euclidean space*, Directions in mathematical quasicrystals, CRM Monogr. Ser., vol. 13, Amer. Math. Soc., Providence, RI, 2000, pp. 329–370, DOI 10.1112/s0024610700008711. MR1798999

Department of Mathematics, University of Florida, Gainesville, Florida
*Email address*: `avince@ufl.edu`

Contemporary Mathematics
Volume **839**, 2026
https://doi.org/10.1090/conm/839/16724

# Statistics of weakly chaotic systems

Leonid A. Bunimovich and Yaofeng Su

ABSTRACT. One of the major breakthroughs in science of the last (20th) century was building a bridge between the worlds of stochastic (random) systems and deterministic (dynamical) systems. It was started by the celebrated 1958 paper by A. N. Kolmogorov [Dokl. Akad. Nauk SSSR (N.S.) 119 (1958), 861–864], who called this new theory (and actually a new way of thinking about deterministic systems) stochasticity of dynamical systems. Later, this name was essentially replaced by a short (sexier but more vague "chaos theory"). Kolmogorov's discovery demonstrated that the time evolution of deterministic systems could be indistinguishable from the evolution of purely random (stochastic) systems. Moreover, it has been later established that typical deterministic systems are chaotic. However, as well as stochastic systems, which could be more or less random (from random processes with independent values to random processes with long memory) , chaotic dynamical systems could be strongly or weakly chaotic. Actually, the majority of chaotic systems are weakly chaotic, especially those that are relevant models for various real-world processes. Naturally, the theory of weakly chaotic systems is less developed because it deals with more complicated problems than the studies of strongly chaotic systems. We present here a brief review of this theory, including recently published and some new results.

## Contents



## 1. Introduction

In this paper we concentrate on weakly chaotic deterministic (dynamical) systems. Therefore, everywhere we just verbally (without exact formulations) describe the results for strongly chaotic systems and present the best (so far) results for weakly chaotic systems, which are more difficult to analyze because they occupy

2020 *Mathematics Subject Classification.* Primary 37A50, 60F15.
The first author was partially supported by NSF grant DMS-2054659.

an intermediate position between strongly chaotic systems and dynamical systems with a non-chaotic, regular behavior. So, neither the methods for studying regular dynamics nor the ones for dealing with strongly chaotic systems are applicable to weakly chaotic systems. We will most often consider examples of billiards and statistical physics systems to demonstrate what the weakly chaotic dynamical systems really are and where they naturally appear as relevant models of various processes in science (mostly in physics). In this review paper, we mean by weak chaos slow mixing rates, that is, if an orbit of the dynamics $f$ can densely fill any subregion of the phase space $\mathcal{M}$ but after a long time. More precisely, let $\phi, \psi$ be smooth functions/observables (e.g., Lipschitz or Hölder) defined in $\mathcal{M}$ and let $\mu_{\mathcal{M}}$ be an invariant measure of dynamics (i.e. $\mu_{\mathcal{M}} \circ f^{-1} = \mu_{\mathcal{M}}$), then

$$\lim_{n\to\infty} \left| \int \phi \circ f^n \psi d\mu_{\mathcal{M}} - \int \psi d\mu_{\mathcal{M}} \int \phi d\mu_{\mathcal{M}} \right| = 0$$

with a slow polynomial decay rate. Such quantitative results give various limit theorems, e.g. central limit theorems, almost sure invariant principles, large deviations, Poisson limit theorems, extreme values theory etc. Relevant references can be found in [**6**–**9**, **11**, **17**, **24**, **27**–**30**]. To obtain the quantitative result for the decay of correlations, readers can refer to various approaches in [**1**, **3**, **14**, **31**, **32**].

We remark that all these results of limit theorems hold for smooth observables only. Our upcoming paper [**4**] proposes a different machinery to address various limit theorems for general observables (e.g., indicator functions).

The organization of the paper is the following. Section 2 deals with the strong law of large numbers (ergodic theorems) and large deviations (with emphasis on maximal-type large deviations, which indicates a core of principal differences between strongly and weakly chaotic systems). In Section 3 we consider the processes of recurrences and related Poisson approximations. In Section 4, we consider central limit theorems and the invariance principles. The stable limit laws are presented in Section 5.

We start by introducing some notations throughout the whole paper.

(1) $C_z$ denotes a constant depending on $z$.
(2) The notation "$a_n \precsim_z b_n$" ("$a_n = O_z(b_n)$") means that there is a constant $C_z \geq 1$ such that (s.t.) $a_n \leq C_z b_n$ for all $n \geq 1$, whereas the notation "$a_n \precsim b_n$" (or "$a_n = O(b_n)$") means that there is a constant $C \geq 1$ such that $a_n \leq C b_n$ for all $n \geq 1$. Next, "$a_n \approx_z b_n$" and $a_n = C_z^{\pm 1} b_n$ mean that there is a constant $C_z \geq 1$ such that $C_z^{-1} b_n \leq a_n \leq C_z b_n$ for all $n \geq 1$. Further, the notations "$a_n = C^{\pm 1} b_n$" and "$a_n \approx b_n$" mean that there is a constant $C \geq 1$ such that $C^{-1} b_n \leq a_n \leq C b_n$ for all $n \geq 1$. Finally, "$a_n = o(b_n)$" means that $\lim_{n\to\infty} |a_n / b_n| = 0$.
(3) The notation $\mathbb{P}$ refers to a probability distribution on the probability space, where a random variable lives, and $\mathbb{E}$ denotes the expectation of the random variable.
(4) $\mu_A$ denotes a measure on a set $A$. If $A$ is a $n$-dimensional (sub)manifold, then $\text{Leb}_A$ denotes the $n$-dimensional Lebesgue measure on $A$.
(5) $\mathbb{N} = \{1, 2, 3, \cdots\}$, $\mathbb{N}_0 = \{0, 1, 2, 3, \cdots\}$.

## 2. Ergodic theorems, large deviations and maximal large deviations

We start with basic theorems in ergodic theory.

DEFINITION 2.1 (Ergodic theorems). *A fundamental Birkhoff's ergodic theorem says that if $\mu_{\mathcal{M}}$ is a probability measure, preserved by an ergodic dynamics $f : \mathcal{M} \to \mathcal{M}$, then almost surely for any function $\phi$ with $\int \phi d\mu_{\mathcal{M}} = 0$,*

$$\lim_{n\to\infty} \frac{\sum_{i\le n} \phi \circ f^i}{n} = 0.$$

Equivalently, for any $\epsilon > 0$, $\mu_{\mathcal{M}}$-a.s. $x \in \mathcal{M}$ there is $N = N_x > 0$, such that

$$\sup_{n\ge N} \left| \frac{\sum_{i\le n} \phi \circ f^i(x)}{n} \right| \le \epsilon.$$

Large deviations theory is concerned with quantifying the convergence rate of the probability of outliers in the convergence of averages of the values of observables along trajectories of dynamical systems, that is,

$$\mu_{\mathcal{M}}\Big(\Big|\frac{\sum_{i\le n} \phi \circ f^i}{n}\Big| \ge \epsilon\Big).$$

We remark that there is no such rate in the context of almost sure convergence.

The paper [**2**] applied large deviations to obtain some geometric properties for certain smooth dynamical systems $f$, while in [**8**], the explicit estimates of constants in the upper bound of the error term in the large deviations principle were obtained.

If the dynamics is sufficiently strongly chaotic (e.g., if it has fast mixing rates or, in other words, correlations decay exponentially) and $\phi$ is sufficiently smooth, then there is a constant $C = C_{\phi,\epsilon} > 0$, such that for any $n, N \ge 1$

$$\mu_{\mathcal{M}}\Big(\Big|\frac{\sum_{i\le n} \phi \circ f^i}{n}\Big| \ge \epsilon\Big) \precsim e^{-Cn},$$

and

$$\mu_{\mathcal{M}}\Big(\sup_{n\ge N}\Big|\frac{\sum_{i\le n} \phi \circ f^i}{n}\Big| \ge \epsilon\Big) \precsim \sum_{n\ge N} e^{-Cn} \precsim e^{-CN}.$$

If the dynamics is weakly chaotic (i.e., slow mixing rates), then this estimate becomes weaker,

$$\mu_{\mathcal{M}}\Big(\Big|\frac{\sum_{i\le n} \phi \circ f^i}{n}\Big| \ge \epsilon\Big) \precsim n^{-a} \text{ where } a > 0.$$

(see [**25**] for more details).

In this situation, in contrast with the case of the exponential estimate, a direct summation over $n$ does not give an analogous estimate

$$\mu_{\mathcal{M}}\Big(\sup_{n\ge N}\Big|\frac{\sum_{i\le n} \phi \circ f^i}{n}\Big| \ge \epsilon\Big) \precsim N^{-a} \text{ where } a > 0.$$

We call this type of estimate a **maximal large deviation**.

One of the motivations to consider maximal large deviations is due to a rather typical approach (called an inducing method) in ergodic theory. This approach is usually aimed at finding a reference set $\Delta_0 \subsetneq \mathcal{M}$ and the first return time to it $R : \Delta_0 \to \mathbb{N}$, in order to prove some "good" properties for the first return map $f^R : \Delta_0 \to \Delta_0$. Then this property is lifted to the original dynamics $f : \mathcal{M} \to \mathcal{M}$. In practice, this lifting technique relies on a quantitative estimate of the time $N$ in

$\sup_{n\geq N} \frac{\sum_{i\leq n} R\circ(f^R)^i}{n}$ (rather than $\frac{\sum_{i\leq N} R\circ(f^R)^i}{N}$). This is related to the presented below maximal large deviation approach.

Let a transfer operators (Perron-Frobenius operator) $P$ associated to $f$ is defined by the duality relations:

$$\int g\cdot P(h)d\mu_{\mathcal{M}} = \int g\circ f\cdot h d\mu_{\mathcal{M}} \text{ for all } h\in L^1(\mu_{\mathcal{M}}),\ g\in L^\infty(\mu_{\mathcal{M}}).$$

Then we have the following.

THEOREM 2.2 (Maximal large deviations, see [**6**]). *Suppose that $\phi\in L^\infty$ has zero mean, and there are constants $p\in\mathbb{N}, C_\phi>0, \beta\in(0,p)$ such that*

$$\|P^n(\phi)\|_p^p \leq C_\phi n^{-\beta} \textit{ for any } n\geq 1.$$

*Then for any $\epsilon>0, N\geq 1$*

$$\mu_{\mathcal{M}}\Big(\sup_{n\geq N}\Big|\frac{\sum_{i\leq n}\phi\circ f^i}{n}\Big|\geq\epsilon\Big)\precsim_{\epsilon,p,\beta}\max\{\|\phi\|_\infty^p, C_\phi\}\cdot\|\phi\|_\infty^p\cdot N^{-\beta},$$

*and*

$$\Big\|\sup_{n\geq N}\Big|\frac{\sum_{i\leq n}\phi\circ f^i}{n}\Big|\Big\|_{2p}^{2p}\precsim_{p,\beta}\max\{\|\phi\|_\infty^p, C_\phi\}\cdot\|\phi\|_\infty^p\cdot N^{-\beta}.$$

*In particular, if the function $\phi\in L^\infty$ satisfies the inequality $\|P^n(\phi)\|_1\leq C_\phi n^{-\beta}$ for some $C_\phi>0$ and any $n\geq 1$, then for any $\epsilon>0$*

$$\|P^n(\phi)\|_{1+\lceil\beta\rceil}^{1+\lceil\beta\rceil}\leq C_\phi\|\phi\|_\infty^{\lceil\beta\rceil}n^{-\beta},$$

$$\mu_{\mathcal{M}}\Big(\sup_{n\geq N}\Big|\frac{\sum_{i\leq n}\phi\circ f^i}{n}\Big|\geq\epsilon\Big)\precsim_{\epsilon,\beta}\max\{\|\phi\|_\infty^{1+\lceil\beta\rceil}, C_\phi\|\phi\|_\infty^{\lceil\beta\rceil}\}\cdot\|\phi\|_\infty^{1+\lceil\beta\rceil}\cdot N^{-\beta}.$$

Now we present applications to some concrete classes of dynamical systems. Denote by $d$ a Riemannian distance in $\mathcal{M}$, and by Leb a Riemannian volume form on $\mathcal{M}$. By $C^{1+}$ we denote the class of continuously differentiable maps with Hölder continuous derivatives. Let $f:\mathcal{M}\to\mathcal{M}$ be a piecewise $C^{1+}$-endomorphism of a Riemannian manifold $\mathcal{M}$.

DEFINITION 2.3 (Gibbs-Markov-Young (GMY) structures, see [**32**] or [**2**]). *Suppose that there exists a ball $\Delta_0\subset\mathcal{M}$, a countable partition $\mathcal{P}$ (mod 0) of $\Delta_0$ into topological balls $U$ with smooth boundaries, and a return time (not necessarily the first return time) function $R:\Delta_0\to\mathbb{N}$, which is constant on each element of $\mathcal{P}$. We say that $f$ admits a Gibbs-Markov-Young (GMY) structure if the following properties hold:*

(1) *Markov property: for each $U\in\mathcal{P}$ and $R=R(U)$, $f^R: U\to\Delta_0$ is a $C^{1+}$-diffeomorphism (and, in particular, it is a bijection). Thus the induced map $F:\Delta_0\to\Delta_0$, $F(x)=f^{R(x)}(x)$, is defined almost everywhere and satisfies a classical Markov property.*
(2) *Uniform expansion: there exists $\rho\in(0,1)$ such that for almost all $x\in\Delta_0$ we have $\|DF(x)^{-1}\|\leq\rho$. In particular the separation time $s(x,y)$, which is equal to the smallest integer $i\geq 0$ such that $F^i(x)$ and $F^i(y)$ lie in distinct elements of the partition $\mathcal{P}$, is defined and is finite for almost all $x,y\in\Delta_0$.*

(3) *Bounded distortion: there exists* $C > 0$ *such that for any two points* $x, y \in \Delta_0$ *with* $s(x,y) \in [1, \infty)$

$$\left| \frac{\det DF(x)}{\det DF(y)} - 1 \right| \leq C\rho^{s(F(x),F(y))},$$

*where* $\det DF$ *is the Radon-Nikodym derivative of* $F$ *with respect to* Leb *on* $\Delta_0$.

Theorem 2.2 implies the following corollary, see a proof in [**6**].

COROLLARY 2.4. *If* $f$ *admits a Gibbs-Markov-Young structure, and* $\text{Leb}(x \in \Delta_0 : R(x) > N) \approx N^{1+\beta}$ *for some* $\beta > 0$, *then there exists an invariant probability measure* $\mu_{\mathcal{M}}$ *and an open and dense set* $H_0$ *in the space of Hölder functions, such that for any zero mean function* $\phi \in H_0$, *and for any* $\epsilon > 0$,

$$\lim_{N\to\infty} \frac{\log \mu_{\mathcal{M}}\Big( \sup_{n \geq N} \left| \frac{\sum_{i \leq n} \phi \circ f^i}{n} \right| \geq \epsilon \Big)}{\log N} = -\beta.$$

## 3. Recurrences and Poisson approximations

The studies of Poisson approximations of the process of recurrences to small subsets in the phase spaces of chaotic dynamical systems are developed now into a large active area in the dynamical systems theory. Another view at this type of problem is a subject of the theory of open dynamical systems where some positive measure subset $A$ of the phase space is named a hole, and hitting and escape through the hole processes are studied. The third view at this type of problems concerns statistics of extreme events ("record values") in the theory of random processes [**15, 18**]. In this section we present some new advances in this area.

In a general set up, one picks a small measure subset $A$ in the phase space $\mathcal{M}$ of hyperbolic (chaotic) ergodic dynamical system and attempts to prove that in the limit, when the measure of $A$ approaches zero, the corresponding process of recurrences to $A$ converges to the Poisson process.

This area received an essential boost in L-S.Young papers [**31,32**], where a new general framework for analysis of statistical properties of hyperbolic dynamical systems was introduced. This approach employs representation of the phase space of a dynamical system as a tower (later called Young tower, Gibbs-Markov-Young tower, etc), which allow to study dynamics by analysing recurrences to the base of this tower. Several developments of this method were proposed later, essentially all focused on the dynamical systems with weak hyperbolicity (slow decay of correlations). For such systems the method of inducing was employed, when the base of the tower is chosen as such subset of the phase space where the induced dynamics, generated by the recurrences to the base, is strongly hyperbolic.

The main results (Theorems 3.11 and 3.31) in this section are dealing with convergence of a random process, generated by the measure preserving dynamics, to the functional Poisson law in the total variation (TV) norm. We also obtain estimates of the corresponding convergence rates in the following form. For almost every $z \in \mathcal{M}$,

$$d_{TV}\left(\mathcal{N}^{r,z,T}, Poisson\right) \precsim_{T,z} r^a, \tag{3.1}$$

where *Poisson* is a Poisson point process and $\mathcal{N}^{r,z,T}$ is a dynamical point process which counts a number of entrances by an orbit to a metric ball $B_r(z)$ with radius

$r$ and the center $z$ in the phase space of a dynamical system during a time interval $[0,T]$ (see their definitions below). The notation $\precsim_{T,z}$ means that a constant in (3.1) depends only on $z$ and $T$.

Definition 3.1 (Dynamical point processes). *Let $(\mathcal{M},d)$ be a Riemannian manifold (with or without boundaries, connected or non-connected, compact or non-compact), $d$ is the Riemannian metric on $\mathcal{M}$ and $B_r(z)$ is a geodesic ball in $\mathcal{M}$ with a radius $r$ and a center $z \in \mathcal{M}$. We assume that dynamics $f:(\mathcal{M},\mu_{\mathcal{M}}) \to (\mathcal{M},\mu_{\mathcal{M}})$ is ergodic with respect to (w.r.t.) some invariant probability measure $\mu_{\mathcal{M}}$.*

*Let $T>0$. Consider a dynamical point process on $[0,T]$, so that for any $t\in[0,T]$*

$$\mathcal{N}_t^{r,T,z} := \sum_{i=0}^{t/\mu_{\mathcal{M}}(B_r(z))} \mathbb{1}_{B_r(z)} \circ f^i.$$

*Thus the dynamical point process $\mathcal{N}^{r,T,z}$ is a random counting measure on $[0,T]$.*

Definition 3.2 (Poisson point processes). *For any $T>0$, we say that Poisson is a Poisson point process on $[0,T]$ if*

1. *Poisson is a random counting measure on $[0,T]$.*
2. *$Poisson(A)$ is a Poisson-distributed random variable for any Borel set $A\subseteq[0,T]$.*
3. *If $A_1,A_2,\cdots,A_n\subseteq[0,T]$ are pairwise disjoint, then $Poisson(A_1),\cdots,Poisson(A_n)$ are independent.*
4. *$\mathbb{E}Poisson(A)=\mathrm{Leb}(A)$ for any Borel set $A\subseteq[0,T]$.*

Definition 3.3 (Total variation norms of point processes). *For any $T>0$ consider the $\sigma$-algebra $\mathcal{C}$ on the space of point processes on $[0,T]$, defined as*

$$\sigma\left\{\pi_A^{-1}B : A\subseteq[0,T], B\subseteq\mathbb{N}\right\}, \tag{3.2}$$

*where $A,B$ are Borel sets and $\pi_A$ is an evaluation map defined on the space of counting measures, so that for any counting measure $N$*

$$\pi_A N := N(A).$$

*Now we can define the total variation norm for the Poisson approximation of a dynamical point process as*

$$d_{TV}\left(\mathcal{N}^{r,T,z}, Poisson\right) := \sup_{C\in\mathcal{C}} \left|\mu_{\mathcal{M}}(\mathcal{N}^{r,T,z}\in C) - \mathbb{P}(Poisson\in C)\right|$$

Definition 3.4 (Poisson approximations for small holes with convergence rates). *Suppose that for any $T>0$ there exists a constant $a>0$ s.t. for almost every $z\in\mathcal{M}$*

$$d_{TV}\left(\mathcal{N}^{r,T,z}, Poisson\right) \precsim_{T,z} r^a \to 0.$$

*Then $a$ is called a convergence rate of a Poisson approximation.*

Remark 3.5. *The convergence rates imply that for any sub-interval $[T_1,T_2]\subseteq[0,T]$,*

$$\sup_{C\subseteq\mathbb{N}} \left|\mu(\mathcal{N}^{r,T,z}[T_1,T_2]\in C) - \mathbb{P}(Poisson[T_1,T_2]\in C)\right| \precsim_{T,z} r^a \to 0.$$

We now turn to hyperbolic Gibbs-Markov-Young structures [**31, 32**]:

DEFINITION 3.6 (Hyperbolic Gibbs-Markov-Young structures). *Introduce at first several notions concerning hyperbolic dynamics $f$ on Riemannian manifolds $(\mathcal{M}, d)$.*

(1) *An embedded disk $\gamma^u$ is called an unstable manifold if for every $x, y \in \gamma^u$*
$$\lim_{n\to\infty} d\Big(f^{-n}(x), f^{-n}(y)\Big) = 0$$
(2) *An embedded disk $\gamma^s$ is called a stable manifold if for every $x, y \in \gamma^s$*
$$\lim_{n\to\infty} d\Big(f^{n}(x), f^{n}(y)\Big) = 0$$
(3) *$\Gamma^u := \{\gamma^u\}$ is called a continuous family of $C^1$-unstable manifolds if there is a compact set $K^s$, a unit disk $D^u$ in some $\mathbb{R}^n$ and a map $\phi^u : K^s \times D^u \to \mathcal{M}$ such that*
   (a) *$\gamma^u = \phi^u(\{x\} \times D^u)$ is an unstable manifold,*
   (b) *$\phi^u$ maps $K^s \times D^u$ homeomorphically onto its image,*
   (c) *$x \to \phi^u|_{\{x\}\times D^u}$ defines a continuous map from $K^s$ to $\mathrm{Emb}^1(D^u, \mathcal{M})$, where $\mathrm{Emb}^1(D^u, \mathcal{M})$ is the space of $C^1$-embeddings of $D^u$ into $\mathcal{M}$.*

*A continuous family of $C^1$-stable manifolds $\Gamma^s := \{\gamma^s\}$ is defined similarly.*

*We say that a compact set $\Lambda \subseteq \mathcal{M}$ has a hyperbolic product structure if there exist continuous families of stable manifolds $\Gamma^s := \{\gamma^s\}$ and of unstable manifolds $\Gamma^u := \{\gamma^u\}$ such that*

(1) $\Lambda = (\bigcup \gamma^s) \bigcap (\bigcup \gamma^u)$,
(2) $\dim \gamma^s + \dim \gamma^u = \dim \mathcal{M}$,
(3) *each $\gamma^s$ intersects each $\gamma^u$ at exactly one point,*
(4) *stable and unstable manifolds are transversal, and the angles between them are uniformly bounded away from 0.*

*A subset $\Lambda_1 \subseteq \Lambda$ is called a s-subset if $\Lambda_1$ has a hyperbolic product structure and, moreover, the corresponding families of stable and unstable manifolds $\Gamma_1^s$ and $\Gamma_1^u$ can be chosen so that $\Gamma_1^s \subseteq \Gamma^s$ and $\Gamma_1^u = \Gamma^u$.*

*Analogously, a subset $\Lambda_2 \subseteq \Lambda$ is called an u-subset if $\Lambda_2$ has a hyperbolic product structure and the families $\Gamma_2^s$ and $\Gamma_2^u$ can be chosen so that $\Gamma_2^u \subseteq \Gamma^u$ and $\Gamma_2^s = \Gamma^s$.*

*For $x \in \Lambda$, denote by $\gamma^u(x)$ (resp. $\gamma^s(x)$) the element of $\Gamma^u$ (resp. $\Gamma^s$) which contains $x$. Also, for each $n \geq 1$, denote by $(f^n)^u$ the restriction of the map $f^n$ to $\gamma^u$-disks, and by $\det D(f^n)^u$ denote the Jacobian of $(f^n)^u$.*

*We say that the set $\Lambda$ with hyperbolic product structure has also a* ***hyperbolic Gibbs-Markov-Young structure*** *if the following properties are satisfied*

(1) *Lebesgue detectability: there exists $\gamma \in \Gamma^u$ such that $\mathrm{Leb}_\gamma(\Lambda \bigcap \gamma) > 0$.*
(2) *Markovian property: there exist pairwise disjoint s-subsets $\Lambda_1, \Lambda_2, \cdots \subseteq \Lambda$ such that*
   (a) $\mathrm{Leb}_\gamma\Big(\Lambda \setminus \Big(\bigcup_{i\geq 1} \Lambda_i\Big)\Big) = 0$ *on each $\gamma \in \Gamma^u$,*
   (b) *for each $i \geq 1$ there exists $R_i \in \mathbb{N}$ such that $f^{R_i}(\Lambda_i)$ is an u-subset, and for all $x \in \Lambda_i$*
$$f^{R_i}(\gamma^s(x)) \subseteq \gamma^s\left(f^{R_i}(x)\right)$$
   *and*
$$f^{R_i}(\gamma^u(x)) \supseteq \gamma^u(f^{R_i}(x)).$$

*Define now a return time function* $R : \Lambda \to \mathbb{N}$ *and a return function* $f^R : \Lambda \to \Lambda$, *so that for each* $i \geq 1$

$$R\big|_{\Lambda_i} = R_i \text{ and } f^R\big|_{\Lambda_i} = f^{R_i}\big|_{\Lambda_i}$$

*Next, the separation time* $s(x, y)$ *for* $x, y \in \Lambda$ *is defined as*

$$s(x, y) := \min\{n \geq 0 : \left(f^R\right)^n(x) \text{ and } \left(f^R\right)^n(y) \text{ belong to different sets } \Lambda_i\}.$$

*We also assume that there are constants* $C > 1, \alpha > 0$ *and* $0 < \beta < 1$, *which depend only on* $f$ *and* $\Lambda$, *such that the following conditions hold*

(3) *Polynomial contraction on stable leaves. For any* $\gamma^s \in \Gamma^s, x, y \in \gamma^s, n \geq 1$,

$$d\Big(f^n(x), f^n(y)\Big) \leq Cn^{-\alpha}.$$

(4) *Backward polynomial contraction on unstable leaves. For any* $\gamma^u \in \Gamma^u$, $x, y \in \gamma^u, n \geq 1$,

$$d\Big(f^{-n}(x), f^{-n}(y)\Big) \leq Cn^{-\alpha}.$$

(5) *Bounded distortion: for any* $\gamma \in \Gamma^u$ *and* $x, y \in \gamma \bigcap \Lambda_i$ *for some* $\Lambda_i$,

$$\log \frac{\det D\left(f^R\right)^u(x)}{\det D\left(f^R\right)^u(y)} \leq C\beta^{s\left(f^R(x), f^R(y)\right)}.$$

(6) *Regularity of the stable foliation. For each* $\gamma, \gamma' \in \Gamma^u$ *denote*

$$\Theta_{\gamma,\gamma'} : \gamma' \bigcap \Lambda \to \gamma \bigcap \Lambda : x \to \gamma^s(x) \bigcap \gamma.$$

*Then the following properties hold*

(a) $\Theta_{\gamma,\gamma'}$ *is absolutely continuous and for any* $x \in \gamma \bigcap \Lambda$

$$\frac{d\left(\Theta_{\gamma,\gamma'}\right)_* \mathrm{Leb}_{\gamma'}}{d\,\mathrm{Leb}_\gamma}(x) = \prod_{n \geq 0} \frac{\det Df^u\left(f^n(x)\right)}{\det Df^u\left(f^n\left(\Theta^{-1}_{\gamma,\gamma'} x\right)\right)},$$

$$\frac{d\left(\Theta_{\gamma,\gamma'}\right)_* \mathrm{Leb}_{\gamma'}}{d\,\mathrm{Leb}_\gamma}(x) = C^{\pm 1},$$

(b) *for any* $x, y \in \gamma \bigcap \Lambda$

$$\log \frac{\frac{d\left(\Theta_{\gamma,\gamma'}\right)_* \mathrm{Leb}_{\gamma'}}{d\,\mathrm{Leb}_\gamma}(x)}{\frac{d\left(\Theta_{\gamma,\gamma'}\right)_* \mathrm{Leb}_{\gamma'}}{d\,\mathrm{Leb}_\gamma}(y)} \leq C\beta^{s(x,y)}.$$

(7) *Aperiodicity:* $\gcd\left(R_i, i \geq 1\right) = 1$.

(8) *Decay rate of the return times* $R$*: there exist* $\xi > 1$ *and* $\gamma \in \Gamma^u$ *such that*

$$\mathrm{Leb}_\gamma\left(R > n\right) \leq Cn^{-\xi}.$$

***SRB measures****: Let the dynamics* $f : \mathcal{M} \to \mathcal{M}$ *has hyperbolic Gibbs-Markov-Young structure. It was proved in* [**31, 32**] *that there exists an ergodic probability measure* $\mu_{\mathcal{M}}$ *such that for any unstable manifold* $\gamma^u$ *(including* $\Gamma^u$*) we have* $\mu_{\gamma^u} \ll \mathrm{Leb}_{\gamma^u}$, *where* $\mu_{\gamma^u}$ *is the conditional measure of* $\mu_{\mathcal{M}}$ *on an unstable manifold* $\gamma^u$. *Such measure* $\mu_{\mathcal{M}}$ *is called Sinai-Ruelle-Bowen measure (SRB measure).*

ASSUMPTION 3.7 (Geometric regularities). *Assume that* $f : \mathcal{M} \to \mathcal{M}$ *has the hyperbolic Gibbs-Markov-Young structure, as described in Definition* 3.6, *and*

(1) $f$ *is bijective and a local* $C^1$*-diffeomorphism on* $\bigcup_{i \geq 1} \bigcup_{0 \leq j < R_i} f^j(\Lambda_i)$.

(2) *The following limit*

$$\dim_H \mu_{\mathcal{M}} := \lim_{r\to 0} \frac{\log \mu_{\mathcal{M}}(B_r(z))}{\log r}$$

*exists for almost every* $z \in \mathcal{M}$. *Then* $dim_H \mu_{\mathcal{M}}$ *is called a Hausdorff dimension of the measure* $\mu_{\mathcal{M}}$.

(3) $\alpha \dim_H \mu_{\mathcal{M}} > 1$, *where* $\alpha$ *is defined in Definition* 3.6.

ASSUMPTION 3.8 (The first returns & interior assumptions on $\Lambda$). *Assume that* $f : \mathcal{M} \to \mathcal{M}$ *has the hyperbolic Gibbs-Markov-Young structure, and there are constants* $C > 1$ *and* $\beta \in (0, 1)$ *(the same as in Definition* 3.6*) such that*

(1) $R : \Lambda \to \mathbb{N}$ *is the first return time and* $f^R : \Lambda \to \Lambda$ *is the first return map for* $\Lambda$.

*This implies that* $f^R$ *is bijective.*

(2) *for any* $\gamma \in \Gamma^s, \gamma_1 \in \Gamma^u, x, y \in \gamma \bigcap \Lambda, x_1, y_1 \in \gamma_1 \bigcap \Lambda$,

$$d\Big( \left(f^R\right)^n (x), \left(f^R\right)^n (y)\Big) \leq C\beta^n,$$

*and*

$$d\Big( \left(f^R\right)^{-n} (x_1), \left(f^R\right)^{-n} (y_1)\Big) \leq C\beta^n d(x_1, y_1).$$

(3) $\mu_{\mathcal{M}}\{\text{int}\,(\Lambda)\} > 0$ *and* $\mu_{\mathcal{M}}(\partial\Lambda) = 0$, *where*

$$\text{int}\,\Lambda := \{x \in \Lambda : \text{ there exists } r_x > 0 \text{ s.t. } \mu_{\mathcal{M}}\left(B_{r_x}(x) \setminus \Lambda\right) = 0\}, \quad \partial\Lambda := \Lambda \setminus \text{int}\,\Lambda.$$

*In other words,* $x \in \text{int}\,\Lambda$ *if and only if* $x \in \Lambda$ *and there is a small ball* $B_{r_x}(x)$ *s.t.* $B_{r_x}(x) \subseteq \Lambda$ $\mu_{\mathcal{M}}$*-almost surely.*

We formulate now the first result in this section.

THEOREM 3.9 (Convergence rates for Poisson laws I, see [**5**]). *Assume that the dynamics* $f : (\mathcal{M}, \mu_{\mathcal{M}}) \to (\mathcal{M}, \mu_{\mathcal{M}})$ *has a first return hyperbolic Gibbs-Markov-Young structure (see Definition* 3.6*) and satisfies Assumptions* 3.7 *and* 3.8. *Then for any* $T > 0$ *the following results hold*

(1) $\dim_H \mu_{\mathcal{M}} \geq \dim \gamma^u$

(2) *if either* $\alpha > \frac{2}{\dim \gamma^u} - \frac{1}{\dim_H \mu}$ *or* $\mu_{\mathcal{M}} \ll \text{Leb}_{\mathcal{M}}$ *with* $\frac{d\mu_{\mathcal{M}}}{d\,\text{Leb}_{\mathcal{M}}} \in L^\infty_{loc}(\mathcal{M})$, *then for almost every* $z \in \mathcal{M}$

$$d_{TV}\left(\mathcal{N}^{r,z,T}, Poisson\right) \precsim_{T,\xi,z} r^a,$$

*where a computable constant* $a > 0$ *depends on* $\xi > 1$, $\dim_H \mu_{\mathcal{M}}, \dim \gamma^u$ *and* $\alpha$, *but it does not depend on* $z \in \mathcal{M}$.

DEFINITION 3.10 (Induced measurable partitions). *We say that a probability measure* $\mu_{\mathcal{M}}$ *for the dynamics* $f : \mathcal{M} \to \mathcal{M}$ *has an induced measurable partition if there are constants* $\beta \in (0, 1), C > 1$ *(the same as in the Definition* 3.6*) and* $b > 0$ *such that*

(1) *There exists a subset* $X \subseteq \mathcal{M}$ *with* $\mu_{\mathcal{M}}\{\text{int}\,(X)\} > 0$, $\mu_{\mathcal{M}}(\partial X) = 0$.

(2) *The subset* $X$ *has a measurable partition* $\Theta := \{\gamma^u(x)\}_{x\in X}$ *(which could be different from* $\Gamma^u$*), such that the elements of* $\Theta$ *are disjoint connected unstable manifolds, so that* $\mu_{\mathcal{M}}$*-almost surely* $X = \bigsqcup_{x\in X} \gamma^u(x)$ *and for any function* $g$

$$\mu_X(g) = \int_X \mu_{\gamma^u(x)}(g) d\mu_X(x),$$

*where* $\mu_X := \frac{\mu_{\mathcal{M}}|_X}{\mu_{\mathcal{M}}(X)}$ *and* $\mu_{\gamma^u(x)}$ *is the conditional probability induced by* $\mu_X$ *on* $\gamma^u(x) \in \Theta$.

(3) *Each* $\gamma^u \in \Theta$ *is (at least* $C^1$*) smooth.*

(4) *All* $\gamma^u \in \Theta$ *have uniformly bounded sectional curvatures and the same dimensions.*

(5) *For any* $\epsilon \in (0,1)$

$$\mu_X\{x \in X : |\gamma^u(x)| < \epsilon\} \leq C\epsilon^b,$$

*where* $|\gamma^u(x)|$ *is the radius of the largest inscribed geodesic ball in* $\gamma^u(x) \in \Theta$*, where a geodesic ball is defined with respect to the distance* $d_{\gamma^u(x)}$ *on* $\gamma^u(x)$*, induced by the Riemannian metric. This property implies that almost every* $\gamma^u(x) \in \Theta$ *is non-degenerate, i.e.,* $|\gamma^u(x)| > 0$ *for almost every* $x \in X$.

(6) *For almost every point* $x \in X$ *we have* $\mu_{\gamma^u(x)} \ll \mathrm{Leb}_{\gamma^u(x)}$, $\mu_{\gamma^u(x)}(\gamma^u(x)) > 0$*, and for any* $y, z \in \gamma^u(x)$

$$\frac{d\mu_{\gamma^u(x)}}{d\,\mathrm{Leb}_{\gamma^u(x)}}(y) = C^{\pm 1}\frac{d\mu_{\gamma^u(x)}}{d\,\mathrm{Leb}_{\gamma^u(x)}}(z).$$

(7) *Denote by* $\overline{R}$ *the first return time to* $X$ *for* $f$*. Then the first return map* $f^{\overline{R}} : X \to X$ *has an exponential u-contraction, i.e., for any* $\gamma^u \in \Theta, x, y \in \gamma^u, n \geq 1$

$$d\Big(\left(f^{\overline{R}}\right)^{-n}(x), \left(f^{\overline{R}}\right)^{-n}(y)\Big) \leq C\beta^n d(x,y),$$

*and an exponential decay of correlations, i.e., for any* $h \in \mathrm{Lip}(X)$

$$\left|\int h \circ \left(f^{\overline{R}}\right)^n h d\mu_X - \left(\int h d\mu_X\right)^2\right| \leq C\beta^n \|h\|_{\mathrm{Lip}}^2.$$

Now we can formulate the second result in this section.

THEOREM 3.11 (Convergence rates for the Poisson laws II, [**5**]). *Assume that the dynamics* $f : (\mathcal{M}, \mu_{\mathcal{M}}) \to (\mathcal{M}, \mu_{\mathcal{M}})$ *has the hyperbolic Gibbs-Markov-Young structure (see Definition* 3.6*), satisfies the Assumption* 3.7*, and* $\mu_{\mathcal{M}}$ *has an induced measurable partition (see Definition* 3.10*). Then for any* $T > 0$*, the following results hold.*

(1) $\dim_H \mu_{\mathcal{M}} \geq \frac{b}{b+\dim\gamma^u}\dim\gamma^u$ *and*

(2) *if either* $\alpha > \frac{2}{\dim\gamma^u}\frac{b+\dim\gamma^u}{b} - \frac{1}{\dim_H \mu_{\mathcal{M}}}$ *or* $\mu_{\mathcal{M}} \ll \mathrm{Leb}_{\mathcal{M}}$ *and* $\frac{d\mu_{\mathcal{M}}}{d\,\mathrm{Leb}_{\mathcal{M}}} \in L^\infty_{loc}(\mathcal{M})$*, then for almost every (a.e.)* $z \in \mathcal{M}$*,*

$$d_{TV}\left(\mathcal{N}^{r,z,T}, Poisson\right) \precsim_{T,\xi,z,b} r^a,$$

*where a constant* $a > 0$ *depends on* $\xi > 1$, $\dim_H \mu_{\mathcal{M}}, \dim\gamma^u, b$ *and* $\alpha$*, but it does not depend on* $z \in \mathcal{M}$.

We will consider now some applications of the above Theorems.

**3.1. Intermittent solenoids.** Following [**1**] let $\mathcal{M} = S^1 \times \mathbb{D}$, $f_\gamma(x,z) = (g_\gamma(x), \theta z + e^{2\pi i x}/2)$, where $g_\gamma : S^1 \to S^1$ is a continuous map of degree $d \geq 2$ and $\gamma \in (0, +\infty)$ such that

(1) $g_\gamma$ is $C^2$ on $S^1 \setminus \{0\}$ and $Dg_\gamma > 1$ on $S^1 \setminus \{0\}$,
(2) $g_\gamma(0) = 0, Dg_\gamma(0+) = 1$ and $xD^2 g_\gamma(x) \sim x^\gamma$ for sufficiently small positive $x$,
(3) $Dg_\gamma(0-) > 1$,
(4) $\theta > 0$ is so small that $\theta \|Dg_\gamma\|_\infty < 1 - \theta$.

By Theorem 3.9 we have the following

COROLLARY 3.12. *Poisson limit laws hold for $f_\gamma$ for any $\gamma \in (0,1)$ with convergence rates* (3.1).

**3.2. Axiom A attractors.** It was well-known that for Axiom A attractors $\Sigma \subsetneq \mathcal{M}$ with $\dim \gamma^u = 1$ the Poisson limit laws hold. We will show that conditions on $\dim_H \mu$ and $\dim \gamma^u$ can be dropped.

DEFINITION 3.13 (Axiom A attractors). *Let $f : \mathcal{M} \to \mathcal{M}$ be a $C^2$-diffeomorphism. A compact set $\Sigma \subseteq \mathcal{M}$ is called an Axiom A attractor if*

(1) *There is a neighborhood $U$ of $\Sigma$, called its basin, such that $f^n(x) \to \Sigma$ for every $x \in U$.*
(2) *The tangent bundle over $\Sigma$ is split into $E^u \oplus E^s$, where $E^u$ and $E^s$ are $df$-invariant subspaces.*
(3) *$df|_{E^u}$ is uniformly expanding, and $df|_{E^s}$ is uniformly contracting.*
(4) *$f : \Sigma \to \Sigma$ is topologically mixing.*

The support of the SRB measure $\mu_\mathcal{M}$ is $\Sigma$. Applying now Theorem 3.9 we obtain the following.

COROLLARY 3.14. *Poisson limit laws hold for Axiom A attractors with convergence rates* (3.1).

**3.3. Hénon attractors.** The Poisson limit laws for certain Hénon attractors (see [**31**]) have been proved in [**10**]. However, convergence rate for this class of dynamical systems was obtained in a weaker form. A stronger rate of convergence was derived (see (3.1)) by using Theorem 3.11.

COROLLARY 3.15. *Poisson limit laws with convergence rates* (3.1) *hold for Hénon attractors which can be modelled by Young towers.*

**3.4. Billiards.**

DEFINITION 3.16 (Billiard tables, billiard maps and phase spaces). *We consider a billiard in a two-dimensional region $Q \subseteq \mathbb{R}^2$ (called a billiard table) with a piecewise smooth (of class $C^3$) boundary $\partial Q$. Each smooth piece has a uniformly bounded curvature. The boundary $\partial Q$ is equipped with a field of inward unit normal vectors $n(q), q \in \partial Q$.*

*A billiard is a dynamical system generated by the motion of a point particle with the unit velocity inside the region $Q$, and being reflected from its boundary according to the law "the angle of incidence equals the angle of reflection". It means that upon reflection the tangent component of the velocity remains the same, while the normal component changes its sign according to the rule $v_+ = v_- - 2\langle n(q), v_-\rangle n(q)$,* *where $v_+$ (resp. $v_-$) is the velocity of the particle immediately after (resp. before) reflection.*

*The phase space of a billiard is the restriction of the unit tangent bundle of $\mathbb{R}^2$ to $Q$. We will use the standard notation for phase points $x = (q, v)$, where $q$ is the point of the configuration space $Q$ and $v$ is the unit velocity vector. The billiard preserves the Liouville measure $d\nu := dqdv$ where $dq$ and $dv$ are Lebesgue measures on $Q$ and on the unit one-dimensional sphere respectively. The corresponding flow will be denoted by $\{S^t\}$. It is customary for billiard-type systems to study instead of $\{S^t\}$ a dynamical system with discrete time, which is called a billiard map $f$. Denote*

$$\mathcal{M} = \{x = (q, v), q \in \partial Q, \langle v, n(q)\rangle > 0\}.$$

*For $x = (q, v) \in \mathcal{M}$ let $\tau(x)$ be the first positive moment of reflection of the billiard orbit defined by $x$ off the boundary. Then the billiard map is defined by $f(x) = (q', v') = S^\tau x$, so that $q'$ is the point of the next reflection and $v'$ is the outcoming velocity vector at that point. We call $\mathcal{M}$ a phase space of the billiard map $f$.*

*Due to the structure of $\partial Q$, the set of singular points of the boundary $\partial Q$ has measure zero. The angle $\phi$ of the velocity vector $v$ varies from $-\pi/2$ to $\pi/2$ at any regular point $q \in \partial Q$. Hence $\mathcal{M} = \partial Q \times [-\pi/2, \pi/2]$ almost surely. In what follows we always identify $\mathcal{M}$ with $\partial Q \times [-\pi/2, \pi/2]$, and denote the phase point $x \in \mathcal{M}$ by $(q, \phi) \in \partial Q \times [-\pi/2, \pi/2]$ throughout this subsection.*

*The phase space $\mathcal{M} = \partial Q \times [-\pi/2, \pi/2]$ is endowed with a natural Riemannian metric $d_{\mathcal{M}}$ and Riemannian volume $\mathrm{Leb}_{\mathcal{M}}$. The billiard map $f$ preserves the measure*

$$d\mu_{\mathcal{M}} := (2\,\mathrm{Leb}_{\partial Q}\,\partial Q)^{-1} \cos\phi d\phi dq = (2\,\mathrm{Leb}_{\partial Q}\,\partial Q)^{-1} \cos\phi\,\mathrm{Leb}_{\mathcal{M}},$$

*where $dq$ is the one-dimensional Lebesgue measure on the boundary $\partial Q$ and $d\phi$ is the one-dimensional Lebesgue (uniform) measure on $[-\pi/2, \pi/2]$.*

Before presenting the results, we introduce some more definitions.

DEFINITION 3.17 (An induced system). *Suppose that there is a fixed subset $X \subseteq \mathcal{M}$ with $\mathrm{Leb}_{\mathcal{M}}(X) > 0$. The first return time to $X$ is denoted as $R : X \to \mathbb{N}$. We assume that $X$ can be partitioned into countably many connected pieces*

$$X = \bigcup_{i \geq 1} X_i, \tag{3.3}$$

*so that $R$ is constant on each $X_i$, and*

$$\mathrm{Leb}_{\mathcal{M}}(\partial X_i) = 0, \quad \mathrm{int}\, X_i \bigcap \mathrm{int}\, X_j = \emptyset \text{ for } i \neq j.$$

*The first return time $R$ induces the first return map $f^R : X \to X$ and a new dynamical system $(X, f^R)$.*

DEFINITION 3.18 (Singularities and (un)stable manifolds). *Denote by $\mathbb{S} \subseteq X$ the singularity set for $f^R$. The set $\mathbb{S}$ has zero Lebesgue measure, and $\mathbb{S}^c \subseteq X$ consists of countably many open connected components. Unstable (resp. stable) manifolds are connected components of $(\bigcup_{i\geq 0}(f^R)^i\mathbb{S})^c$ (resp. $(\bigcup_{i\geq 0}(f^R)^{-i}\mathbb{S})^c$). A closed and connected part of the unstable (resp. stable) manifold will be called an unstable (resp. stable ) disk. We denote each unstable (resp. stable) manifold/disk by $\gamma^u$ (resp. $\gamma^s$), and its tangent vectors by $v^u$ (resp. $v^s$).*

REMARK 3.19. *The singularity set $\mathbb{S}$ consists of the points in $X$ which are not "well-behaved". It includes the discontinuities and the points where the map $f^R$ is not differentiable. (It may also include other points in $X$ with some "bad" properties).*

DEFINITION 3.20 (Chernov-Markarian-Zhang (CMZ) structures). *We say that an induced system $(X, f^R)$ in Definition 3.17 is a CMZ structure on $(\mathcal{M}, f)$ if there are constants $C > 0$ and $\beta \in (0,1)$ such that the following conditions hold*

(1) *Hyperbolicity. For any $n \in \mathbb{N}$, $v^u$ and $v^s$,*
$$|D(f^R)^n v^u| \geq C\beta^{-n}|v^u|, \quad |D(f^R)^n v^s| \leq C\beta^n |v^s|,$$
*where $|\cdot|$ is the Riemannian metric induced from $\mathcal{M}$ to (un)stable manifolds.*

(2) *SRB measures and u-SRB measures. The conditional probability measure $\mu_X := \frac{\mu_{\mathcal{M}}|_X}{\mu_{\mathcal{M}}(X)} \ll \mathrm{Leb}_X$ with $\frac{d\mu_X}{d\,\mathrm{Leb}_X} \in L^\infty$ and $\frac{d\mu_X}{d\,\mathrm{Leb}_X} > 0$ $\mathrm{Leb}_X$-a.s. on $X$. Moreover, conditional distributions on $\gamma^u$ (say $\mu_{\gamma^u}$) are absolutely continuous w.r.t. Lebesgue measure $\mathrm{Leb}_{\gamma^u}$ on $\gamma^u$.*

(3) *Distortion bounds. Let $d_{\gamma^u}(\cdot,\cdot)$ be the distance measured along $\gamma^u$. By $\det D^u f^R$ we denote the Jacobian of $Df^R$ along the unstable manifolds. Then, if $x, y \in X$ belong to a $\gamma^u$, and $(f^R)^n$ is $C^2$-smooth on $\gamma^u$, then the following relation holds*
$$\log \frac{\det D^u(f^R)^n(x)}{\det D^u(f^R)^n(y)} \leq \psi\left[d_{\gamma^u}\left((f^R)^n x, (f^R)^n y\right)\right],$$
*where $\psi(\cdot)$ is some function, which does not depend on $\gamma^u$, and $\lim_{s\to 0^+} \psi(s) = 0$.*

(4) *Bounded curvatures. The curvatures of all $\gamma^u$ are uniformly bounded by $C$.*

(5) *Absolute continuity. Consider a holonomy map $h : \gamma_1^u \to \gamma_2^u$, which maps a point $x \in \gamma_1^u$ to the point $h(x) \in \gamma_2^u$, such that both $x$ and $h(x)$ belong to the same $\gamma^s$. We assume that the holonomy map satisfies the following relation*
$$\frac{\det D^u(f^R)^n(x)}{\det D^u(f^R)^n(h(x))} = C^{\pm 1} \text{ for all } n \geq 1 \text{ and } x \in \gamma_1^u,$$

(6) *Growth lemmas. For any sufficiently small $\delta > 0$ and for any disk on a $C^1$-smooth unstable manifold $\gamma^u$ with $\mathrm{diam}\,\gamma^u \leq \delta_0$ denote by $U_\delta \subseteq \gamma^u$ a $\delta$-neighborhood of the subset $\gamma^u \bigcap \bigcup_{0\leq i\leq N} (f^R)^{-i}\mathbb{S}$ within the set $\gamma^u$. Then there exist $N \in \mathbb{N}$, sufficiently small $\delta_0 > 0$ and constants $\kappa, \sigma > 0$ which satisfy the following condition. There exists an open subset $V_\delta \subseteq \gamma^u \setminus U_\delta$, such that $\mathrm{Leb}_{\gamma^u}\left(\gamma^u \setminus (U_\delta \bigcup V_\delta)\right) = 0$, and for any $\epsilon > 0$*
$$\mathrm{Leb}_{\gamma^u}(r_{V_\delta,N} < \epsilon) \leq 2\epsilon\beta + \epsilon C\delta_0^{-1}\,\mathrm{Leb}_{\gamma^u}(\gamma^u),$$
$$\mathrm{Leb}_{\gamma^u}(r_{U_\delta,0} < \epsilon) \leq C\delta^{-\kappa}\epsilon,$$
$$\mathrm{Leb}_{\gamma^u}(U_\delta) \leq C\delta^\sigma,$$
*where*
$$r_{U_\delta,0}(x) := d_{\gamma^u}(x, \partial U_\delta), \qquad r_{V_\delta,N}(x) := d_{(f^R)^N\gamma^u}\left((f^R)^N x, \partial (f^R)^N V_\delta(x)\right),$$
*and $V_\delta(x)$ is a connected component of $V_\delta$, which contains $x$.*

(7) *Mixing:* $\gcd\{R\} = 1$.

REMARK 3.21. *In the growth lemmas a positive integer $N$ is usually chosen as a sufficiently large number.The conditions that* $\gcd R = 1$ *and K-mixing of* $f^R$ *guarantee that* $f$ *is also K-mixing. This condition is not included in the original papers* [**12**, **13**]. *Here we include it because this can be easily verified in applications.*

Consider now the first return tower

$$\Delta := \{(x, n) \in X \times \{0, 1, 2, \cdots\} : n < R(x)\}.$$

Dynamics $F : \Delta \to \Delta$ is defined as $F(x, n) = (x, n + 1)$ if $n + 1 \leq R(x) - 1$ and $F(x, n) = (f^R x, 0)$ if $n = R(x) - 1$. The projection $\pi : \Delta \to \mathcal{M}$ is defined by $\pi(x, n) := f^n(x)$ as $\pi \circ F = f \circ \pi$. Finally we introduce projections $\pi_X : \Delta \to X$ and $\pi_{\mathbb{N}} : \Delta \to \mathbb{N}_0$, so that for any $(x, n) \in \Delta$

$$\pi_X(x, n) = x, \quad \pi_{\mathbb{N}}(x, n) = n. \tag{3.4}$$

Extend now $\mu_X$ from $X$ to $\Delta$ as

$$\mu_\Delta := (\int R d\mu_X)^{-1} \sum_j (F^j)_* (\mu_X|_{\{R>j\}}),$$

which reproduces the invariant probability measure on $\mathcal{M}$

$$\mu_{\mathcal{M}} = (\pi)_* \mu_\Delta.$$

We identify $\Delta_0 := (X \times \{0\}) \bigcap \Delta$ with $X$, $\mu_{\Delta_0}$ with $\mu_X$ and $F^R$ with $f^R$. Therefore $\pi : X \to X$ is the identity map.

Note that $\pi : \Delta \to \mathcal{M}$ is bijective. Thus $(\Delta, F)$ is identical to $(\mathcal{M}, f)$, and $(X, f^R) = (\Delta_0, F^R)$ is a CMZ structure on $(\Delta, F)$.

REMARK 3.22.

(1) *If* $R = 1$, *then* $X = \mathcal{M}$, *and thus* $(\mathcal{M}, f, \mu_{\mathcal{M}})$ *has a CMZ structure.*
(2) *It follows from* [**12**, **13**] *that* $(X, f^R, \mu_X)$ *can be modelled by a hyperbolic Young tower* [**31**].
(3) $(\mathcal{M}, f, \mu_{\mathcal{M}})$ *is a K-system (and therefore mixing) because of the condition that* $f^R$ *is K-mixing and* $\gcd\{R\} = 1$.

DEFINITION 3.23 (Holes and dynamical point processes). *Throughout this subsection, a hole within the boundary* $\partial Q$ *is an open disk* $B_r(q)$ *with radius* $r$ *and the center at a regular point* $q \in \partial Q$ *of the boundary of a billiard table.*

*Similar to Definition 3.1, define now a different dynamical point process* $\mathcal{N}^{r,q}$ *on* $\mathbb{R}^+ \bigcup \{0\}$. *For any measurable* $A \subseteq \mathbb{R}^+ \bigcup \{0\}$, *and any* $x \in \mathcal{M}$.

$$\begin{aligned} \mathcal{N}^{r,q}(x)(A) : &= \#\{i \geq 0 : f^i(x) \in B_r(q) \times [-\pi/2, \pi/2], \\ &\qquad\qquad i \cdot \mu_{\mathcal{M}}(B_r(q) \times [-\pi/2, \pi/2]) \in A\} \\ &= \sum_{i \cdot \mu_{\mathcal{M}}(B_r(q) \times [-\pi/2, \pi/2]) \in A} \mathbb{1}_{B_r(q) \times [-\pi/2, \pi/2]} \circ f^i(x). \end{aligned}$$

*We will usually drop the symbol* $x$ *and write*

$$\mathcal{N}^{r,q}(A) = \sum_{i \cdot \mu_{\mathcal{M}}(B_r(q) \times [-\pi/2, \pi/2]) \in A} \mathbb{1}_{B_r(q) \times [-\pi/2, \pi/2]} \circ f^i.$$

*By using* $\mu_{\mathcal{M}} := (\pi)_* \mu_\Delta$, *we get*

$$\mathcal{N}^{r,q}(A) = \sum_{i \cdot \mu_\Delta(A_r) \in A} \mathbb{1}_{A_r} \circ F^i,$$

*where* $A_r := \pi^{-1}(B_r(q) \times [-\pi/2, \pi/2])$.

REMARK 3.24. *Following the theory of point processes we have that*

$$\mathcal{N}^{r,q}(x) = \sum_{i\geq 0: f^i(x)\in B_r(q)\times[-\pi/2,\pi/2]} \delta_{i\cdot\mu_{\mathcal{M}}(B_r(q)\times[-\pi/2,\pi/2])},$$

*where $\delta$ is a Dirac measure. Hence, $\mathcal{N}^{r,q}$ is a random counting measure, e.g., $\mathcal{N}^{r,q}(x)[0,1]$ counts the number of $i \in [0, \frac{1}{\mu_{\mathcal{M}}\{B_r(q)\times[-\pi/2,\pi/2]\}}]$, such that $f^i(x)$ belongs to $B_r(q) \times [-\pi/2, \pi/2]$. $B_r(q) \times [-\pi/2, \pi/2]$ is a large hole in the phase space $\mathcal{M}$.*

DEFINITION 3.25 (Poisson approximations for large holes). *We say that $\mathcal{N}^{r,q} \to_d$ Poisson (see Definition 3.2) if $\mathcal{N}^{r,q}g \to_d$ Poisson$(g)$ for any $g \in C_c^+(\mathbb{R}^+\bigcup\{0\})$, i.e.,*

$$\lim_{r\to 0}\int \exp(-t\cdot\mathcal{N}^{r,q}g)d\mu_{\mathcal{M}} = \int \exp[-t\cdot Poisson(g)]d\mathbb{P} \text{ for all } t > 0,$$

*where $C_c^+(\mathbb{R}^+\bigcup\{0\})$ is the space consisting of positive continuous functions with a compact support, defined on $[0,\infty)$.*

*This is equivalent to the relation*

$$(\mathcal{N}^{r,q}I_1, \mathcal{N}^{r,q}I_2, \cdots, \mathcal{N}^{r,q}I_k) \to_d (Poisson(I_1), Poisson(I_2), \cdots, Poisson(I_k))$$

*for any $k \in \mathbb{N}$ and any bounded intervals $I_1, \cdots I_k \subseteq R^+\bigcup\{0\}$.*

*In other words, the limit distribution of $\mathcal{N}^{r,q}$ is Poissonian, when the disk $B_r(q)$ shrinks to a regular point on the boundary $\partial Q$ of a billiard table.*

REMARK 3.26. *If $I_1, \cdots, I_k$ are restricted inside an interval $[0,T]$, the convergence in this definition is equivalent to the $d_{TV}$-convergence in Definition 3.4 without convergence rates.*

DEFINITION 3.27 (Sections and quasi-sections). *Recall that $\pi_X : \Delta \to X$ is defined as $\pi_X(x,n) = x$. We say that $B_r(q) \times [-\pi/2, \pi/2] \subseteq \mathcal{M}$ is a section if $\pi_X : \pi^{-1}(B_r(q) \times [-\pi/2, \pi/2]) \to X$ is injective for any sufficiently small $r > 0$. Further, $B_r(q) \times [-\pi/2, \pi/2] \subseteq \mathcal{M}$ is a quasi-section if for any sufficiently small $r > 0$ there is a measurable set $S_r \subseteq B_r(q) \times [-\pi/2, \pi/2]$, such that $\mu_{\mathcal{M}}[(B_r(q) \times [-\pi/2, \pi/2]) \setminus S_r] = o(\mu_{\mathcal{M}}[B_r(q)\times[-\pi/2,\pi/2]])$, and $\pi_X : \pi^{-1}S_r \to X$ is injective. In this case, we also refer to $S_r$ as a section in $B_r(q) \times [-\pi/2, \pi/2]$.*

REMARK 3.28. *In applications to two-dimensional billiards, $B_r(q)\times[-\pi/2,\pi/2]$ is a strip in $\mathcal{M}$, and $B_r(q) \times [-\pi/2, \pi/2] \setminus S_r$ is usually a union of finitely many rectangles, whose measures are of order $O(r^2)$. To avoid unnecessary complications, we always assume that $B_r(q)\times[-\pi/2,\pi/2]\setminus S_r$ has a regular shape, e.g., as a union of finitely many rectangles.*

ASSUMPTION 3.29 (Geometric assumptions).

(1) *For a.e. $q \in \partial Q$ the set $B_r(q) \times [-\pi/2, \pi/2]$ is a quasi-section.*

(2) *$\bigcup_{i\geq 1}\partial X_i \subseteq \mathbb{S}$ (see the definition of $X_i$ in (3.3)).*

(3) *There are constants $C > 0$ and $\alpha \in (0,1]$, such that for any $\gamma^k$, where $k = u$ or $s$, and for any $x, y \in \gamma^k$,*

$$d_{f^j\gamma^k}(f^jx, f^jy) \leq Cd_{\gamma^k}(x,y)^\alpha \text{ for all } j \in [0, R(x)).$$

*(Observe that the condition $\bigcup_{i\geq 1}\partial X_i \subseteq \mathbb{S}$ implies that $\gamma^k \subseteq X_i$ for some $i \geq 1$).*

(4) *There exist two cones (i.e. affine subspaces) $C^u, C^s \subseteq \mathcal{T}(\mathcal{M})$, such that*

$$\dim(\operatorname{int} C^u \bigcap \operatorname{int} C^s) < 1, \quad (Df)C^u \subseteq C^u, \quad (Df)^{-1}C^s \subseteq C^s,$$

*and for all $n \geq 1$ and $\mathrm{Leb}_{\partial Q}$-a.e. $q \in \partial Q$*

$$\dim\left(\mathcal{T}(\{q\}\times[-\pi/2,\pi/2])\bigcap \operatorname{int} C^u\right)<1, \quad \dim\left(\mathcal{T}(\{q\}\times[-\pi/2,\pi/2])\bigcap \operatorname{int} C^s\right)<1,$$

$$(Df)^n\mathcal{T}(\{q\} \times [-\pi/2, \pi/2]) \subseteq \operatorname{int} C^u, \quad (Df)^{-n}\mathcal{T}(\{q\} \times [-\pi/2, \pi/2]) \subseteq \operatorname{int} C^s,$$

*where $\operatorname{int} C^u$ (resp. $\operatorname{int} C^s$) is the interior of $C^u$ (resp. $C^s$) and $\mathcal{T}(A)$ means a tangent space of a (sub)manifold $A$.*

REMARK 3.30.

(1) *The existence of $C^u, C^s$ always holds for two dimensional hyperbolic billiards.*
(2) *The Hölder condition is natural, and it is traditionally used for hyperbolic systems, (particularly, for billiards).*

Now we can present the main results in this subsection.

THEOREM 3.31 (Poisson limit laws, see [**7**]). *Suppose that dynamical system $(\mathcal{M}, f)$ has a CMZ structure $(X, f^R)$ (see Definitions 3.20 and 3.17), and the Assumption 3.29 holds. Then, when $r \to 0$, we have $\mathcal{N}^{r,q} \to_d$ Poisson (see Definition 3.25) for $\mathrm{Leb}_{\partial Q}$-a.s. $q \in \partial Q$.*

REMARK 3.32. *The convergence rate (see Definition 3.4) for this approximation in Theorem 3.31 can be obtained by maximal large deviations [**6**] and assuming polynomial mixing rates, while in Theorem 3.31 the convergence can be proven without specifying mixing rates.*

COROLLARY 3.33 (First hitting times, see [**7**]). *Under the same conditions as in Theorem 3.31 consider the moment of time when the first hitting of (passage through) the hole occurs, i.e.,*

$$\tau_{r,q}(x) := \inf\{n \geq 1 : f^n(x) \in B_r(q) \times [-\pi/2, \pi/2]\}$$

*for any $x \in \mathcal{M}$. Then for any $t > 0$ and almost every $q \in \partial Q$, the following relation for the first hitting probability holds*

$$\lim_{r\to 0} \mu_{\mathcal{M}}\Big\{\tau_{r,q} > t/\mu_{\mathcal{M}}\Big(B_r(q) \times [-\pi/2, \pi/2]\Big)\Big\} = e^{-t}.$$

We present now applications of Theorem 3.31 to some weakly chaotic billiards.

3.4.1. *Squashes or Stadium-type billiards.* A billiard table $Q$ of a squash billiard is a convex domain bounded by two circular arcs and two straight (flat) segments tangent to the arcs at their common endpoints. A squash billiard is called a stadium if flat sides are parallel, see Figure 1. Initially called squash billiards, they were later sometimes called "skewed" stadia, drive-belt billiards, etc, see Figure 2. Note that squashes contain a boundary arc, which is longer than a half circle. Applying Theorem 3.31 we have the following.

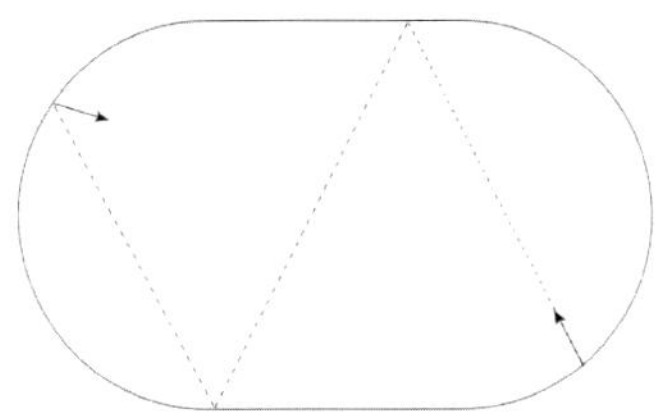

FIGURE 1. Stadium billiard

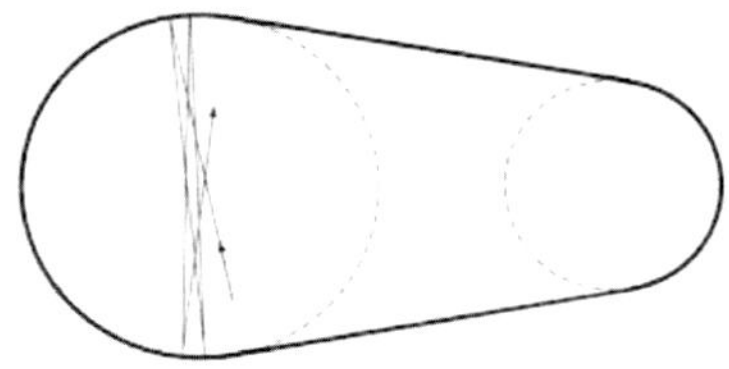

FIGURE 2. Squash billiard

COROLLARY 3.34. *Theorem* 3.31 *holds for squash billiards and stadiums.*

3.4.2. *Another class of billiards with focusing components.* In this section we consider billiard tables $Q$ for which each smooth component $\Gamma_i \subseteq \partial Q$ of the boundary is either dispersing, i.e., convex inwards, or focusing, i.e., convex outwards. A curvature of every dispersing component is bounded away from zero and infinity. We assume that every focusing component is an arc of a circle, and that there are no points of $\partial Q$ on that circle or inside it, other than the arc itself (the strong focusing or SFC-condition). We assume also that two dispersing components intersect (if they do) transversally (i.e., there are no cusps) and, besides, every focusing arc is not longer than a half of the corresponding circle, e.g. see the Figure 3. Applying Theorem 3.31 we have the following.

FIGURE 3. Flower billiard

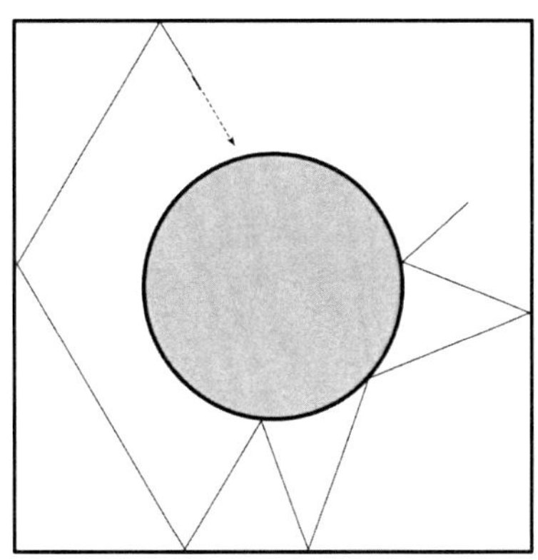

FIGURE 4. Semidispersing billiard

COROLLARY 3.35. *Theorem* 3.31 *holds for the class of billiards considered in this subsection.*

3.4.3. *Semi-dispersing billiards.* In this subsection we consider billiard tables of the following type. Let $R_0 \subseteq \mathbb{R}^2$ be a rectangle, and scatterers $B_1, \cdots, B_r \subseteq \operatorname{int} R_0$ are open strictly convex sub-domains with smooth, (at least $C^3$)), or piece-wise smooth boundaries whose curvature is bounded away from zero, and such that $B_i \bigcap B_j = \emptyset$ for $i \neq j$. The boundary of a billiard table $Q = R_0 \setminus \bigcup_i B_i$ is partially dispersing (convex inwards) and partially neutral (flat), e.g. see Figure 4. The flat part is $\partial R_0$. Applying Theorem 3.31 we have the following.

COROLLARY 3.36. *Theorem 3.31 holds for the class of semi-dispersing billiards considered in this subsection.*

## 4. Central limit theorems and invariance principles

In this section we consider random dynamical systems. A special subclass of random systems is formed by stationary systems.

DEFINITION 4.1 (Random and stationary dynamical systems). *A random dynamical system is defined as $(\mathcal{M}, (T_\omega)_{\omega\in\Omega}, \Omega, \mathbb{P}, \sigma, (\mu_\omega)_{\omega\in\Omega})$. Here $\sigma : (\Omega, \mathbb{P}) \to (\Omega, \mathbb{P})$ preserves a probability measure $\mathbb{P}$, and it is ergodic, $T_\omega : \mathcal{M} \to \mathcal{M}$ is a map, $\mu_\omega$ is a probability measure on $X$ satisfying $(T^k_\omega)_*\mu_\omega = \mu_{\sigma^k\omega}, k \geq 0$ where $T^k_\omega := T_{\sigma^{k-1}\omega} \circ \cdots \circ T_{\sigma\omega} \circ T_\omega$.*

*If $\Omega$ is a single point, then we call this system a stationary dynamical system.*

DEFINITION 4.2 (Vector-valued Almost Sure Invariance Principle (VASIP)). *For a random dynamical system $(\mathcal{M}, (T_\omega)_{\omega\in\Omega}, \Omega, \mathbb{P}, \sigma, (\mu_\omega)_{\omega\in\Omega})$, consider observables (functions on the phase space of all states of the system) $\{\phi_\omega \in L^\infty(\mathcal{M}, \mu_\mathcal{M}; \mathbb{R}^d) : \omega \in \Omega\}$ which satisfy*

$$\sup_{\omega\in\Omega} \|\phi_\omega\|_\infty < \infty, \qquad \int \phi_\omega d\mu_\omega = 0 \textit{ for all } \omega \in \Omega.$$

*We say that $(\phi_{\sigma^k\omega} \circ T^k_\omega)_{k\geq 1,\omega\in\Omega}$ satisfies a quenched vector-valued almost sure invariance principle (quenched VASIP) if, for a.e. $\omega \in \Omega$, there exists a constant $\epsilon \in (0,1)$ and a family of zero mean d-dimensional independent Gaussian random vectors $(G^\omega_k)_{k\geq 1}$ defined on some extended probability space of $(\mathcal{M}, \mu_\omega)$ such that*

$$\sum_{k=1}^{n} \phi_{\sigma^k\omega} \circ T^k_\omega - \sum_{k=1}^{n} G^\omega_k = o(n^{\frac{1-\epsilon}{2}}) \textit{ a.s.} \tag{4.1}$$

$$\sigma^2_n(\omega) = \sum_{k=1}^{n} \tilde{\mathbb{E}}^\omega[(G^\omega_k)(G^\omega_k)^T] + o(n^{1-\epsilon}), \tag{4.2}$$

$$\sigma^2_n(\omega) \approx_\omega nI_{d\times d}, \tag{4.3}$$

*where $\sigma^2_n(\omega) := \int (\sum_{k=1}^n \phi_{\sigma^k\omega} \circ T^k_\omega)(\sum_{k=1}^n \phi_{\sigma^k\omega} \circ T^k_\omega)^T d\mu_\omega$ and $\tilde{\mathbb{E}}^\omega$ in (4.2) is the expectation w.r.t. the probability $\tilde{P}^\omega$ on an extended probability space of $(\mathcal{M}, \mu_\omega)$. Here the extended probability space is $(\mathcal{M}, \mu_\omega) \otimes \otimes_{i\geq 1}(I_i, \mathrm{Leb})$ where $I_i$ is a copy of $[0,1]$.*

REMARK 4.3. *If $d = 1$, then the quenched VASIP implies the quenched central limit theorem (CLT):*

$$\lim_{n\to\infty} \mu_\omega\Big(\frac{\sum_{k=1}^n \phi_{\sigma^k\omega} \circ T^k_\omega}{\Sigma\sqrt{n}} \leq z\Big) = \frac{1}{\sqrt{2\pi}} \int_{-\infty}^{z} e^{-s^2/2} ds \textit{ for all } z \in \mathbb{R},$$

*(in other words, $\frac{\sum_{k=1}^n \phi_{\sigma^k\omega} \circ T^k_\omega}{\Sigma\sqrt{n}}$ converges in distribution to a Gaussian variable $\mathcal{N}(0,1)$) and the quenched iterated logarithm law(LIL):*

$$\limsup_{n\to\infty} \frac{\sum_{k=1}^n \phi_{\sigma^k\omega} \circ T^k_\omega}{\sqrt{n \log\log n}} = \Sigma \quad \mu_\omega\text{-a.s.}$$

*for a.e. $\omega \in \Omega$ and a constant $\Sigma > 0$, which does not depend on $\omega$.*

REMARK 4.4. *If $\Omega$ is a single point $\{\omega_0\}$, then the random system becomes a stationary system. Then VASIP, CLT and LIL have the following forms, respectively:*

$$\sum_{k=1}^{n} \phi_{\omega_0} \circ T_{\omega_0}^k - \sum_{k=1}^{n} G_k = o(n^{\frac{1-\epsilon}{2}}) \ a.s.$$

$$\lim_{n\to\infty} \mu_{\omega_0}\Big(\frac{\sum_{k=1}^n \phi_{\omega_0} \circ T_{\omega_0}^k}{\Sigma\sqrt{n}} \le z\Big) = \frac{1}{\sqrt{2\pi}} \int_{-\infty}^{z} e^{-s^2/2} ds \text{ for all } z \in \mathbb{R},$$

$$\limsup_{n\to\infty} \frac{\sum_{k=1}^n \phi_{\omega_0} \circ T_{\omega_0}^k}{\sqrt{n \log\log n}} = \Sigma \quad \mu_{\omega_0}\text{-}a.s.$$

The following is a general VASIP result for random dynamical systems.

THEOREM 4.5 (Random Dynamical Systems, see [**27**]). *Consider a random dynamical system $(\mathcal{M}, (T_\omega)_{\omega\in\Omega}, \Omega, \mathbb{P}, \sigma, (\mu_\omega)_{\omega\in\Omega})$, where $d\mu_\omega = h_\omega d\mu_{\mathcal{M}}$ is a probability distribution on $\mathcal{M}$ and $\mu_{\mathcal{M}}$ is a measure on $\mathcal{M}$. Let the transfer operators (Perron-Frobenius operators) $P_\omega$ associated to $T_\omega$ are defined by the following duality relations*

$$\int g \cdot P_\omega(h) d\mu_{\mathcal{M}} = \int g \circ T_\omega \cdot h d\mu_{\mathcal{M}} \text{ for all } h \in L^1(\mu_{\mathcal{M}}),\ g \in L^\infty(\mu_{\mathcal{M}}).$$

*Let the observables $\{\phi_\omega \in L^\infty(\mathcal{M}, \mu_{\mathcal{M}}; \mathbb{R}^d) : \omega \in \Omega\}$ be the ones from the Definition 4.2. Suppose that they satisfy the conditions* (A1')–(A3') *below, and there is a constant $C > 0$ such that for any $i, j \ge 0, n \ge 1$, and for almost every $\omega \in \Omega$,*

(A1') $$\int |P_{\sigma^i\omega}^n(\phi_{\sigma^i\omega} h_{\sigma^i\omega})| d\mu_{\mathcal{M}} \le C n^{1-1/\alpha},$$

(A2') $$\int \Big|P_{\sigma^i\omega}^n\Big[(\phi_{\sigma^i\omega}\phi_{\sigma^i\omega}^T - \int \phi_{\sigma^i\omega}\phi_{\sigma^i\omega}^T d\mu_{\sigma^i\omega}) h_{\sigma^i\omega}\Big]\Big| d\mu_{\mathcal{M}} \le C n^{1-1/\alpha},$$

(A3')
$$\int \Big|P_{\sigma^{i+j}\omega}^n\Big[P_{\sigma^i\omega}^j(\phi_{\sigma^i\omega} h_{\sigma^i\omega})\phi_{\sigma^{i+j}\omega}^T - h_{\sigma^{i+j}\omega}\int P_{\sigma^i\omega}^j(\phi_{\sigma^i\omega} h_{\sigma^i\omega})\phi_{\sigma^{i+j}\omega}^T d\mu_{\mathcal{M}}\Big]\Big| d\mu_{\mathcal{M}} \le C n^{1-1/\alpha},$$

*where $\alpha < 1/2$, $|\cdot|$ is the Euclidean norm for vectors or the spectral norm for matrices.*

*Then there are two linear subspaces (which do not depend on $\omega$): $W_1, W_2 \subseteq \mathbb{R}^d$, $\mathbb{R}^d = W_1 \bigoplus W_2$ with projections $\pi_1 : W_1 \bigoplus W_2 \to W_1, \pi_2 : W_1 \bigoplus W_2 \to W_2$ such that*

(1) *$(\pi_1 \circ \phi_{\sigma^k\omega} \circ T_\omega^k)_{k\ge1,\omega\in\Omega}$ satisfies the quenched VASIP (see Definition 4.2).*
(2) *$(\pi_2 \circ \phi_{\sigma^k\omega} \circ T_\omega^k)_{k\ge1,\omega\in\Omega}$ is a coboundary, i.e., there is $\psi \in L^1(\Omega\times X, d\mu_\omega d\mathbb{P})$ such that*

$$\pi_2 \circ \phi_{\sigma\omega}(T_\omega x) = \psi(\sigma\omega, T_\omega x) - \psi(\omega, x) \ a.e.\ (\omega, x) \in \Omega \times X.$$

The paper [**23**] considers the so called Pomeau-Manneville type maps, which are expanding everywhere besides at an indifferent (parabolic) point. Existence of just one such point may essentially change dynamics. This class of maps is defined in the following way. For $\beta \in [0,1)$, define the map

$$T_\beta(x) = \begin{cases} x + 2^\beta x^{1+\beta}, & 0 \le x \le 1/2 \\ 2x-1, & 1/2 < x \le 1 \end{cases}. \tag{4.4}$$

According to [**23**, **32**], its mixing rate is $O(n^{1-1/\beta})$ and this map preserves a probability measure absolutely continuous respect to the Lebesgue measure on $[0,1]$. If we randomly perturbed the parameters $\beta \in [0,1/2)$, and apply Theorem 4.5 to the class of these random systems, then the following result holds.

COROLLARY 4.6 (Polynomially mixing random systems, see [**27**]). *Let* $\mu_{\mathcal{M}} :=$ Leb *be the Lebesgue measure on* $\mathcal{M} := [0,1]$, $\Omega := [0,1/2)^{\mathbb{Z}}$, *and* $T_\omega := T_{\omega_{[0]}}$ *are the Pomeau-Manneville type maps* (4.4) *which are picked from* $\{T_\beta : \beta \in [0,1/2)\}$. *Let also* $\sigma : \Omega \to \Omega$ *is a* $\mathbb{P}$*-preserving ergodic and invertible left shift, and* $\omega_{[0]}$ *is the 0th entry of* $\omega$.

*Then there are functions* $h_\omega \in L^1(\text{Leb})$ *and probabilities* $d\mu_\omega := h_\omega d\,\text{Leb}$ *such that* $(T_\omega)_*\mu_\omega = \mu_{\sigma\omega}$ *for a.e.* $\omega \in \Omega$. *Moreover, suppose that the observables* $(\phi_\omega)_{\omega\in\Omega} \subseteq \text{Lip}([0,1];\mathbb{R}^d)$ *satisfy the relations* $\sup_\omega \|\phi_\omega\|_{\text{Lip}} < \infty$ *and* $\int \phi_\omega d\mu_\omega = 0$. *Then there are two linear subspaces (which do not depend on* $\omega$*)* $W_1, W_2 \subseteq \mathbb{R}^d$, $\mathbb{R}^d = W_1 \bigoplus W_2$ *with projections* $\pi_1 : W_1 \bigoplus W_2 \to W_1, \pi_2 : W_1 \bigoplus W_2 \to W_2$ *such that*

(1) $(\pi_1 \circ \phi_{\sigma^k\omega} \circ T^k_\omega)_{k\ge 1,\omega\in\Omega}$ *satisfies the quenched VASIP.*
(2) $(\pi_2 \circ \phi_{\sigma^k\omega} \circ T^k_\omega)_{k\ge 1,\omega\in\Omega}$ *is a coboundary, which means that there is* $\psi \in L^1(\Omega \times [0,1], d\mu_\omega d\mathbb{P})$ *such that*

$$\pi_2 \circ \phi_{\sigma\omega}(T_\omega x) = \psi(\sigma\omega, T_\omega x) - \psi(\omega, x) \text{ a.e. } (\omega, x) \in \Omega \times [0,1].$$

REMARK 4.7. $\sigma : \Omega \to \Omega$ *is ergodic, and* $1/2$ *is optimal, see also* [**16**]. *If* $\sigma$ *is Bernoulli, then the range of* $\beta$ *can be larger than* $[0,1/2)$ *(see* [**28**]*).*

REMARK 4.8. *By Remark 4.4, when* $\beta \in [0,1/2), \Omega = \{\beta\}$ *and* $\phi$ *has zero mean, then we obtain the corresponding results for the single dynamics* $T_\beta$ *and a zero mean function* $\phi$

$$\sum_{k=1}^n \phi \circ T^k_\beta - \sum_{k=1}^n G_k = o(n^{\frac{1-\epsilon}{2}}) \text{ a.s.}$$

$$\frac{\sum_{k=1}^n \phi \circ T^k_\beta}{\Sigma\sqrt{n}} \to_d \mathcal{N}(0,1)$$

*(Gaussian distribution) or there is a function* $\chi$ *such that* $\phi = \chi \circ T_\beta - \chi$ *a.s. Different statistics (called escape rates and orbits preferences) of this stationary dynamical system can be found in the papers* [**9**, **29**].

If $\beta \in (1/2,1)$, then weakly chaotic stationary dynamics $T_\beta$ does not satisfy the standard CLT or VASIP, but the following different limit theorems hold [**16**].

## 5. Stable limit laws

Let $1/2 < \beta < 1$, and the corresponding stationary dynamics $T_\beta : [0,1] \to [0,1]$ be the Liverani-Saussol-Vaienti map described by (4.4).

THEOREM 5.1 (see [**16**]). *Let h be the density of its unique absolutely continuous invariant probability measure $\mu_\beta$, and $\phi : [0,1] \to \mathbb{R}$ is Hölder with $\int \phi d\mu_\beta = 0$.*

*If $\phi(0) \neq 0$, then the following stable limit law holds:*

$$\frac{\sum_{i\le n} \phi \circ T_\beta^i}{n^\beta} \to_d X_{1/\beta, c, sgn(\phi(0))}$$

*whose characteristic function is*

$$\mathbb{E}(e^{itX_{1/\beta,c,sgn(\phi(0))}}) = e^{-c|t|^{1/\beta}(1-isgn(\phi(0))sgn(t)\tan(\pi/(2\beta)))}$$

*with* $c = \frac{h(1/2)}{4(\beta/|\phi(0)|)^{1/\beta}}\Gamma(1-1/\beta)\cos(\pi/(2\beta))$.

*If $\phi(0) = 0$, assume also that $|\phi(x)| \le Cx^\gamma$ with $\gamma > \beta - 1/2$. Then there exists a constant $\sigma^2 \ge 0$ such that*

$$\frac{\sum_{i\le n} \phi \circ T_\beta^i}{\sqrt{n}} \to_d \mathcal{N}(0, \sigma^2).$$

*Moreover, $\sigma^2 = 0$ if and only if there exists a measurable function $\chi$ such that $\phi = \chi \circ T_\beta - \chi$.*

Besides expanding maps $T_\beta$, we next consider two-dimensional dispersing billiards with a cusp at a flat point. More precisely, for any fixed constant $\eta > 2$ (which will determine the sharpness of the cusp), we take a dispersing billiard table $Q = Q_\eta$ with a boundary consisting of a finite number of $C^3$ smooth curves $\Gamma_i$, $i = 1 \cdots, n_0$, where $n_0 \ge 3$, where a cusp formed by two curves $\Gamma_1, \Gamma_2$ such that cusp at $P = \Gamma_1 \bigcap \Gamma_2$. Moreover, both $\Gamma_1$ and $\Gamma_2$ have zero derivatives up to $\eta - 1$ order at $P$, and the $\eta$-order derivative is not zero. We also assume that all other boundary components are dispersing and have curvature bounded away from zero. Choose now a Cartesian coordinate system $(s, z)$ with the origin at $P$, and with the horizontal $s$-axis being the tangent line to both $\Gamma_1$ and $\Gamma_2$. Assume also that $\Gamma_1$ and $\Gamma_2$ can be represented as

$$z_1(s) = \eta^{-1}s^\eta, z_2(s) = -\eta^{-1}s^\eta \tag{5.1}$$

for $s \in [0, \epsilon_0]$ with $\epsilon_0 > 0$ being a small fixed number. The mixing rate for this billiard system is $O(n^{-1/(\eta-1)})$. A stable limit law for this billiard system $(\mathcal{M}, f)$ (see Definition 3.16) is proved by Jung and Zhang [**21**].

THEOREM 5.2 (see [**21**]). *Let $Q_\eta$, where $\eta \in (2, \infty)$, is a billiard table with a cusp defined by (5.1), $\alpha = \eta/(\eta - 1)$ and suppose that a mean zero function $\psi$ is Hölder and defined on $\mathcal{M}$. Let also*

$$I_\psi := 1/4 \int_0^\pi (\psi(q'', \phi) + \psi(q', \phi)) \sin^{1/\alpha} \phi d\phi \neq 0,$$

*where $q', q''$ are the $q$-coordinates of the cusps point $P$ in the phase space $\mathcal{M}$ with respect to $\Gamma_1$ and $\Gamma_2$, respectively. Then as $n \to \infty$,*

$$\frac{\sum_{i\le n} \psi \circ f^i}{n^{1/\alpha}} \to_d S_{\alpha,\sigma}$$

*where* $\sigma = \frac{2I_\psi}{\eta \,\mathrm{Leb}_{\partial Q}\, \partial Q}$ *and the characteristic function of $S_{\alpha,\sigma}$ is*

$$\mathbb{E}(e^{itS_{\alpha,\sigma}}) = e^{-|\sigma|^\alpha |t|^\alpha (1-isgn(t)\tan(\alpha\pi/2))}.$$

REMARK 5.3. *We note that when the mixing rate is slower and slower, the Poisson limit laws (Theorem* 3.31*) still hold, while the central limit theorems (Remark* 4.8*) will change to stable limit laws (Theorem* 5.1*).*

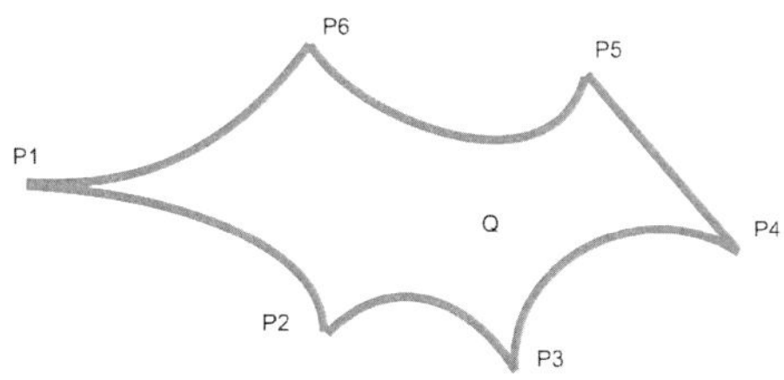


FIGURE 5. A billiard table with several cusps at flat-points [**20**]

REMARK 5.4. *In a slightly more complicated case, a billiard table has, possibly non-isometric and asymmetric cusps, at flat points (see Figure* 5*). It was shown in*[**19**, **20**, **26**] *that in such situation a functional limit law with Hölder observables holds in Skorokhod* $\mathcal{M}_1$*-topology (or the weaker* $\mathcal{M}_2$*-topology) if*

$$I_\psi(s) := 1/4 \int_0^s (\psi(q'', \phi) + \psi(q', \phi)) \sin^{1/\alpha} \phi d\phi, s \in [0, \pi]$$

*is monotone (or the image of* $I_\psi$ *is contained in* $[0, I_\psi(\pi)]$*). We refer the readers to* [**19**, **20**, **26**] *for more details.*

## Acknowledgments

The second author thanks Professor Ian Melbourne and the anonymous referees for useful comments.

## References

[1] José F. Alves and Davide Azevedo, *Statistical properties of diffeomorphisms with weak invariant manifolds*, Discrete Contin. Dyn. Syst. **36** (2016), no. 1, 1–41, DOI 10.3934/dcds.2016.36.1. MR3369212

[2] José F. Alves, Jorge M. Freitas, Stefano Luzzatto, and Sandro Vaienti, *From rates of mixing to recurrence times via large deviations*, Adv. Math. **228** (2011), no. 2, 1203–1236, DOI 10.1016/j.aim.2011.06.014. MR2822221

[3] Viviane Baladi, *Positive transfer operators and decay of correlations*, Advanced Series in Nonlinear Dynamics, vol. 16, World Scientific Publishing Co., Inc., River Edge, NJ, 2000, DOI 10.1142/9789812813633. MR1793194

[4] Leonid A. Bunimovich and Yaofeng Su, *Limit theorems for general and irregular observables in slowly mixing hyperbolic systems.*

[5] Leonid A. Bunimovich and Yaofeng Su, *Poisson approximations and convergence rates for hyperbolic dynamical systems*, Comm. Math. Phys. **390** (2022), no. 1, 113–168, DOI 10.1007/s00220-022-04309-w. MR4381186

[6] Leonid A. Bunimovich and Yaofeng Su, *Maximal large deviations and slow recurrences in weakly chaotic systems*, Adv. Math. **432** (2023), Paper No. 109267, 58, DOI 10.1016/j.aim.2023.109267. MR4630843

[7] Leonid A. Bunimovich and Yaofeng Su, *Back to boundaries in billiards*, Comm. Math. Phys. **405** (2024), no. 6, Paper No. 140, 74. MR4751722

[8] Leonid A. Bunimovich and Yaofeng Su, *Estimates of constants in the limit theorems for chaotic dynamical systems*, Stoch. Dyn. **24** (2024), no. 1, Paper No. 2450004, 15, DOI 10.1142/S0219493724500047. MR4745238

[9] Leonid A. Bunimovich and Yaofeng Su, *Which subsets and when orbits of non-uniformly hyperbolic systems prefer to visit: operator renewal theory approach*, arXiv:2412.04615, 2024

[10] J.-R. Chazottes and P. Collet, *Poisson approximation for the number of visits to balls in non-uniformly hyperbolic dynamical systems*, Ergodic Theory Dynam. Systems **33** (2013), no. 1, 49–80, DOI 10.1017/S0143385711000897. MR3009103

[11] Jean-René Chazottes, *Fluctuations of observables in dynamical systems: from limit theorems to concentration inequalities*, Nonlinear dynamics new directions, Nonlinear Syst. Complex., vol. 11, Springer, Cham, 2015, pp. 47–85, DOI 10.1007/978-3-319-09867-8_4. MR3379397

[12] N. Chernov, *Decay of correlations and dispersing billiards*, J. Statist. Phys. **94** (1999), no. 3-4, 513–556, DOI 10.1023/A:1004581304939. MR1675363

[13] N. Chernov and H.-K. Zhang, *Billiards with polynomial mixing rates*, Nonlinearity **18** (2005), no. 4, 1527–1553, DOI 10.1088/0951-7715/18/4/006. MR2150341

[14] Mark F. Demers, Niloofar Kiamari, and Carlangelo Liverani, *Transfer operators in hyperbolic dynamics—an introduction*, 33° Colóquio Brasileiro de Matemática, Instituto Nacional de Matemática Pura e Aplicada (IMPA), Rio de Janeiro, 2021. MR4472820

[15] Ana Cristina Moreira Freitas, Jorge Milhazes Freitas, and Mário Magalhães, *Complete convergence and records for dynamically generated stochastic processes*, Trans. Amer. Math. Soc. **373** (2020), no. 1, 435–478, DOI 10.1090/tran/7922. MR4042881

[16] Sébastien Gouëzel, *Central limit theorem and stable laws for intermittent maps*, Probab. Theory Related Fields **128** (2004), no. 1, 82–122, DOI 10.1007/s00440-003-0300-4. MR2027296

[17] Sébastien Gouëzel, *Almost sure invariance principle for dynamical systems by spectral methods*, Ann. Probab. **38** (2010), no. 4, 1639–1671, DOI 10.1214/10-AOP525. MR2663640

[18] Nicolai Haydn and Sandro Vaienti, *Limiting entry and return times distribution for arbitrary null sets*, Comm. Math. Phys. **378** (2020), no. 1, 149–184, DOI 10.1007/s00220-020-03795-0. MR4124984

[19] Paul Jung, Ian Melbourne, Françoise Pène, Paulo Varandas, and Hong-Kun Zhang, *Necessary and sufficient condition for $\mathcal{M}_2$-convergence to a Lévy process for billiards with cusps at flat points*, Stoch. Dyn. **21** (2021), no. 5, Paper No. 2150024, 8, DOI 10.1142/S0219493721500246. MR4297947

[20] Paul Jung, Françoise Pène, and Hong-Kun Zhang, *Convergence to $\alpha$-stable Lévy motion for chaotic billiards with several cusps at flat points*, Nonlinearity **33** (2020), no. 2, 807–839, DOI 10.1088/1361-6544/ab5148. MR4048780

[21] Paul Jung and Hong-Kun Zhang, *Stable laws for chaotic billiards with cusps at flat points*, Ann. Henri Poincaré **19** (2018), no. 12, 3815–3853, DOI 10.1007/s00023-018-0726-y. MR3877517

[22] A. N. Kolmogorov, *A new metric invariant of transient dynamical systems and automorphisms in Lebesgue spaces* (Russian), Dokl. Akad. Nauk SSSR (N.S.) **119** (1958), 861–864. MR103254

[23] Carlangelo Liverani, Benoît Saussol, and Sandro Vaienti, *A probabilistic approach to intermittency*, Ergodic Theory Dynam. Systems **19** (1999), no. 3, 671–685, DOI 10.1017/S0143385799133856. MR1695915

[24] Valerio Lucarini, Davide Faranda, Ana Cristina Gomes Monteiro Moreira de Freitas, Jorge Miguel Milhazes de Freitas, Mark Holland, Tobias Kuna, Matthew Nicol, Mike Todd, and Sandro Vaienti, *Extremes and recurrence in dynamical systems*, Pure and Applied Mathematics (Hoboken), John Wiley & Sons, Inc., Hoboken, NJ, 2016, DOI 10.1002/9781118632321. MR3558780

[25] Ian Melbourne, *Large and moderate deviations for slowly mixing dynamical systems*, Proc. Amer. Math. Soc. **137** (2009), no. 5, 1735–1741, DOI 10.1090/S0002-9939-08-09751-7. MR2470832

[26] Ian Melbourne and Paulo Varandas, *Convergence to a Lévy process in the Skorohod $\mathcal{M}_1$ and $\mathcal{M}_2$ topologies for nonuniformly hyperbolic systems, including billiards with cusps*, Comm. Math. Phys. **375** (2020), no. 1, 653–678, DOI 10.1007/s00220-019-03501-9. MR4082176

[27] Yaofeng Su, *Vector-valued almost sure invariance principles for (non)stationary and random dynamical systems*, Trans. Amer. Math. Soc. **375** (2022), no. 7, 4809–4848, DOI 10.1090/tran/8609. MR4439492

[28] Yaofeng Su, *Random Young towers and quenched limit laws*, Ergodic Theory Dynam. Systems **43** (2023), no. 3, 971–1003, DOI 10.1017/etds.2021.164. MR4544151

[29] Yaofeng Su, *Polynomial escape rates via maximal large deviations*, J. Stat. Phys. **192** (2025), no. 5, Paper No. 73, 7, DOI 10.1007/s10955-025-03455-4. MR4905601

[30] Yaofeng Su, *Almost surely invariance principle for non-stationary and random intermittent dynamical systems*, Discrete Contin. Dyn. Syst. **39** (2019), no. 11, 6585–6597, DOI 10.3934/dcds.2019286. MR4026993

[31] Lai-Sang Young, *Statistical properties of dynamical systems with some hyperbolicity*, Ann. of Math. (2) **147** (1998), no. 3, 585–650, DOI 10.2307/120960. MR1637655

[32] Lai-Sang Young, *Recurrence times and rates of mixing*, Israel J. Math. **110** (1999), 153–188, DOI 10.1007/BF02808180. MR1750438

School of Mathematics, Georgia Institute of Technology, Atlanta, Georgia
*Email address*: `leonid.bunimovich@math.gatech.edu`

Southern University of Science and Technology, Shenzhen, People's Republic of China.
*Email address*: `suyaofengmath@gmail.com, suyf@sustech.edu.cn`

Contemporary Mathematics
Volume **839**, 2026
https://doi.org/10.1090/conm/839/16725

# Fixed flowers in hyperbolic laminations

Md. Abdul Aziz, Brittany E. Duncan, and John C. Mayer

Abstract. Laminations are a combinatorial and topological way to study Julia sets. Laminations give information about the structure of parameter space of degree $d$ polynomials with connected Julia sets. We first study fixed point portraits in laminations and their respective global count. Then, we investigate the correspondence between locally unicritical laminations and locally maximally critical laminations with rotational polygons. The global correspondence has been shown in Burdette, Falcione, Hale, and Mayer [Contemp. Math., 797, 2024].

## 1. Introduction

**1.1. Motivation.** What are "rotational sets" of laminations of the closed unit disk $\overline{\mathbb{D}}$ under the action of the *angle d-tupling map*, $\sigma_d$ on the unit circle $\mathbb{S}$ and why try to understand their relationship to fixed objects in a lamination? Laminations are a topological and combinatorial model of the connected Julia sets of polynomials considered as functions of the complex plane $\mathbb{C}$. Such models are used both to understand specific types of polynomials and their Julia sets and to study the parameter spaces of polynomials. For example, the well-known Mandelbrot set $\mathcal{M}_2$, also called the quadratic connectedness locus, is the parameter space of quadratic polynomials (of the form $z \mapsto z^2 + c$ for parameter $c \in \mathbb{C}$) with connected Julia set. The hyperbolic domains of that parameter space, where the critical point is attracted to an attracting periodic orbit, are of interest, including how they are connected to each other, how they are arranged in the Mandelbrot set, and how many domains there are that are associated with attractive orbits (of the associated polynomials) of a given period, fixed points of a given rotation number, and the like. These terms are all defined below.

As explicated by Thurston [**21**], the Mandelbrot set has an underlying combinatorial tree structure. We suggest this tree structure in Figure 1.

The *principal hyperbolic domain* $PHD_2$ of the Mandelbrot set $\mathcal{M}_2$, where the critical point is attracted to an attractive fixed point, is the central region bounded by a cardioid in Figure 1. Immediately attached to $PHD_2$ at their respective boundaries are hyperbolic domains, appearing as attached disks, in each of which the critical point is attracted to an attracting periodic orbit. For example, the rabbit Julia set in Figure 1 has the critical point attracted to an orbit of period 3 which moves around a repelling fixed cut point of the Julia set. This cut point

2020 *Mathematics Subject Classification.* Primary 37F46.

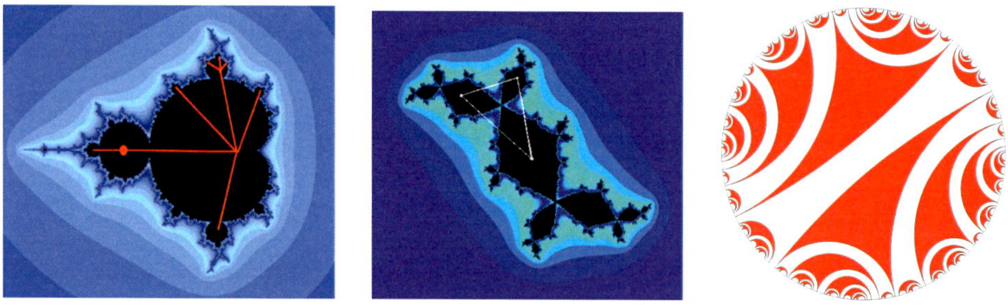

FIGURE 1. On left, the Mandelbrot set with indication of underlying combinatorial tree structure of hyperbolic domains. In the center, the quadratic rabbit Julia set with attractive orbit around a repelling fixed cut point. The parameter $c = -0.12 + 0.78i$ is in the hyperbolic domain attached to the main cardioid at the top. On the right, the corresponding rabbit lamination; each triangle corresponds to a cut point of the Julia set.

corresponds in the lamination to the large triangle in the upper part of the disk. The other triangles correspond to cut points pre-periodic it. The closures of the hyperbolic domains attached to the main cardioid do not meet each other, hence, the combinatorial tree structure. Our goal is to understand how the principal hyperbolic domain $PHD_d$ in the connectedness locus $\mathcal{M}_d$ of degree $d > 2$ polynomials, where all critical points are attracted to the same attractive fixed point, meets other hyperbolic domains at the boundary of $PHD_d$, and how the other hyperbolic domains meet each other, if they do. In **[7]** and **[8]** the authors study the boundary of the principal hyperbolic domain of cubic parameter space and the hyperbolic domains that meet that boundary. Their tool is cubic laminations. Our tool will be laminations of higher degree ($d > 2$) and rotational polygons, like the rabbit triangle, in them. Our focus on fixed point portraits, rotational sets, and flower-like sets follows.

Laminations are composed of *leaves* (chords of the closed unit disk $\overline{\mathbb{D}}$) which form a closed collection of non-crossing line segments that are forward and backward invariant under a natural extension of the map $\sigma_d$ (defined below) on the unit circle $\mathbb{S}$. A *rotational set* (defined below) is a set of points of $\mathbb{S}$ carried forward by $\sigma_d$ preserving their circular order. Leaves connecting points of a rotational set in circular order form polygons in the lamination. There is a correspondence between rotational polygons in laminations and fixed cut points of polynomials that have an infinitesimal rotation number $\frac{p}{q} \in [0, 1)$ (determined by the argument $\frac{2\pi p}{q}$ of the derivative of the polynomial at the fixed point). Such polygons are in correspondence to a fundamental class of hyperbolic domains of the parameter space of degree $d$ polynomials with connected Julia set, those directly attached to the principal hyperbolic domain. Fixed points of polynomials corresponding to such rotational sets were studied in the polynomial setting by Goldberg in **[14]** and by Goldberg and Milnor in **[15]**. By studying them in the laminational setting, we do not first assume that there is a corresponding polynomial.

The connection between the Julia set of a polynomial and its corresponding lamination is made by Böttcher's version of the Riemann Mapping Theorem. If the Julia set of the polynomial $P(z)$ of degree $d$ is connected, then the unbounded component $B_\infty$ of the complement of the Julia set is topologically a disk about the point $\infty$. Let $\mathbb{D}_\infty$ denote the complement of the closed unit disk in the compactified plane, the Riemann sphere $\mathbb{C}_\infty$. By Böttcher's Theorem (see Figure 2), there is a complex analytic homeomorphism $\phi : \mathbb{D}_\infty \to B_\infty$ that conjugates the power function $z \mapsto z^d$ to the action of $P$ on $B_\infty$. In $\mathbb{D}_\infty$ there are polar coordinates, concentric circles and radial lines from the boundary $\mathbb{S}$ of the unit disk $\mathbb{D}$ out toward $\infty$. Each radial line is associated with an angle in the usual way. The map $\phi$ carries this system of coordinates onto $B_\infty$. In $B_\infty$, it is customary to call the images of the radial lines *external rays* encoded by their angle. We should distinguish between the Julia set, which is the boundary of $B_\infty$ and the *filled* Julia set, which is that boundary together with what it bounds.

In case the Julia set of $P$ is locally connected, then by a theorem of Carathéodory all these external rays "land" at points of the Julia set, and every point of the Julia set is a landing point of such a ray. Often more than one external ray lands at the same point. If two rays land at the same point of the Julia set, then in the corresponding lamination, those two points on $\mathbb{S}$ are joined by a chord of $\overline{\mathbb{D}}$, and thus a leaf of the lamination is born. By studying laminations in the abstract, without having a particular polynomial or Julia set in mind, we aim to reverse this process and move from a lamination to a topological Julia set to a specific polynomial and its actual Julia set. Thus, we aim to understand what is possible for laminations and use that information to constrain what is possible for locally connected Julia sets of polynomials. This in turn will provide information about the combinatorial structure of the hyperbolic domains of the corresponding parameter space.

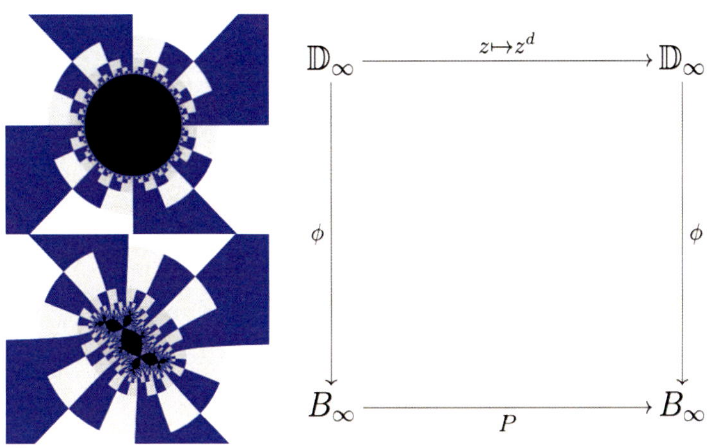


FIGURE 2. On the left, we see trivial Julia set and the famous Douady Rabbit Julia set. We pinch the disk at certain angles to get the Douady Rabbit Julia set corresponding to $P(z) = z^2 + (-0.12 + 0.78i)$. The checkerboards correspond to binary coordinates. On the right, we have the Böttcher's Diagram mentioned above.

**1.2. Preliminary definitions.** We cite the following preliminary definitions from [**9**].

DEFINITION 1.1 ($\sigma_d$ map). We use "d-nary" coordinates on the circle $\mathbb{S}$. The map $\sigma_d : \mathbb{S} \to \mathbb{S}$ is defined to be $\sigma_d(t) = dt \pmod 1$. For example, $\sigma_2$ has binary coordinates, $\sigma_3$ has ternary, $\sigma_4$ has quartenary, and so on.

REMARK 1.2. As we are in binary in $\sigma_2$, the digits 0 and 1 are used. For example _001 represents $\frac{1}{7}$. The digits following a "_" are repeated. For an arbitrary $d$-nary coordinate, a geometric series calculation can be performed to retrieve its fractional equivalent.

DEFINITION 1.3 (Lamination). A *lamination*, $\mathcal{L}$, is a collection of chords of the closed unit disk, $\overline{\mathbb{D}}$, which we call *leaves*, such that:

(1) Any two leaves of $\mathcal{L}$ meet, if at all, in a point of the circle $\mathbb{S}$.
(2) $\mathcal{L}^* := \mathbb{S} \cup \{\cup \mathcal{L}\}$ is a closed subset of $\overline{\mathbb{D}}$.

In the case that condition (1) only is met, we call our collection of chords, a *prelamination.*

DEFINITION 1.4 (Leaf Mapping). We denote the image of a leaf, $\ell = [a, b]$, with endpoints $a$ and $b$, under the map $\sigma_d$ by

$$\sigma_d(\ell) = [\sigma_d(a), \sigma_d(b)]$$

The map $\sigma_d$ is combinatorial; it tells where the endpoints of leaves map but not where points on the leaves map. It is straightforward to show that there is a continuous linear extension $\sigma_d^*$ of $\sigma_d$ to $\mathcal{L}^*$. We introduce an extension $\sigma_d^\#$ of $\sigma_d^*$ as a branched covering map to the entire plane in Section 2.

DEFINITION 1.5 (Critical Chords). A chord $\ell = [a, b]$, is called *critical* when both its endpoints map to a single point $\sigma_d(a) = \sigma_d(b) \in \mathbb{S}$.

DEFINITION 1.6 (All Critical Polygon). A polygon whose endpoints all map to a single point is called an *all critical polygon.*

For example, in Figure 3, the two laminations have critical chords colored blue with an all critical polygon made of three critical chords.

DEFINITION 1.7 (Sibling Leaves). Let $\ell_1 \in \mathcal{L}$ be a leaf and suppose $\sigma_d(\ell_1) = \ell'$, for some non-degenerate leaf $\ell' \in \mathcal{L}$. A leaf $\ell_2 \in \mathcal{L}$, disjoint from $\ell_1$, is called a *sibling* of $\ell_1$ provided $\sigma_d(\ell_2) = \ell' = \sigma_d(\ell_1)$. A collection $\mathcal{S} = \{\ell_1, \ell_2, \dots, \ell_d\} \subset \mathcal{L}$ is called a *full sibling collection* provided that for each $i$, $\sigma_d(\ell_i) = \ell'$ and for all $i \neq j$, $\ell_i \cap \ell_j = \emptyset$.

DEFINITION 1.8 (Sibling Invariant Lamination). A lamination $\mathcal{L}$ is said to be *sibling d-invariant* (or simply *invariant* if no confusion will result) provided that the following three statements hold:

(1) (Forward Invariant) For every $\ell \in \mathcal{L}$, $\sigma_d(\ell) \in \mathcal{L}$.
(2) (Backward Invariant) For every non-degenerate $\ell' \in \mathcal{L}$, there is a leaf $\ell \in \mathcal{L}$ such that $\sigma_d(\ell) = \ell'$.
(3) (Sibling Invariant) For every $\ell_1 \in \mathcal{L}$ with $\sigma_d(\ell_1) = \ell'$, a non-degenerate leaf, there is a full sibling collection $\{\ell_1, \ell_2, \dots, \ell_d\} \subset \mathcal{L}$ such that $\sigma_d(\ell_i) = \ell'$.

REMARK 1.9. A sibling $d$-invariant lamination induces an equivalence relation. Two points on $\mathbb{S}$ are equivalent if they are joined by a finite concatenation of leaves. We consider laminations for which this results in a closed equivalence relation. Thus, the sibling invariant laminations we will be considering have a fourth condition from **[6]** not listed in the definition:

(4) $\mathcal{L}$ has finite equivalence classes, and all leaves are boundary chords of the convex hulls of equivalence classes.

DEFINITION 1.10 (Gap). A *gap* in a lamination, $\mathcal{L}$, is the closure of a component of $\overline{\mathbb{D}} \setminus \mathcal{L}^*$. A gap is *critical* if and only if the degree of $\sigma_d^*$ restricted to the boundary of the gap is greater than one. A gap with finitely many leaves in its boundary is usually called a *polygon*. The leaves bounding a finite gap are called the *sides* of the polygon.

DEFINITION 1.11 (Fatou Gap). A *Fatou gap* in a lamination is a gap whose boundary intersected with $\mathbb{S}$ contains a Cantor set.

In a Julia set corresponding to a lamination via the Böttcher coordinates, a *Fatou domain* corresponds to a Fatou gap. For example, in Figure 1 the white regions in the lamination correspond to the black regions immediately between cut points in the filled Julia set. There is a period 3 Fatou domain orbit, with one domain critical, forming a "flower" around the fixed cut point.

PROPOSITION 1.12 (Order Preserving **[6]**). *On both finite and infinite gaps of a d-invariant lamination, the map $\sigma_d$ preserves circular order.*

DEFINITION 1.13 (Critical Portrait). A maximal collection of critical chords for $\sigma_d$ is called a *critical portrait* (maximal here meaning at least $d-1$ non-crossing critical chords). When a collection of critical chords meet at endpoints, the number may exceed $d-1$ by forming all critical polygons.

DEFINITION 1.14 (Critical Sectors). A critical sector is a region $C$ in the closed unit disk that is bounded by critical chords and arcs of the circle such that the boundary of $C$ maps onto the circle with degree 1. There are $d$ critical sectors.

In Figure 3, as we are in $\sigma_4$, we have a maximal collection of 4 critical chords. By Definition 1.13, we have at least $d-1 = 4-1 = 3$ critical chords. Since we have formed an all critical polygon (triangle), we have an additional critical chord added in this lamination.

By **[12]**, a critical chord has length $i/d$ for $i \in \{1, 2, \dots j | j = \lfloor \frac{d}{2} \rfloor\}$. In the same figure, we can see four critical sectors. There are three critical sectors formed by a single critical chord of length $\frac{1}{4}$. The remaining critical sector is between the critical chord of length $\frac{1}{2}$ and of length $\frac{1}{4}$.

DEFINITION 1.15 (Forward Invariant Set). A *forward invariant set* consists of periodic or pre-periodic leaves or polygons that map forward preserving circular order without intersecting.

DEFINITION 1.16 (Canonical Critical Portrait). Given a forward invariant set, a critical portrait is called *canonical* provided all critical polygons/chords touch some end point of a leaf of the forward invariant set.

All examples in this paper have a canonical critical portrait. Canonical critical portraits force certain leaves of the forward invariant set to be major leaves; see Definition 1.22.

DEFINITION 1.17 (Compatible). A critical portrait $C$ is said to be *compatible* with a forward invariant set $F$ provided that if $\cup C$ meets $F$, it is only at endpoints of leaves or vertices of polygons contained in $F$.

DEFINITION 1.18 (Branches of the Inverse). Let $C$ be a critical portrait. Every critical sector $S$ in $\overline{\mathbb{D}}$ defined by $C$ will have a function $\tau$ defined on $\partial\mathbb{D}$ onto $\partial S$ that is one to one, order-preserving, and $\sigma_d \circ \tau$ is the identity on $\partial\mathbb{D}$.

Combining the previous definitions we can now define a pullback scheme.

DEFINITION 1.19 (Pullback Scheme). Let $C$ be a critical portrait and $F$ a compatible forward invariant set. The corresponding collection $PB(F, C)$ of the branches of the inverse determined by $C$ as in Definition 1.18 gives us a *pullback scheme* for $F$.

There can be multiple critical portraits compatible with $F$ that define the pullback scheme differently. There are some critical portraits that are compatible with $F$, but the related pullback lamination does not satisfy condition (4) of Definition 1.8. The following theorem is well known.

THEOREM 1.20 ([**9**]). *Let $F$ be a periodic forward invariant set under $\sigma_d$, $C$ a compatible critical portrait, and the pullback scheme $PB(F, C) = \{\tau_1, \tau_2, \ldots, \tau_d\}$. Let $F_0 = F$, and $F_1 = F_0 \cup \tau_1(F_0) \cup \tau_2(F_0) \cup \cdots \cup \tau_d(F_0)$. In general, for any given stage $n$ of the pull back $F_n = F_{n-1} \cup \tau_1(F_{n-1}) \cup \cdots \cup \tau_d(F_{n-1})$. Let $F_\infty = \bigcup_{n=0}^{\infty} F_n$, and let $\mathcal{L} = \overline{F_\infty}$. Then, $\mathcal{L}$ is a sibling $d$-invariant lamination.*

DEFINITION 1.21 (Grand Orbit). Let $P$ be an arbitrary subset of a lamination (for example a point, leaf, polygon, gap, etc.) in a $d$-invariant lamination. The *grand orbit* $\mathcal{GO}(P)$ is defined to be

$$\mathcal{GO}(P) = \{\sigma_d^{-n}(\sigma_d^k(P)) | n, k \in \mathbb{Z}^+ \cup 0\}.$$

DEFINITION 1.22 (Major and Minor Leaves). For the orbit of a periodic leaf in a lamination, a leaf, $M$, closest to critical length is called a *major*, and the image $\sigma_d(M)$ of a major is called the *minor* usually denoted $m$.

In Figure 3, there are two leaf orbits, a fixed leaf and a rotating triangle. The major leaf for the triangle is the leaf that is closest to a blue critical chord in the lamination. The minor leaf is that leaf's image and is the smallest of the three leaves that form the red triangle. The major leaf for the fixed black leaf connecting two fixed points is itself, and it is also the minor.

DEFINITION 1.23 (Hyperbolic Lamination). A $d$-invariant lamination is said to be *hyperbolic* if and only if all compatible critical chords are interior to periodic or pre-periodic Fatou gaps.

DEFINITION 1.24 (Sub-lamination). A *sub-lamination* is a $d$-invariant lamination that is a subset of another $d$-invariant lamination.

DEFINITION 1.25 (Rotational Set). Consider $\sigma_d\colon \mathbb{S} \to \mathbb{S}$ for a particular $d \geq 2$.

Let $P = \{x_i | 0 < x_1 < x_2 < x_3 < \cdots < x_k < 1\}$ be a finite set of consecutive points on $\mathbb{S}$ in anticlockwise order. We say that $P$ is a *rotational set* (for $\sigma_d$) provided

(1) $\sigma_d(P) = P$, and
(2) For $1 \leq j \leq k$, if $\sigma_d(x_j) = x_i$, set $i(j) = i$. Then for all $j$, $j - i(j) (\mathrm{mod}\ k)$ is the same.

If (2) holds (but possibly not (1)), we say $\sigma_d$ is circular order-preserving on $P$. We call a rotational set which is a single periodic orbit, a *rotational orbit.*

A rotational set is a basic example of a forward invariant set. There are infinite rotational sets ([**17**] and [**4**]), but we do not consider them in this paper. As we define below, a rotational set will always have a unique rotation number. See Theorem 1.28.

DEFINITION 1.26 (Rotation Number). To each rotational set we can assign a *rotation number*, a rational number $0 \leq \frac{p}{q} < 1$ in lowest terms. That is, let $\mathcal{O} = \{x_1 < x_2 < x_3 \cdots < x_q\}$ be a rotational periodic orbit. Suppose that $\sigma_d(x_1) = x_j$. Set $p = j - 1$. The rotation number of $\mathcal{O}$ is $\frac{p}{q}$. Our notation is $\rho(\mathcal{O}) = \rho(x_1) = \rho(x_i) = \frac{p}{q}$.

DEFINITION 1.27 (Rotational Polygon). A polygon (or a leaf), $P = P_0$ is said to be *rotational* if $\sigma_d|_{P_0}$ maps $P_0$ back to itself preserving circular order and $\forall i$, $\exists j \neq i, \sigma_d(x_i) = x_j$ according to the same nonzero rotation number.

The red triangle in Figure 3 is a rotational polygon with rotation number $\frac{1}{3}$. The black leaf is a leaf with rotation number 0. The vertices of a rotational polygon satisfy the conditions of a rotational set.

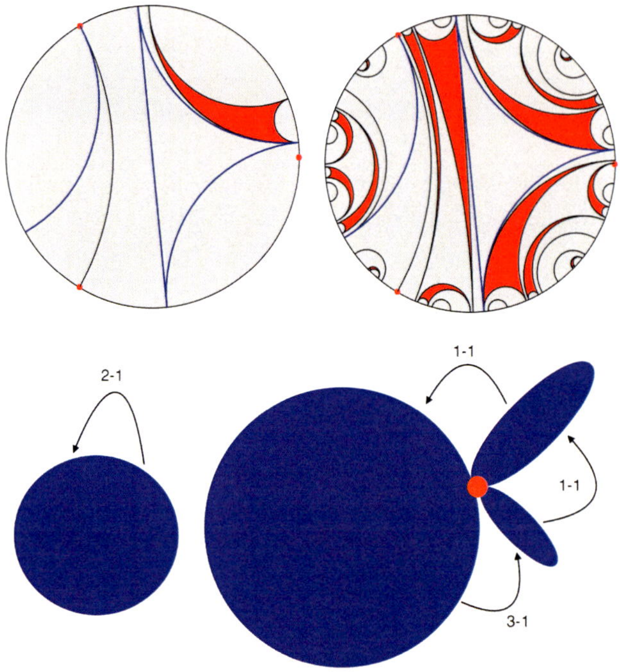


FIGURE 3. On the top left, we have a rotational polygon [_001, _010, _100] in $\sigma_4$ where only a portion of the criticality is associated with the polygon. On the top right, we have the first two pullbacks of the polygon. We can see the Fatou gaps beginning to form in the lamination. We also notice that pre-images of the polygon limit to the fixed leaf [_1,_2]. Thus, the forward orbit of the two critical Fatou gaps will not meet. On the bottom, we have a sketch of the Fatou domains that are associated with criticality and how they map forward. We can see the "flower-like" orbit on the bottom right. The red point, where three Fatou gaps meet, is the rotational polygon that was collapsed to a point in the quotient space. The pullback laminations are generated by Lamination Builder [**13**].

### 1.3. Uniqueness of rotation numbers in a critical sector.

THEOREM 1.28 (Uniqueness of Rotation Numbers [**17**]). *Each critical sector's boundary maps onto the boundary of the disk one-to-one except on finitely many points. Consequently, the boundary of each critical sector on the circle contains a closed invariant set. This invariant set has a unique rotation number. In other words, a non-zero rotation set and a fixed point cannot exist in the same critical sector.*

DEFINITION 1.29 (Terminal and Non-Terminal Critical Sectors). We will say a critical sector is *terminal* if it is formed by a singular critical chord of length $\frac{1}{d}$. Otherwise a critical sector is formed by two or more critical chords and will be called *non-terminal*.

COROLLARY 1.30. *There cannot be non-zero rotational sets in two adjacent critical sectors.*

PROOF. Assume we have two adjacent critical sectors, $S_1$ and $S_2$, both with a rotational set within it. We have three cases: two terminal adjacent sectors, one terminal and one non-terminal adjacent sectors, and two non-terminal adjacent sectors.

We observe that this result follows immediately in the easiest case that two critical sectors form a singular arc of the circle. The sectors are length $\frac{1}{d}$; the arc they form together is length $\frac{2}{d}$. We will have a fixed point in that arc as the fixed points are distributed $\frac{1}{d-1} < \frac{2}{d}$ apart contradicting Theorem 1.28.

In the mixed case, we have two sub-cases: one where the non-terminal critical sector has two arcs on the circle and the other where it has three or more. When the non-terminal sector has two arcs on the circle and is adjacent to a terminal critical sector, its longest critical chord will be of length $\frac{2}{d}$. Thus, by the previous argument, we will be able to fit a fixed point underneath this non-simple sector again contradicting Theorem 1.28. Now we consider the case when the non-terminal sector has three or more arcs on the circle. Our two rotational sectors take up $\frac{2}{d}$ of the circle. In the remaining portion of the circle, we can fit up to $d-2$ fixed points in the other critical sectors. This leaves one fixed point left to place. Since our other critical sectors already have as many fixed points as allowed, this forces a fixed point in one of our rotational sectors contradicting Theorem 1.28.

The final case of two non-terminal adjacent sectors is the same as the previous sub-case. □

**1.4. Flower-like sets.** We cite [**8**] for specific motivation for this paper. The authors introduced the concept of "flower-like sets" which gives motive to expand the global correspondence in [**9**] to a more "local" correspondence. The following definitions are from [**8**] and give additional motivation to studying the correspondence between locally unicritical and locally maximally critical rotational polygons (defined in Section 4).

DEFINITION 1.31 (Lap). A lap of $\mathcal{L}$ is either a finite gap of $\mathcal{L}$ or a non-degenerate leaf of $\mathcal{L}$ not on the boundary of a finite gap.

DEFINITION 1.32 (Flower-Like Sets). Suppose that $\mathcal{L}$ has an infinite invariant gap $U$ or an invariant lap $G$ and an infinite gap $U$ that shares an edge with $G$ (in the latter case it follows that $U$ is periodic). Then U (in the former case) or the set consisting of $G$ and all periodic Fatou gaps attached to it (in the latter case) is said to be a *flower-like set*.

Thus, a flower-like set is a certain set of gaps or laps. An example which pertains to our desired correspondence is that of a rotational polygon in a fixed sector of a lamination as they will have an associated invariant gap. Fixed objects and sectors are defined in the following section. In Figure 3, we show a lamination with a rotational polygon and a fixed leaf in the lamination. This fixed leaf separates the unit disk into two fixed sectors which forces our polygon not to be associated with the full degree of criticality thus creating what we call "fixed flowers."

**1.5. Outline of subsequent sections.** In Section 2 we introduce fixed point portraits (FPP) for laminations and provide examples of degree 5 fixed point portraits and the corresponding Julia sets of some corresponding degree 5 polynomials. We also prove that the count of fixed point portraits for a given degree $d$ is the Catalan numbers for $n = d - 1$.

In Section 3 we introduce the degree of a fixed sector of a FPP, and show that there is a finite selection of canonical critical portraits for a given FPP. From this we derive properties of canonical fixed point laminations. One conclusion is that in any hyperbolic lamination, there is a uniquely determined invariant object inside each fixed sector of the given FPP.

In Section 4 we distinguish locally unicritical from globally unicritical laminations and extend the characterizations of [**9**] to the locally unicritical case. This is the result which motivated the term "fixed flowers." The locally unicritical flower in a fixed sector bifurcates to a locally maximally critical flower within the fixed sector.

## 2. Fixed point portraits for laminations

Fixed point portraits in the polynomial setting were studied by Goldberg in [**14**] and by Goldberg and Milnor in [**15**]. Our aim in starting with laminations is to provide more detail about the association of fixed points with zero rotation and fixed points with non-zero rational rotation, including counting all the possibilities. In this paper we restrict ourselves to the hyperbolic case.

### 2.1. Fixed point portraits.

DEFINITION 2.1 (Fixed Point). We say a point $x$ is *fixed* by a map $f$, if and only if, $f(x) = x$.

PROPOSITION 2.2. *For the map $\sigma_d$ on the circle, the fixed points are $\frac{i}{d-1}$, where $0 \leq i < d - 1$, for $i \in \mathbb{Z}^+$.*

DEFINITION 2.3 (Fixed Point Portrait (FPP)). We call a partition of $\mathbb{D}$, obtained by connecting any number of fixed points with leaves in a manner that two leaves only meet at a fixed point, a *fixed point portrait.* We call a sector of $\mathbb{D}$ bounded by fixed leaves and arcs of $\mathbb{S}$ a *fixed sector* and a polygon inside $\mathbb{D}$ bounded by fixed leaves a *fixed polygon.* If two fixed leaves do meet at a point, then they must be part of the boundary of a fixed polygon.

REMARK 2.4. For us, a fixed point portrait is defined to contain only the $d-1$ fixed points that are "externally visible" via Böttcher's uniformization, see Figure 2. The "missing" fixed points are then shown necessarily to exist in the laminational setting at the end of Section 3 (Theorem 3.9).

The only fixed point portrait for degree 2 is the empty portrait. For degree 3, we have the empty portrait and the portrait containing the fixed leaf $[\_0, \_1]$. For degree 4, we have the empty portrait, one portrait (up to rotation by a fixed point) containing a fixed leaf (see Figure 3, and one portrait containing a fixed triangle joining all three fixed points. Figures 4, 5, 6, 7, 8, and 9 are all the degree 5 fixed point portraits (up to rotation by a fixed point), a finite pullback of the lamination for each, and a corresponding Julia set for each. The fixed points on the circle in the lamination are marked with a dot. The degree 5 polynomials in Figures 4-8 are

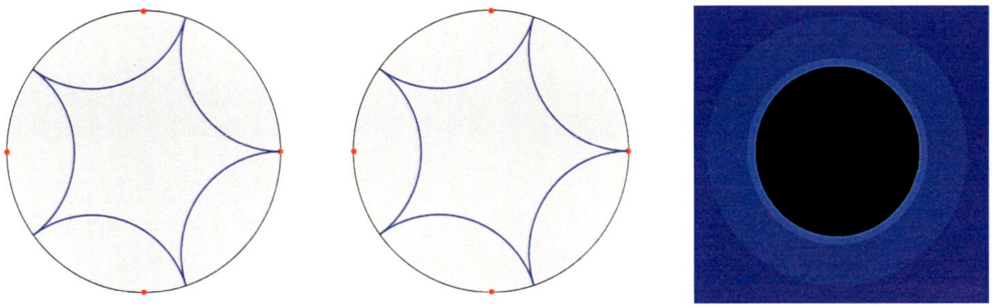


FIGURE 4. This is the simplest fixed point portrait as it has no fixed leaves only fixed points on the boundary. Here, the pullback lamination is the same as the fixed point portrait as there are no leaves to pull back. We call this the *empty lamination*. The Julia set is just a topological circle since there are no "pinching" leaves.

from the two-parameter family $z \mapsto z^5/5 + az^4 + bz^3$, all having 0 as a fixed critical point (*super-attracting* fixed point) of degree $> 2$. We give the degree 5 fixed point portraits as examples as this is the first degree where more "interesting" portraits occur.

**2.2. Extending $\sigma_d$ to the disk.** In [**10**] and [**1**], the authors extend $\sigma_d$ on a hyperbolic lamination $\mathcal{L}$ continuously, linearly to the leaves and then to the plane as a branched covering map $\sigma_d^{\#}$ carrying the unit disk to itself. Branch points (critical points) are exclusively inside Fatou gaps by the definition of hyperbolicity. We will refer to them and their pre-images in pre-image gaps as the *centers* of the Fatou gaps. The hyperbolic lamination induces a *topological Julia set* $J_{\mathcal{L}}$ and a branched covering map $P_{\mathcal{L}}$ on the plane. The first author then showed that such a branched covering map satisfies the Thurston criterion for the existence of a polynomial $P$ Thurston equivalent to $P_{\mathcal{L}}$. Moreover, the lamination induced by the Julia set of $P$ is $\mathcal{L}$. The reader is referred to the cited dissertation and thesis for definitions and details. In Subsection 2.3, for each fixed point portrait for $\sigma_5$ we find a Julia set corresponding to the fixed point portrait shown with its degree 5 polynomial equation.

**2.3. Examples of fixed points portraits of degree 5.** With the fixed point portraits, we can see the relation between laminations and Julia sets: leaves in the laminations correspond to cut points in the Julia sets. As these fixed point portraits have compatible fixed leaves added to the empty lamination, the principal hyperbolic domain in the connectedness locus for degree 5 polynomials will have a hyperbolic domain attached to it for each different fixed point portrait with the critical points being attracted to fixed attracting points. We leave rigorously establishing that statement about parameter space to a future paper. Here we provide substantial evidence.

**2.4. Counting fixed point portraits.** Recall that we defined the map $\sigma_d$ in Definition 1.1. Also, recall that Proposition 2.2 states that there are $d-1$ fixed points on the boundary of $\mathbb{S}$ for $\sigma_d$.

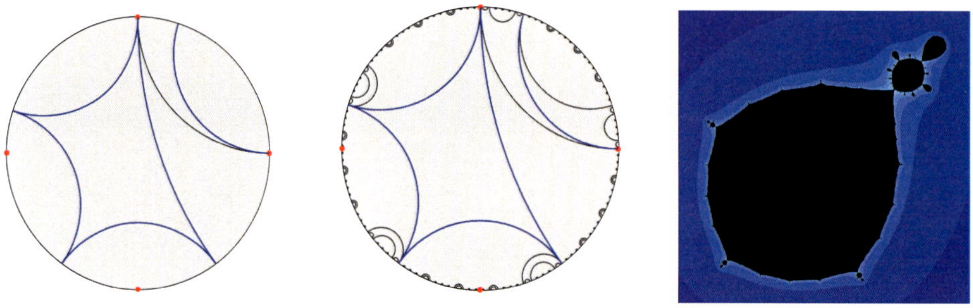

$z \mapsto z^5/5 + (-0.373837 - 0.373837i)z^4$

FIGURE 5. This is the fixed point portrait with a fixed leaf of length $\frac{1}{4}$ made by connecting $\_0$ and $\_1$. This single fixed leaf forms two fixed sectors. One Fatou gap contains an all-critical quadrilateral. In the corresponding Julia set we expect this gap to correspond to a rather large Fatou domain which we see on the right. The pinching point between the large Fatou domain and the "middle" Fatou domain is the initial fixed leaf in our lamination. In the filled Julia set, 0 is a degree 4 critical point that is fixed.

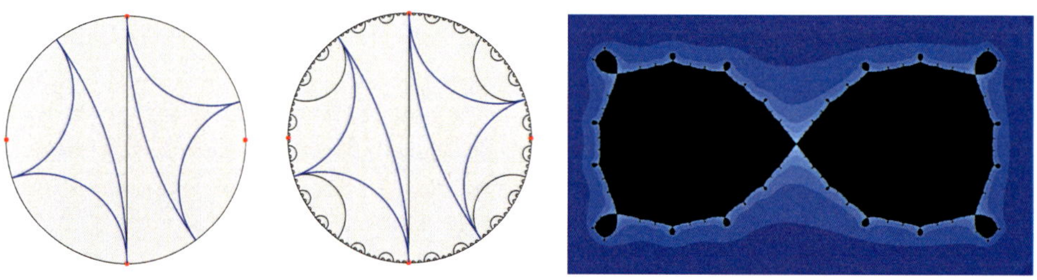

$z \mapsto z^5/5 + (1.17017)z^4 + (1.82574)z^3$

FIGURE 6. This is the fixed point portrait with a fixed diameter (length $\frac{1}{2}$) made by connecting $\_1$ and $\_3$ forming a rotationally symmetric FPP and the associated Julia set with two symmetric main Fatou domains. In the filled Julia set, there are two degree three critical points, one of which is 0, fixed.

THEOREM 2.5 (Counting Theorem). *The number of fixed point portraits for $\sigma_d$ are the Catalan numbers $\frac{(2n)!}{(n+1)!(n)!}$, where $n = d-1$, the number of fixed points of $\sigma_d$.*

PROOF. Let $d \geq 2$. By Proposition 2.2, the map $\sigma_d$ has $d-1$ fixed points $\{\_0, \_1, \ldots, \_(n-1)\}$. We will show the number of fixed point portraits satisfies the recurrence relation of Catalan numbers.

The Catalan numbers are defined by the recurrence

$$C_0 = 1$$

and

$$C_{n+1} = \sum_{i=0}^{n} C_i C_{n-i}$$

for $n \geq 0$.

$z \mapsto z^5/5 + (-0.912871i)z^3$

FIGURE 7. This is the fixed point portrait with two fixed leaves both of length $\frac{1}{4}$, one from $\_0$ to $\_1$ and another from $\_2$ to $\_3$ forming a rotationally symmetric FPP and the associated Julia set. Zero is a degree 3 fixed critical point.

$z \mapsto z^5/5 + (1.01236)z^4 + (1.49071)z^3$

FIGURE 8. This is the fixed point portrait with a fixed triangle connecting $\_1$, $\_2$, and $\_3$. Here the fixed triangle has two sides of length $\frac{1}{4}$ and one side of length $\frac{1}{2}$ which has the associated Julia set having two invariant Fatou domains of the same size and one much larger main Fatou domain. Zero is a degree three critical fixed point.

By way of induction, let $f(n)$ be the number of fixed point portraits for $\sigma_d$ using the fixed points $\{\_0, \_1, \ldots, \_(n-1)\}$. We need to show $f(n)$ satisfies the above recurrence relation. We define $f(0) = 1$.

Now, consider a portrait for $\sigma_{d+1}$ using fixed points $\{\_0, \_1, \ldots, \_n\}$ that contains $\_n$. Also, let $\_k$ be the least fixed point in this portrait, i.e., $\_k$ and $\_n$ are connected and no fixed points of $\{\_0, \_1, \ldots, \_(k-1)\}$ is connected with any fixed points of $\{\_k, \ldots, \_n\}$. Therefore, $\{\_0, \_1, \ldots, \_(k-1)\}$ can form portraits on their own and from the induction hypothesis, there are $f(k)$ such portraits. Also, there are $(n-k+1)$ fixed points from $\_k$ to $\_n$. All of these fixed points lie on the arc of the circle that is bounded by the chord $\overline{\_k\_n}$. Therefore, any portraits among $\{\_k, \ldots, \_n\}$ will not cross the the chord $\overline{\_k\_n}$. Furthermore, since $\_k$ and $\_n$ are already connected there are $(n-k)$ fixed points to form portraits from $\_k$ to $\_n$. Hence, by the induction hypothesis, there are $f(n-k)$ portraits.

$z \mapsto z^5/5 + 1.25z$

FIGURE 9. This is the maximal fixed point portrait: all fixed points are in a single fixed quadrilateral. The associated Julia set has 90° rotational symmetry. Zero is a fixed non-critical point, and the 4 critical fixed points of degree 2 are symmetrically located in Fatou domains around the cut point 0.

Now, the set $\{\_0, \ldots, \_n\}$ has $(n+1)$ fixed points and $\_k$ ranges from $\_0$ to $\_n$. Thus we have,

$$f(n+1) = \sum_{k=0}^{n} f(k) f(n-k)$$

Therefore, by induction, we can conclude that the number of fixed point portraits of $\sigma_d$ satisfies the recurrence relation of the Catalan numbers. □

## 3. Canonical laminations for fixed point portraits

In this section, we formalize the concept of a "canonical critical portrait" for a fixed point portrait (FPP) and the associated canonical lamination for a given fixed point portrait. We conclude by exploring structural properties of these laminations and confirming the existence of invariant objects within each fixed sector.

### 3.1. Canonical critical portraits for fixed point portraits.

DEFINITION 3.1 (Degree of a fixed sector). The *degree of a fixed sector* is the cardinality of the maximal number of disjoint critical chords that is compatible with the region.

THEOREM 3.2. *The degree of a fixed sector is one more than the number of arcs of the circle $\mathbb{S}$ between the adjacent fixed points in the boundary of that region.*

PROOF. Each arc between two adjacent fixed points (length $\frac{1}{d-1}$) can contain only one critical chord (length $\frac{1}{d}$). See Figures 5 and 7 for examples. □

The definition below for canonical critical portraits for fixed point portraits is a special case of to Definition 1.16.

DEFINITION 3.3 (Canonical Critical Portraits for Fixed Point Portrait). Let $\mathcal{P}$ be a fixed point portrait. For each fixed sector in $\mathcal{P}$ a maximal all-critical polygon that touches a fixed point in the boundary of that sector is called a *canonical critical portrait* for $\mathcal{P}$.

In the next theorems, we will show that it does not matter which fixed point in a given fixed sector the all-critical polygon touches. This justifies calling the critical portrait canonical.

THEOREM 3.4. *There are infinitely many critical portraits compatible with the initial FPP data of leaves connecting fixed points but there are only finitely many critical portraits that are canonical, i.e., touch the fixed points.*

PROOF. Given any initial laminational data (including a FPP), there are an infinite number of compatible critical chords. However, if we restrict to critical portraits that touch the fixed points, there are only a finite number of ways for a chord of critical length to touch one of the finite number of fixed points. □

THEOREM 3.5. *Canonical critical portraits compatible with a given fixed point portrait result in the same sibling portrait in the first pullback and thus in all subsequent pullbacks.*

PROOF. Suppose for a fixed $d$, we are given a fixed point portrait. By Proposition 2.2, the map $\sigma_d$ has $d-1$ fixed points on $\mathbb{S}$. Also, each of the $d-1$ fixed points pulls back to $d$ pre-images spaced evenly on $\mathbb{S}$, one of them the fixed point itself. Since fixed points on $\mathbb{S}$ are spaced evenly and pre-images of individual fixed points are also spaced evenly on $\mathbb{S}$, between any two consecutive fixed points there is exactly one pre-image of each of the fixed points.

First, we consider a simple case where the fixed point portrait is just a leaf connecting any two consecutive fixed points. Then the canonical critical portrait compatible with this fixed point portrait will consist of a critical chord and an all-critical $(d-1)$-gon. The single critical chord lies in the smaller fixed sector of $\mathbb{S}$ bounded by the fixed leaf, and one of its ends touches one of the two fixed points (see Figure 10). Therefore, the critical chord will connect the fixed endpoint and the pre-image of that fixed point in this fixed sector in the first pullback. The $(d-1)$-gon lies in the other fixed sector of $\mathbb{S}$ and one of its vertices touches one of the $d-1$ fixed points in this sector. Therefore, the sides of the $(d-1)$-gon will connect the consecutive pre-images of the vertex fixed point. We call these closed intervals of $\mathbb{S}$ bounded by the end points of a critical chord a critical interval. Every critical interval contains pre-images of each fixed point exactly once except for the touching fixed point, which has two pre-images in every interval.

Now, in the critical interval bounded by the critical chord, we have only one choice of pre-image leaf. Thus, if we change the touching fixed point of the critical chord, it would not affect the pullback leaf and the leaf will remain the same. In the critical intervals bounded by the sides of the $(d-1)$-gon, we have exactly one pre-image of each fixed point except for the touching fixed point, which has two pre-images; we have exactly one option to pullback the fixed leaf in every critical interval.

Moreover, since the pullback leaves are unique in every interval, shifting the critical polygon to a different touching fixed point will not affect the pullback leaves. Therefore, we can conclude that we will have the same siblings irrespective of the choice of the touching fixed point. Since we have the same sibling portrait in the first pullback and the branches of inverse are also fixed, we will have the same sibling portrait in every pullback. Therefore, the lamination is canonical.

In each fixed sector of local degree $d'$, there is an all critical $d'$-gon. The number of the pre-images of the fixed points in each critical interval of that fixed sector allow

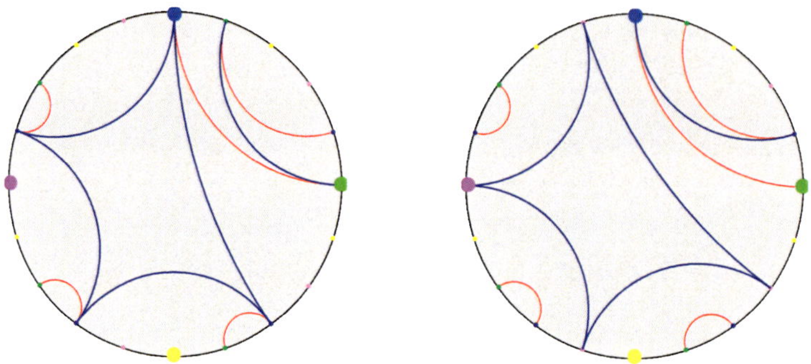

FIGURE 10. Here we have fixed leaf [_0, _1] connecting the blue and green fixed points. We choose two canonical critical portraits: one that touches at the blue fixed point, _1 and one that touches at the pink fixed point, _2. The first pullback lamination with either critical portrait results in the same sibling portrait.

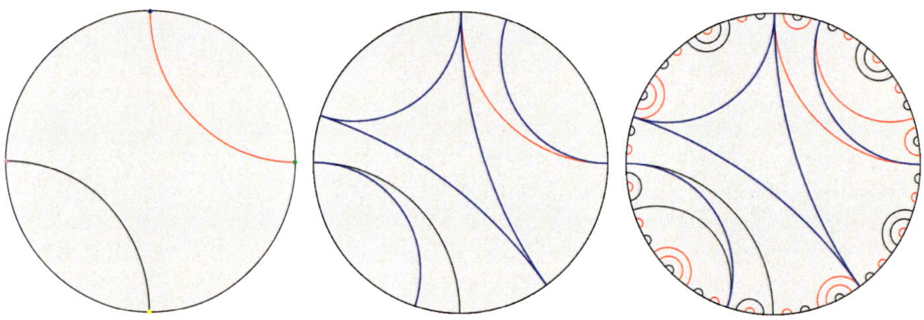

FIGURE 11. This is a canonical lamination for $\sigma_5$ with two fixed leaves, [_0, _1] and [_2, _3]. On the left we have the two fixed leaves. In the middle, a canonical critical portrait is added in. On the right, we have the first few pullbacks of the lamination.

for only one way to connect pre-images of the initial FPP data. Details are left to the reader. □

### 3.2. Canonical laminations for fixed point portraits.

DEFINITION 3.6 (Canonical Lamination for Fixed Point Portrait). Let $\mathcal{P}$ be a fixed point portrait. Choose for each fixed sector in $\mathcal{P}$ a maximal all-critical polygon touching a fixed point in the boundary of that sector. Then the pullback lamination $\mathcal{L}$ with respect to the chosen critical portrait is the *canonical lamination* for $\mathcal{P}$.

It follows from Theorems 3.4 and 3.5 that Definition 3.6 is well-defined. That is, it does not matter which canonical critical portrait is used in the pullback scheme. See Figures 10 and 11.

DEFINITION 3.7 (Center of a Gap). The center of a gap in a lamination is the unique center determined in the extension $\sigma_d^{\#}$ of $\sigma_d$ to the closed unit disk in Subsection 2.2.

THEOREM 3.8 (Properties of Canonical Fixed Point Lamination). *Let $\mathcal{P}$ be a fixed point portrait and $\mathcal{L}$ be its canonical pullback lamination. Then $\mathcal{L}$ has the following properties:*

(1) *Every pullback leaf is a pre-image of a leaf of $\mathcal{P}$.*
(2) *Pullback leaves converge to points of $\mathbb{S}$.*
(3) *There are no limit leaves.*
(4) *Every fixed sector $F$ in $\mathcal{P}$ contains a unique invariant critical Fatou gap $G(F)$ whose center is a fixed point.*
(5) *The boundary leaves of $G(F)$ are pre-images of boundary leaves of $F$.*

PROOF.

(1) Every point on $\mathbb{S}$ for the pullback lamination $\mathcal{L}$ is a pre-image of the endpoints of the leaves of $\mathcal{P}$. So in $\mathcal{L}$, if we connect two pre-images of two endpoints of a leaf of $\mathcal{P}$ by a leaf, then this leaf in $\mathcal{L}$ maps forward to the corresponding leaf of $\mathcal{P}$. Therefore, every pullback leaf is a pre-image of a leaf of $\mathcal{P}$.
(2) Let $\mathcal{L}$ be a canonical lamination of the fixed point portrait $\mathcal{P}$, i.e., $\mathcal{L}$ touches some fixed points of $\mathcal{P}$. We can see the initial pullback of a fixed leaf is trapped under a critical chord. Since the maximum length of a fixed leaf is $\frac{1}{2}$, the maximum length of pullback leaf is $\frac{1}{2d}$. Thus, $n$ pullbacks will be bounded in length by $\frac{1}{2d^n}$.
(3) Since from (2) we have pullback leaves converge to points, there are no limit leaves.
(4) Let $F$ be a fixed sector for $\mathcal{P}$. Let the degree of the sector $F$ be $n$. Then, there is a canonical all-critical $n$-gon (Definition 3.3), touching one of the fixed points on the boundary of $F$. The chords of the all-critical polygon determine the critical sectors of $F$. When we pull back, each of the critical sectors has a pre-image of every fixed point inside it. In the first pullback, we will see a pre-image of each fixed leaf in that critical sector. The boundary leaves of $F$ subtend any other fixed leaves of $\mathcal{P}$. Since the pullback of the fixed points preserves counterclockwise circular order, the pullbacks of other fixed leaves that are not boundary leaves of $F$ will be subtended by pullbacks of the boundary leaves of $F$. If we continue pulling back, we will have a gap $G$ in $F$ that meets the circle in a Cantor set. By Definition 1.11, $G$ is a Fatou gap. The pullback process uniquely determines the Fatou gap.
(5) From the proof of (4) we have seen that in the first pullback inside $F$, the pullback leaves that are not pullbacks of boundary leaves of $F$ are subtended by the pullbacks of the boundary leaves of $F$. Therefore, the boundary leaves of the Fatou gap are pre-images of the boundary leaves of that fixed sector. □

**3.3. Fixed points in a lamination.** Every hyperbolic lamination of degree $d$ is supposed to correspond to a polynomial of degree $d$. Every polynomial of degree $d$ has $d$ fixed points. But on the boundary of $\mathbb{S}$ we have exactly $d-1$ fixed points. If in the lamination two fixed points are joined by a leaf, then in the corresponding Julia set this will be a multiple fixed point which will count as only one fixed point for the polynomial. For example, in Figure 7 four peripheral fixed points in the lamination have become two multiple fixed points in the Julia set. In the canonical case, there is an invariant Fatou gap in each fixed sector as in Theorem 3.8, with an attractive fixed point in each invariant Fatou gap. In this subsection, we will show that there exists a fixed object in the interior of every fixed sector in all non-canonical hyperbolic cases as well.

FIGURE 12. Left to right cases I, II, and III of Theorem 3.9. Note that the degrees of the laminations illustrated are 5, 4, and 4. On the left, we have Case I where no fixed object, say in the central degree 3 fixed sector, is subtended by the leaves of the lamination. In the middle, we have Case II where every fixed object in the degree 3 fixed sector is subtended by leaves of the lamination. On the right, we have Case III where in the degree 3 fixed sector only some fixed objects are subtended by leaves of the lamination and some are not.

THEOREM 3.9. *Let $\mathcal{L}$ be a hyperbolic lamination containing a particular FPP $\mathcal{P}$. In each fixed sector $F$ of $\mathcal{P}$ there is an invariant object of one of the following types:*

(1) *The invariant object is a fixed point at the center of an invariant Fatou gap $G(F)$.*
(2) *The invariant object is a rotational polygon.*

PROOF. Let $F$ be a fixed sector of $\mathcal{P}$. The proof is done in three cases; the first and third case are of Type (1). The cases divide on whether leaves of the lamination subtend fixed points or fixed leaves of the fixed point portrait.

CASE 1. No fixed point or fixed leaf of $F$ is subtended by the leaves of the lamination. The left-most lamination in Figure 12 is a $\sigma_5$ example: there are two fixed leaves and three canonical invariant gaps. $F$ is the central fixed sector.

Since no fixed point or fixed leaf is subtended by leaves of the lamination, the fixed points and leaves must be on the boundary of a gap, $G$, in $F$. Since the fixed points and leaves in the boundary of $G$ are disjoint, $G$ is a Fatou gap. Since $G$ is an invariant Fatou gap, there are points $x$ and $y$ that map to the same point [**4**]. Thus, $G$ is critical. Since the lamination is hyperbolic, the criticality is interior to the Fatou gap. As the extension $\sigma_d^{\#}$ [**10**] of $\sigma_d$ to the plane is defined, the center of $G$ is a fixed point in $G$. In this case, the fixed point is of Type (1).

CASE 2. Every fixed object of $F$ is subtended by leaves of the lamination. The middle lamination of Figure 12 is a $\sigma_4$ example: there is one fixed leaf and the degree three fixed sector $F$ contains a degree 3 rotational triangle. The fixed leaf and the fixed point in the degree three fixed sector are subtended by pullbacks of the triangle.

For each fixed object, we pick a sufficiently close leaf $l$ of the lamination $\mathcal{L}$ so that $l$ is repelled from the fixed object, but the image of $l$ remains in the fixed sector $F$. Let $\widetilde{\mathbb{D}} \subset F$ be the disk bounded by our chosen leaves, the intervals of $\mathbb{S}$, and the endpoints of adjacent leaves in $F$. We retract every point of $\mathbb{D} \setminus \widetilde{\mathbb{D}}$ to the nearest point of the boundary of $\widetilde{\mathbb{D}}$ by a retraction mapping $r$. $\widetilde{\mathbb{D}}$ does not contain any fixed point of $\sigma_d$ on $\mathbb{S}$. We have, $r(\sigma_d^{\#}(\widetilde{\mathbb{D}})) \subset \widetilde{\mathbb{D}}$. Since $r \circ \sigma_d^{\#}$ is continuous, Brouwer's fixed point theorem shows at least one fixed point exists in $\widetilde{\mathbb{D}}$. However, these fixed points can not occur on $\partial\widetilde{\mathbb{D}}$, so they must occur inside $F$ as $\widetilde{\mathbb{D}} \subset F$.

In a hyperbolic lamination, fixed points are either the center of an invariant Fatou gap or an invariant equivalence class. In this case, we do not have an invariant Fatou gap. Therefore, the fixed point must occur in an invariant finite gap. These are rotational polygons.

CASE 3. Some fixed objects of $F$ are subtended by leaves of the lamination, and some are not. The rightmost lamination in Figure 12 is a $\sigma_4$ example: there is one fixed leaf and in the degree three fixed sector, one critical chord touches a fixed point, but the other touches a period two leaf in $F$.

As in Case I, the fixed objects not subtended by the leaves of the lamination are on the boundary of an invariant critical gap. As we have defined $\sigma_d^{\#}$, the center of this invariant critical Fatou gap is fixed. □

## 4. Unicritical and maximally critical correspondence

In [**9**], a correspondence between unicritical and maximally critical laminations was established. In this section, we show a similar correspondence between locally unicritical laminations and locally maximally critical laminations.

### 4.1. Locally unicritical laminations.

DEFINITION 4.1 (Unicritical Lamination). A $d$-invariant lamination which is compatible with an all critical $d$-gon is called a *unicritical lamination.* We also refer to this as a *globally unicritical lamination.*

The following corollary follows immediately from Theorem 1.30.

COROLLARY 4.2. *A unicritical rotational polygon is in a terminal critical sector, and thus has all fixed points subtended by one of its sides.*

DEFINITION 4.3 (Maximally Critical Lamination). A $d$-invariant lamination which has a polygon with $d-1$ degree 2 Fatou gaps on sides of the polygon in separate orbits is called a *maximally critical lamination.*

DEFINITION 4.4 (Locally Unicritical Lamination). A $d$-invariant lamination with a rotational polygon which is in a terminal critical sector of an all critical $k$-gon where $k < d$ is called a *locally unicritical lamination.*

DEFINITION 4.5 (Locally Maximally Critical Lamination). A $d$-invariant lamination which has a rotational polygon with $k-1$ degree 2 Fatou gaps in separate orbits on the sides of the polygon where $k < d$ is called a *locally maximally critical lamination.*

REMARK 4.6. We will refer to our local laminations as *canonical* if the critical portrait intersects the vertices of the polygon when possible. See Figure 14 for examples of both canonical locally unicritical and locally maximally critical laminations.

LEMMA 4.7. *The length of the major of a unicritical polygon is within* $\frac{1}{d(d+1)}$ *of critical length* $\frac{1}{d}$.

PROOF. We know the length of any leaf in a lamination must get at least as long as $\frac{1}{d+1}$ ([**12**]). Since the critical length for the critical chord associated with a unicritical major is $\frac{1}{d}$, our major leaf must be within $\frac{1}{d} - \frac{1}{d+1} = \frac{1}{d(d+1)}$ of critical length. □

The following definition and theorem are adapted from [**19**].

DEFINITION 4.8 (Co-root). A *co-root* is a point, other than an endpoint of the major, in the boundary of the central gap of the unicritical lamination that is fixed under the first return map.

THEOREM 4.9 (Co-root Theorem). *Given a degree $d$ unicritical lamination, there will be $d-2$ co-roots compatible with our unicritical polygon. The distance between co-roots is greater than $1/d$.*

The following algorithm gives rise to the existence of co-roots for all unicritical laminations.

REMARK 4.10 (The Generalized Lavaur's Algorithm, Section 6 of [**3**]). For all unicritical laminations of any degree, we can uniquely find the corresponding minor leaf that generates the lamination. In this process of finding these minors, the algorithm "skips" over certain points of every period. These points that are skipped over are the desired co-roots. However, the Lavaur's Algorithm identifies the minor leaf rather than the major. Thus, the "co-roots" it identifies are actually the images of the points described in the co-root definition. These points are in the same orbit so we will use the term interchangeably when no confusion will arise. We will identity a co-root with the leaf with which it is associated in its orbit.

The following theorem follows from Definitions 4.1 and 4.4.

THEOREM 4.11. *A polygon (or leaf) is compatible with a locally unicritical lamination if, and only if it is compatible with a globally unicritical lamination.*

The global correspondence between unicritical rotational laminations and maximally critical rotational laminations was shown in [**9**]. The language used between that paper and this paper differ. In the language of this paper, an SCM lamination is a maximally critical lamination, and a MAC lamination is a (globally) unicritical lamination. The global correspondence is stated in the following two theorems:

THEOREM 4.12 (Maximally Critical to Unicritical Lamination). [**9**] *Let $\mathcal{S}$ be a canonical maximally critical lamination containing a rotational maximally critical polygon, $P$. Then, there exists a unique corresponding unicritical lamination, $\mathcal{L}$, with a leaf $M(P)$ as its major leaf.*

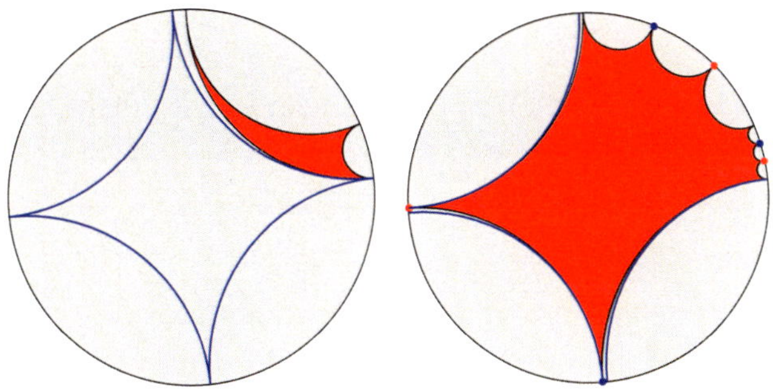

FIGURE 13. A globally unicritical rotational polygon (the [_0, _1] triangle) (left) and a globally maximally critical rotational polygon (right) for $\sigma_4$ with co-roots marked.

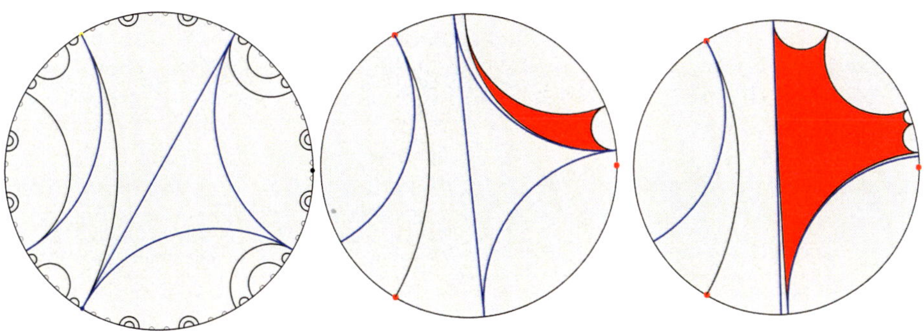

FIGURE 14. On left, pullback 3 of the fixed leaf [_1, _2] alone. In middle, the same rotational triangle as seen in Figure 13, but now locally unicritical (of local degree 3). On right, the corresponding locally maximally critical polygon.

THEOREM 4.13 (Unicritical to Maximally Critical Lamination). **[9]** *Let $\mathcal{L}(M)$ be a canonical unicritical lamination with major leaf $M$. Then there is a canonical maximally critical lamination $\mathcal{S}(M)$ that contains a maximally critical rotational $k(d-1)$-gon $P(M)$ with adjacent majors in which $k$ is the number of vertex orbits of $P(M)$.*

Now, we wish to establish the local correspondence. Comparing Figures 13 and 14, we can visually see what changes when we are in the local case.

In the global case, the Generalized Lavaur's algorithm **[3]** gives $d-2$ co-roots for globally unicritical rotational polygons. These co-roots and the original vertices of the rotational polygon give the vertices for the maximally critical rotational polygon. See Figure 13.

In the local case, we can again use the Generalized Lavaur's algorithm to find $d-2$ co-roots. However, not all will be compatible with our locally unicritical polygon. There will be $d'-2$ compatible co-roots for the locally unicritical rotational polygon, where $d'$ is the local degree of the fixed point sector containing the locally unicritical polygon. These compatible co-roots for the local case can be used to find an associated locally maximally critical polygon. See Figure 14.

THEOREM 4.14 (Locally Maximally Critical to Locally Unicritical). *Given a canonical lamination, $\mathcal{L}$, with a locally maximally critical polygon $P'$ with adjacent majors, there exists a canonical lamination with a unique locally unicritical rotational polygon with major leaf $M(P')$.*

PROOF. This follows from Theorems 4.11 and 4.12. □

THEOREM 4.15 (Locally Unicritical to Locally Maximally Critical). *Let $\mathcal{L}$ be a canonical lamination with a locally unicritical rotational polygon with major $M$. Then, there exists a canonical lamination with a locally maximally critical $k(d'-1)$-gon $P'(M)$ with adjacent majors where $k$ is the number of vertex orbits of $P'(M)$.*

PROOF. This follows from Theorems 4.11 and 4.13. □

### 4.2. Remarks about future work.

REMARK 4.16. It seems quite reasonable to conclude that not only are the hyperbolic domains corresponding to the Julia sets with the laminations in Figure 14 attached to the principal hyperbolic domain of the degree 4 connectedness locus, but those hyperbolic domains are attached to each other. Rigorous justification is left to a future paper.

REMARK 4.17. The global count of fixed point portraits in Subsection 2.4 can be extended to a local count given a locally unicritical rotational polygon whose canonical all critical polygon separates the fixed points. Details are left to the reader.

QUESTION 1. With reference to Theorem 3.9, we raise the question: Under what conditions is the canonical pullback lamination of the FPP, $\mathcal{P}$, a sublamination of $\mathcal{L}$?

## References


[1] M. Aziz. Canonical Laminations for Fixed Point Portraits. Masters Thesis UAB 2023. https://digitalcommons.library.uab.edu/etd-collection/93.

[2] Brandon L. Barry, *On the simplest lamination of a given identity return triangle*, ProQuest LLC, Ann Arbor, MI, 2015. Thesis (Ph.D.)–The University of Alabama at Birmingham. MR3419326

[3] Sourav Bhattacharya, Alexander Blokh, and Dierk Schleicher, *Unicritical laminations*, Fund. Math. **258** (2022), no. 1, 25–63, DOI 10.4064/fm18-2-2022. MR4422233

[4] Alexander Blokh, James M. Malaugh, John C. Mayer, Lex G. Oversteegen, and Daniel K. Parris, *Rotational subsets of the circle under $z^d$*, Topology Appl. **153** (2006), no. 10, 1540–1570, DOI 10.1016/j.topol.2005.04.010. MR2216120

[5] Alexander M. Blokh, Clinton P. Curry, and Lex G. Oversteegen, *Locally connected models for Julia sets*, Adv. Math. **226** (2011), no. 2, 1621–1661, DOI 10.1016/j.aim.2010.08.011. MR2737795

[6] Alexander M. Blokh, Debra Mimbs, Lex G. Oversteegen, and Kirsten I. S. Valkenburg, *Laminations in the language of leaves*, Trans. Amer. Math. Soc. **365** (2013), no. 10, 5367–5391, DOI 10.1090/S0002-9947-2013-05809-6. MR3074377

[7] Alexander Blokh, Lex Oversteegen, Ross Ptacek, and Vladlen Timorin, *Laminations from the main cubioid*, Discrete Contin. Dyn. Syst. **36** (2016), no. 9, 4665–4702, DOI 10.3934/dcds.2016003. MR3541500

[8] Alexander Blokh, Lex Oversteegen, Vladlen Timorin, and Yimin Wang, *A model of the cubic connectedness locus*, Nonlinearity **38** (2025), no. 7, Paper No. 075014, 37, DOI 10.1088/1361-6544/addfa3. MR4919956

[9] Brittany E. Burdette, Caleb S. Falcione, Cameron G. Hale, and John C. Mayer, *Unicritical and maximally critical laminations*, Recent developments in fractal geometry and dynamical systems, Contemp. Math., vol. 797, Amer. Math. Soc., [Providence], RI, 2024, pp. 147–172, DOI 10.1090/conm/797/15952. MR4730767

[10] Adam Carty, *Realization of Hyperbolic Laminations and Properties of Limit Laminations*, ProQuest LLC, Ann Arbor, MI, 2023. Thesis (Ph.D.)–The University of Alabama at Birmingham. MR4675584

[11] Douglas K. Childers, *Wandering polygons and recurrent critical leaves*, Ergodic Theory Dynam. Systems **27** (2007), no. 1, 87–107, DOI 10.1017/S0143385706000526. MR2297088

[12] David J. Cosper, Jeffrey K. Houghton, John C. Mayer, Luka Mernik, and Joseph W. Olson, *Central strips of sibling leaves in laminations of the unit disk*, Topology Proc. **48** (2016), 69–100. MR3340275

[13] C. Falcione. Lamination Builder. `https://csfalcione.github.io/lamination-builder/`.

[14] Lisa R. Goldberg, *Fixed points of polynomial maps. I. Rotation subsets of the circles*, Ann. Sci. École Norm. Sup. (4) **25** (1992), no. 6, 679–685. MR1198093

[15] Lisa R. Goldberg and John Milnor, *Fixed points of polynomial maps. II. Fixed point portraits*, Ann. Sci. École Norm. Sup. (4) **26** (1993), no. 1, 51–98. MR1209913

[16] Jan Kiwi, *Wandering orbit portraits*, Trans. Amer. Math. Soc. **354** (2002), no. 4, 1473–1485, DOI 10.1090/S0002-9947-01-02896-3. MR1873015

[17] J. Malaugh. Rotational subsets of the circle under $z^d$. PhD Dissertation UAB. ProQuest Dissertations Publishing, 2003. 3101495.

[18] John Milnor, *Dynamics in one complex variable*, 3rd ed., Annals of Mathematics Studies, vol. 160, Princeton University Press, Princeton, NJ, 2006. MR2193309

[19] Dierk Schleicher, *On fibers and local connectivity of Mandelbrot and Multibrot sets*, Fractal geometry and applications: a jubilee of Benoît Mandelbrot. Part 1, Proc. Sympos. Pure Math., vol. 72, Amer. Math. Soc., Providence, RI, 2004, pp. 477–517, DOI 10.1090/pspum/072.1/2112117. MR2112117

[20] Dierk Schleicher (ed.), *Complex dynamics*, A K Peters, Ltd., Wellesley, MA, 2009. Families and friends, DOI 10.1201/b10617. MR1500163

[21] Dierk Schleicher (ed.), *Complex dynamics*, A K Peters, Ltd., Wellesley, MA, 2009. Families and friends, DOI 10.1201/b10617. MR1500163

Department of Mathematics, University of Alabama at Birmingham, Birmingham, Alabama

*Email address*: `azizm@uab.edu`

Department of Mathematics, University of North Georgia, Dahlonega, Georgia

*Email address*: `brittany.duncan@ung.edu`

Department of Mathematics, University of Alabama at Birmingham, Birmingham, Alabama

*Email address*: `jcmayer@uab.edu`

Contemporary Mathematics
Volume **839**, 2026
https://doi.org/10.1090/conm/839/16726

# Maximal inequalities for the ergodic Hilbert transform along sequences with cone condition and a local moving ergodic theorem

Doğan Çömez

Abstract. The local ergodic Hilbert transform along sequences with cone condition (CC) is known to exist a.e. with respect to invertible measure preserving flows. In this article we obtain maximal inequalities for ergodic Hilbert transform along sequences satisfying CC with respect to invertible measure preserving flows (continuous parameter case) as well as with respect to invertible measure preserving transformations (discrete case) in the setting of additive and admissible processes. As an application, we extend the existence theorem to bounded symmetric admissible processes relative to such flows. On the other hand, the discrete ergodic Hilbert transform along sequences with CC fails to exist.

## 1. Introduction

The operator $\mathcal{H}^{\tau} f := \lim_{\varepsilon\to 0^+} \mathcal{H}^{\tau}_{\varepsilon} f$, where $f \in L^p(X)$, $\tau = \{T_s\}_{s\in\mathbb{R}}$ be an invertible measure-preserving flow, and $\mathcal{H}^{\tau}_{\varepsilon} f(x) = \int_{\varepsilon<|s|<1/\varepsilon} \frac{f(T_s x)}{s}\, ds$, is known as the ergodic Hilbert transform. Its discrete version is defined as $\mathcal{H}^T f := \lim_{n\to\infty} \mathcal{H}^T_n f$,where $f \in L^p(X)$, $T$ is an invertible measure-preserving transformation, and $\mathcal{H}^T_n f(x) = \sum_{0<|i|\le n} \frac{T^i f(x)}{i}$.

When T is induced by an invertible measure preserving transformation, the a.e. existence of the discrete ergodic Hilbert transform was shown by M. Cotlar [**C**]. This result, both in the local and discrete case, has been extended to several settings ever since [**CaP**, **CaJRW**, **FJR**, **P1**, **P2**]. Among them, Petersen [**P1**, **P2**] gave a direct proof of this result, in [**CaJRW**] oscillation and variation inequalities of Hilbert transform was studied. The study of Hilbert transform is strongly connected to the study of singular integrals and on the approach regions for the existence of such integrals. In 1984, Nagel and Stein [**NS**] gave a necessary and sufficient

---

2020 *Mathematics Subject Classification.* Primary 60Exx, 28A80; Secondary 94A34.

*Key words and phrases.* Local ergodic Hilbert transform, admissible processes, moving averages, sequences with cone condition.

Professor Doğan Çömez passed away on June 7, 2025. All correspondence regarding this paper should be directed to Professor Mrinal K. Roychowdhury at mrinal.roychowdhury@utrgv.edu.

condition, known as the *cone condition* (CC), on an approach region $\Omega$ that ensures the existence of singular integrals. The cone condition has led many authors to obtain several ergodic theorems, both in the ordinary ergodic averages setting and in the Hilbert transform setting [**AD**, **BJR**, **CaJRW**, **FJR**]. The results in [**FJR**] has the distinctive flavor of giving the a.e. existence of *local* ergodic Hilbert transform for an approach region $\Omega$ satisfying the CC. All the results above deal with additive processes in their respective settings. This naturally leads to the question of whether the analogous results can be obtained for discrete ergodic Hilbert transform for additive processes, and for not-neccessarily additive processes in both local and discrete cases. By an example in [**Ç3**], the answer to the first question negative. In [**Ç2**] and [**Ç3**] the existence of ergodic Hilbert transform and discrete singular transforms for admissible processes were studied. In this article, utilizing the tools developed in [**Ç1**] and [**Ç2**], we will extend the local result in [**FJR**] to the setting of a class of bounded superadditive processes, namely, for *admissible* processes. This type of processes include additive processes $\{T^k f\}_{k\in\mathbb{Z}}$ as a special case.

## 2. Preliminaries and statements of the maximal inequalities

A sequence $\mathbf{w} = \{(v_q, r_q)\}_{q\in\mathbb{Q}^+} \subset \mathbb{R}\times\mathbb{R}^+$ satisfies the *cone condition* if there exists $C > 0$ such that $|\Omega_\alpha(s)| \leq Cs$, for all $s \in \mathbb{R}^+$, where

$$\Omega_\alpha = \{(t,s) : \exists q, |t - v_q| \leq \alpha(s - r_q)\}, \quad \Omega_\alpha(s) = \{x \in \mathbb{R} : (x,s) \in \Omega_\alpha\}.$$

Let $(X, \Sigma, \mu)$ be a probability space, and let $\tau = \{T_s\}_{s\in\mathbb{R}}$ be a group of measure-preserving transformations, where $T_t : X \to X$ for every $t \in \mathbb{R}$, with $T_0 = I$, the identity map.

Given a sequence $\mathbf{w} = \{(v_q, r_q)\}$ satisfyting CC and $f \in L_p(X)$, the ordinary *local moving ergodic averages along* $\mathbf{w}$ is defined as

$$A^\tau_{(q,\mathbf{w})} f(x) = \frac{1}{r_q}\int_0^{r_q} f(T_{v_q+t}x)\,dt,$$

Analogously, if

$$H^\tau_{(q,w)} f(x) = \int_{r_q\leq|t|\leq 1/r_q} \frac{f(T_{v_q+t}x)}{t}\,dt.$$

then the *local moving ergodic Hilbert transform along* $\mathbf{w}$ is defined as $\mathcal{H}^\tau_{\mathbf{w}} f = \lim\limits_{q\to 0} \mathcal{H}^\tau_{(q,w)} f$.

The Proposition 2.1 below, which is the basis for the moving Hilbert transform results in this article, was obtained in [**FJR**] in the setting of general singular integral operators (which are generalizations of continuous parameter Hilbert transform). Here, we restate it in terms of moving Hilbert transform along a sequence $\mathbf{w}$ satisfying CC for easy reference. With this, for $f \in L_p(\mathbb{R})$, the continuous parameter moving Hilbert transform of $f$ is defined as

$$\mathcal{H}_{\mathbf{w}} f = \lim_{q\to 0}\mathcal{H}_{(q,w)} f \quad \text{where } \mathcal{H}_{(q,w)} f \int_{r_q\leq|t|\leq 1/r_q} \frac{f(t_s + v_q)}{t} dt.$$

PROPOSITION 2.1 ([**FJR**]). *Let* $\mathbf{w} = \{(v_q, r_q)\}$ *satisfy CC and* $f \in L_p(\mathbb{R})$. *Then there exists constants* $C_p$ *such that*

(a) $|\{s \in \mathbb{R} : \mathcal{H}^*_{\mathbf{w}} f(s)| > \lambda\}| \leq \dfrac{C_1}{\lambda}\|f\|_1$ *for every* $\lambda > 0$;
(b) $\|\mathcal{H}^*_{\mathbf{w}} f\|\|_p \leq C_p\|f\|_p$, *for* $1 < p < \infty$,

*where* $\mathcal{H}^*_{\mathbf{w}} f(s) = \sup\limits_q |\mathcal{H}_{(q,\mathbf{w})} f(s)|$ *is the associated maximal operator.*

The ergodic analog of Proposition 2.1 is obtained via the transference principle [**FJR**]:

THEOREM 2.2. *Let* $\mathbf{w} = \{(v_q, r_q)\}$ *satisfy CC, and* $f \in L^p(X)$. *Then there exists constants* $C_p$ *such that*

(a) $\mu\left(\left\{x \in X : \sup\limits_q |\mathcal{H}^*_{\mathbf{w}} f(x)| > \lambda\right\}\right) \leq \dfrac{C}{\lambda}\|f\|_1$ *for all* $\lambda > 0$;
(b) $\|\sup\limits_q |\mathcal{H}^*_{\mathbf{w}} f|\|_p \leq C_p\|f\|_p$, *for* $1 < p < \infty$,

*where* $\mathcal{H}^*_{\mathbf{w}} f(x) = \sup\limits_q |H^\tau_{(q,\mathbf{w})} f(s)|$. *Furthermore, if* $v_q, r_q \to 0$ *as* $q \to 0$ *(along rational numbers) then* $\mathcal{H}^\tau_{\mathbf{w}} f(x)$ *exists a.e.*

A family $F = \{f_t\}_{t\in\mathbb{R}} \subset L_p(X)$ is a $\tau$-*superadditive process* on $\mathbb{R}$ if it satisfies

$$F_{s+t} \geq F_s + T_s F_t \quad \text{and} \quad F_{-(s+t)} \geq F_{-s} + T_{-s} F_{-t}, \quad \text{for all } s, t \geq 0$$

where $F_t = \displaystyle\int_0^t f_u du$, if $t \geq 0$ and $F_t = \displaystyle\int_t^0 f_u du$, it $t < 0$, $\tau = \{T_s\}_{s\in\mathbb{R}}$ is an invertible measure preserving flow, and abusing notation, $T_s F_t = F_t \circ T_s = \displaystyle\int_0^t f_u \circ T_s du$. These integrals are defined in the Riemann sense. $F$ is called bounded if $\gamma_F := \sup\limits_{t>0} \dfrac{1}{t}\|F_t\|_1 < \infty$. When the reverse inequalities hold, $F$ is called $\tau$-*subadditive*. Note that if $F$ is both superadditive and subadditive, it is called $\tau$-*additive process* and is necessarily of the form $F = \{T_t f\}_{t\in\mathbb{R}}$, for some function $f$.

A process $F = \{f_t\} \subset L_p$ is called $\tau$-*admissible process on* $\mathbb{R}$ if

$$T_s f_t \leq f_{s+t} \quad \text{and} \quad f_{-s} f_{-t} \leq f_{-s-t} \quad \text{for } t, s \geq 0.$$

Obviously, any $\tau$-admissible process is $\tau$-superadditive. The family $F = \{f_t\}$ is called *positive* if $f_t \geq 0$ for all $t \in \mathbb{R}$, and is called *strongly bounded* if $\sup\limits_{t\in\mathbb{R}} \|f_t\|_p < \infty$.

The local ergodic Hilbert transform of a $\tau$-superadditive process $F = \{f_t\}$ is defined as

$$\mathcal{H}^\tau F(x) = \lim_{\epsilon\to 0} \mathcal{H}^\tau_\epsilon F(x), \quad \text{where } \mathcal{H}^\tau_\epsilon F(x) = \int_{\epsilon \leq |t| \leq 1/\epsilon} \frac{f_s(x)}{s} ds.$$

When $\mathbf{w} = \{(v_q, r_q)\}$ is a sequence satisfying CC, then the (local) moving Hilbert transform of $F$ along $\mathbf{w}$ is defined as

$$\mathcal{H}^\tau_{\mathbf{w}} F(x) = \lim_{q\to 0} \mathcal{H}^\tau_{(q,\mathbf{w})} F(x), \quad \text{where } \mathcal{H}^\tau_{(q,\mathbf{w})} F(x) = \int_{r_q \leq |s| \leq 1/r_q} \frac{f_s(T_{v_q} x)}{s} ds.$$

In what follows, all families $F = \{f_t\}_{t\in\mathbb{R}}$ defining a $\tau$-admissible process will be assumed to satisfy the symmetry condition:

$$T_{2t}f_{-t} = f_t \quad \text{for all } t \in \mathbb{R}^+.$$

The theorem below is the analogue of the maximal inequality in Theorem 2.2 extended to the admissible processes setting.

THEOREM 2.3. *Let $F = \{f_t\} \subset L_1$ be a symmetric, strongly bounded $\tau$-admissible process with exact dominant $\delta$ and $\mathbf{w} = \{(v_q, r_q)\}$ be a sequence satisfying CC. Then, there exists a constant $C$ such that, for any $\lambda > 0$,*

$$\mu\left\{x : \mathcal{H}^*_{\mathbf{w}}F(x) > \lambda\right\} \le \frac{C}{\lambda}\|\delta\|_1,$$

*where* $H^*_{\mathbf{w}}F(x) = \sup_{q>0}\left|\int_{r_q\le|s|\le 1/r_q} \frac{f_s(T_{v_q}x)}{s}ds\right|$.

Having the results in [**FJR**] for the local (ergodic) Hilbert transform, the maximal inequalities for the discrete (ergodic) Hilbert transform along sequences satisfying CC follows via "discretization". Before proceeding further, we will restate some of the terminology introduced above in the discrete setting. In this case, we will consider sequences $\mathbf{w} = \{(v_k, r_k)\}_{k\ge1} \subset \mathbb{Z}\times\mathbb{Z}^+$, with $r_k > 0$, satisfying CC.

If $T$ is an invertible measure preserving transformation, then the operator $\mathcal{H}^T f := \lim_{n\to\infty}\mathcal{H}^T_n f$, where $\mathcal{H}^T_n f = \sum_{0<|i|\le n}\frac{T^i f}{i}$, $f \in L_p(X)$, is the classical ergodic Hilbert transform. When $\mathbf{w} = \{(v_k, r_k)\}$ is a sequence satisfying CC, the moving ergodic Hilbert transform along $\mathbf{w}$ is

$$\mathcal{H}^T_{\mathbf{w}}f := \lim_{n\to\infty} H^T_{(n,\mathbf{w})}f, \quad \text{where } H^T_{(n,\mathbf{w})}f = \sum_{0<|k|\le r_n}\frac{T^{v_n+k}f}{k}.$$

A *$T$-admissible process on* $\mathbb{Z}$ is a family of functions $F = \{f_i\}_{i\in\mathbb{Z}} \subset L_p$ satisfying the inequalities

$$Tf_i \le f_{i+1} \text{ for } i \ge 0, \quad \text{and } T^{-1}f_i \le f_{i-1} \text{ for } i \le 0.$$

$F = \{f_i\} \subset L_p$ is called *strongly bounded* if $\sup_{n\in\mathbb{Z}}\|f_n\|_p < \infty$. When $p = 1$, a $T$-superadditive process $F \subset L_1$ on $\mathbb{Z}$ is called *bounded* if $\gamma_F := \sup_{n\in\mathbb{Z}}\frac{1}{|n|}\|F_n\|_1 < \infty$. Then, analogously to the continuous parameter setting, the ergodic Hilbert transform of a $T$-admissible process $F$ is defined as

$$\mathcal{H}^T F := \lim_n \mathcal{H}^T_n F, \quad \text{where } \mathcal{H}^T_n F = \sum_{0<|i|\le n}\frac{f_i}{i},$$

and, for a sequence $\mathbf{w} = \{(v_n, r_n)\}$ with CC,

$$H^T_{\mathbf{w}}F := \lim_n H^T_{(n,\mathbf{w})}F, \quad \text{where } H^T_{(n,\mathbf{w})}F = \sum_{0<|i|\le r_n}\frac{f_i(T^{v_n}x)}{i},$$

is the *moving Hilbert transform of $F$ along* $\mathbf{w}$.

In what follows, $T$ will always be an invertible measure preserving transformation, and all families $F = \{f_i\}_{i\in\mathbb{Z}}$ defining a T-admissible process will be assumed to satisfy the *symmetry condition*:

$$T^{2i}f_{-i} = f_i \quad \text{for all } i \in \mathbb{Z}.$$

The statement below provides the discrete version of Proposition 2.1.

THEOREM 2.4. *Let* $\mathbf{w} = \{(v_n, r_n)\}$ *be a sequence satisfying CC such that* $r_n \to \infty$ *as* $n \to \infty$, *and* $a \in l_p(\mathbb{Z})$. *Then there exists constants* $C_p$ *such that*

(1) $|\{j \in \mathbb{Z} : \mathcal{H}^*_{\mathbf{w}}a(j) > \lambda\}| \le C_1\dfrac{\|a\|_1}{\lambda}$, *for every* $\lambda > 0$,
(2) $\|\mathcal{H}^*_{\mathbf{w}}a\|_p \le C_p\|a\|_p$, *for* $1 < p < \infty$,

*where* $\mathcal{H}^*_{\mathbf{w}}a(j) = \sup\limits_n \left|\sum\limits_{0<|k|\le r_n} \dfrac{a(k + v_n + j)}{k}\right|$.

For any $f \in L_p()$, $1 \le p < \infty$, and for a sequences $\mathbf{w} = \{(v_n, r_n)\}$ satisfying CC we will define the maximal operator as $\mathcal{H}^*_{\mathbf{w}}f = \sup\limits_n \left|\sum\limits_{0<|k|\le r_n} \dfrac{f(T^{v_n+k})}{k}\right|$, and its truncated version as $\mathcal{H}^*_{\mathbf{w},N}f = \sup\limits_{n\le N} \left|\sum\limits_{0<|k|\le r_n} \dfrac{f(T^{v_n+k})}{k}\right|$. The statements below are the discrete versions of Theorem 2.2 and Theorem 2.3, respectively.

THEOREM 2.5. *Let* $\mathbf{w} = \{(v_n, r_n)\}$ *be a sequence satisfying CC such that* $r_n \to \infty$ *as* $n \to \infty$, *and* $f \in L_p(X)$. *Then there exists constants* $C_p$ *such that*

(1) $\mu(x \in X : \mathcal{H}^*_{\mathbf{w}}f(x) > \lambda| \le C_1\dfrac{\|f\|_1}{\lambda}$, *for every* $\lambda > 0$,
(2) $\|\mathcal{H}^*_{\mathbf{w}}f\|_p \le C_p\|f\|_p$, *for* $1 < p < \infty$.

THEOREM 2.6. *Let* $F = \{f_t\} \subset L_1$ *be a symmetric, strongly bounded* $\tau$-*admissible process with exact dominant* $\delta$ *and* $\mathbf{w} = \{(v_q, r_q)\}$ *be a sequence satisfying CC. Then, there exists a constant* $C$ *such that, for any* $\lambda > 0$,

$$\mu\{x : \mathcal{H}^*_{\mathbf{w}}F(x) > \lambda\} \le \frac{C}{\lambda}\|\delta\|_1,$$

*where* $H^*_{\mathbf{w}}F(x) = \sup\limits_{n\ge 1} \left|\sum\limits_{0<|k|\le r_n} \dfrac{f_k(T_{v_q}x)}{k}\right|$.

Since $F = \{f_t\}$ is a family of equivalence classes of functions with t ranging through all reals, statements like $\lim\limits_t \dfrac{1}{t}F_t$ are meaningless, unless one either selects a suitable representation for each $F_t$ or let $t$ range through a countable set of reals. In our case, when we are dealing with limits as $t \to a$, $t$ will run through a countable set, which will be the set of rationals if it is not clear from the context.

## 3. Proof of maximal inequality for the local moving ergodic Hilbert transform of admissible processes

In this section we will prove the maximal inequality for the local ergodic Hilbert transform of symmetric strongly bounded admissible processes along sequences satisfying CC. First, we need to observe some properties of symmetric, strongly bounded admissible processes. As the Proposition 3.1 below indicates, these processes admit a simpler representation, enabling us to utilize some well-known tools for the a.e. existence of the local ergodic Hilbert transform.

When $F$ is a $\tau$-admissible process on $\mathbb{R}$, we have $f_t - T_t f_0 \geq 0$, and consequently, if a result holds for $\mathcal{H}^\tau f_0\,(\mathcal{H}^\tau_{\mathbf{w}} f_0)$, then the same is also valid for $\mathcal{H}^\tau F\,(\mathcal{H}^\tau_{\mathbf{w}} F)$ if and only if it is valid for $\mathcal{H}^\tau G\,(\mathcal{H}^\tau_{\mathrm{w}} G)$, where $g_t = f_t - T_t f_0$. Therefore, in such a case, we can assume that $f_t \geq 0$, for each $t \in \mathbb{R}$.

PROPOSITION 3.1. *Let $\tau$ be a measure preserving flow and $F = \{f_t\} \subset L_p$ be a positive symmetric strongly bounded $\tau$-admissible process. Then there exists a sequence of non-negative functions $\{u_t\} \in L_p$ such that $f_t = T_t u_{|t|}$ for all $t \in \mathbb{R}$. Furthermore, there exists $\delta \in L_p$ such that $f_t \leq T_t \delta$ for all $t \in \mathbb{R}$ and $\|\delta\|_p = \sup_{t \in \mathbb{R}} \|f_t\|_p$.*

PROOF. Define $\{u_t\}_{t\in\mathbb{R}}$ by

$$u_t = T_{-t} f_t \text{ and } u_{-t} = T_t f_{-t} \text{ for } t \geq 0.$$

Then, by the symmetry property, $u_t = T_{-t} f_t = T_{-t}\left(T_{2t} f_{-t}\right) = T_t f_{-t} = u_{-t}$; hence, $\{u_t\}$ is symmetric and satisfies $f_t = T_t u_{|t|}$ for all $t \in \mathbb{R}$. Now, for $t, s \geq 0$, by admissibility

$$u_{t+s} = T_{-(t+s)} f_{t+s} \geq T_{-(t+s)}\left(T_s f_t\right) = T_{-t} f_t = u_t,$$

thus $u_t$ is increasing. By Monotone Convergence Theorem and strong boundedness of $F = \{f_t\}$, we see that there exists $\delta \in L_p^+$such that $u_t \uparrow \delta$ and $\|\delta\|_p = \sup_t \|u_t\|_p = \sup_t \|f_t\|_p$. Since $u_t \leq \delta$, it we have $f_t \leq T_t \delta$, for all $t$. □

REMARK. Proposition 3.1 is the continuous parameter analogue of Proposition 2.1 in [C$_1$]. In the discrete setting, if $p = 1$, the function $\delta$ is called the exact dominant for $F$ and $\gamma_F := \|\delta\|_1$ is called the time constant for $F$.

The following statement, proved in [**AK**], is one of the most important properties of (continuous parameter) additive processes.

THEOREM 3.2. *Let $F = \{f_t\}$ be a positive symmetric, bounded $\tau$-superadditive process. Given $\epsilon > 0$, there exists a $\tau$-additive process $G^\epsilon = \{T_t g_\epsilon\}$ such that*
(1) $G^\epsilon_t \leq F_t$ *for all* $t \in \mathbb{R}$,
(2) $\|g_\epsilon\|_1 > \gamma_F - \epsilon$, *and,*
(3) $\displaystyle \lim_t \frac{1}{t} \int_X G^\epsilon_t d\mu = \lim_t \frac{1}{t} \int_X F_t d\mu.$

PROOF OF THEOREM 2.3. We can assume that $f_t \geq 0$ for all $t \in \mathbb{R}$. Since this is a local a.e. convergence result, we have $(v_q, r_q) \to (0, 0)$; hence we can also assume that $r_q \leq 1, |v_q| \leq 1$. We will work in a discrete setting; so, let $\{(v_n, r_n)\}$ be a a subsequence of $\{(v_q, r_q)\}$ such that $(v_n, r_n) \to (0, 0)$. For $N$ a large positive integer, consider $\{(v_n, r_n) : 1 \leq n \leq N\}$. Pick another integer $M' > 0$ with the property $\dfrac{1}{2M'} < \min_{n \leq N} \{r_n, |v_n|\}$, and let $M$ be an integer with $M > M'$. Define

$$H^*_N F(x) = \sup_{1 \leq n \leq N} \left| \int_{r_n \leq |s| \leq 1/r_n} \frac{f_s\left(T_{v_n} x\right)}{s} ds \right| \text{ and } E_N = \left\{x \in X : H^*_N F(x) > \lambda\right\}.$$

Then $E_N = E_N^1 \cup E_N^2$, where

$$E_N^1 = \left\{ x \in X : \sup_{1 \le n \le N} \int_{r_n \le |s| \le 1/r_n} \frac{f_s\left(T_{v_n} x\right)}{s} ds > \lambda \right\} \text{ and}$$

$$E_N^2 = \left\{ x \in X : \inf_{1 \le n \le N} \int_{r_n \le |s| \le 1/r_n} \frac{f_s\left(T_{v_n} x\right)}{s} ds < -\lambda \right\}.$$

Now, consider $E_N^1$. By Proposition 3.1, $f_t = T_t u_{|t|}$ for all $t \in \mathbb{R}$, where $u_p \uparrow \delta$, and $u_p \le u_{p'}$ for all $0 \le p \le p'$. Hence, we have

$$T_t u_0 \le f_t \le T_t \delta, \qquad \text{for all } t \in \mathbb{R} \tag{1}$$

By the measure preserving property of $\tau$, for any $t \in \mathbb{R}$,

$$\begin{aligned} \mu\left(E_N^1\right) = \mu\left(T_t E_N^1\right) &= \mu \left\{ x : \sup_n \int_{r_n \le |s| \le 1/r_n} \frac{T_{t+v_n} f_s(x)}{s} ds > \lambda \right\} \\ &= \mu \left\{ x : \sup_n \int_{r_n \le |s| \le 1/r_n} \frac{T_{t+v_n+s} u_{|s|}(x)}{s} ds > \lambda \right\}. \end{aligned}$$

By (1), for any $t$, in particular, for any $|t| \le M$,

$$\begin{aligned} \lambda &< \int_{r_n \le |s| \le 1/r_n} \frac{T_{t+v_n+s} u_{|s|}(x)}{s} ds \\ &= \int_{-1/r_n}^{-r_n} \frac{T_{t+v_n+s} u_{|s|}(x)}{s} ds + \int_{r_n}^{1/r_n} \frac{T_{t+v_n+s} u_{|s|}(x)}{s} ds \\ &\le \int_{-1/r_n}^{-r_n} \frac{T_{t+v_n+s} u_0(x)}{s} ds + \int_{r_n}^{1/r_n} \frac{T_{t+v_n+s} \delta(x)}{s} ds \end{aligned}$$

Hence,

$$\mu\left(E_N^1\right) \le \mu\left(\left\{ x : \sup_n \left[ \int_{-1/r_n}^{-r_n} \frac{T_{t+v_n+s} u_0(x)}{s} ds + \int_{r_n}^{1/r_n} \frac{T_{t+v_n+s} \delta(x)}{s} ds \right] > \lambda \right\}\right)$$

Define a function $\mathbf{a}_x = a_x(p)$ by

$$a_x(p) = \begin{cases} \delta(T_p x) & \text{if } 0 \le p \le 3M, \\ u_0(T_p x) & \text{if } -3M \le p < 0, \\ 0 & \text{otherwise.} \end{cases}$$

Therefore, for $x \in X$, not belonging a set of measure zero, we have $\mathbf{a}_x \in L_1(\mathbb{R})$. Thus, since $0 < |t| + |v_n| + |s| < 3M$,

$$\int_{r_n \le |s| \le 1/r_n} \frac{T_{t+v_n+s} u_{|s|}(x)}{s} ds \le \int_{-1/r_n}^{-r_n} \frac{a\left(t + v_n + s\right)}{s} ds + \int_{r_n}^{1/r_n} \frac{a\left(t + v_n + s\right)}{s} ds$$

which implies that

$$E_N^1 \subset \left\{ x : \sup_n \int_{r_n \le |s| \le 1/r_n} \frac{a\left(t + v_n + s\right)}{s} ds > \lambda \right\}.$$

For $E_N^2$, similarly, by (1), we have

$$\int_{r_n \le |s| \le 1/r_n} \frac{T_{t+v_n+s} u_{|s|}(x)}{s} ds \ge \int_{-1/r_n}^{-r_n} \frac{T_{t+v_n+s}\delta(x)}{s} ds + \int_{r_n}^{1/r_n} \frac{T^{t+v_n+s} u_0(x)}{s} ds$$

Hence, define a function $\mathbf{b}_x = b_x(p)$ by

$$\mathbf{b}_x(p) = \begin{cases} \delta(T_p x) & \text{if } -3M \le p < 0, \\ u_0(T_p x) & \text{if } 0 \le p \le 3M, \\ 0 & \text{otherwise,} \end{cases}$$

it follows that, for $x \in X$, not belonging a set of measure zero, we have $\mathbf{b}_x \in L_1(\mathbb{R})$. Therefore, as before,

$$\begin{aligned} E_N^2 &\subset \left\{ x : \inf_n \int_{r_n \le |s| \le 1/r_n} \frac{b\,(t+v_n+s)}{s} ds < -\lambda \right\} \\ &= \left\{ x : \sup_n \int_{r_n \le |s| \le 1/r_n} \frac{b\,(t+v_n-s)}{s} ds > \lambda \right\}. \end{aligned}$$

Consequently, since $|t| + |v_n| + |s| < 3M$, we observe that

$$\begin{aligned} &\left\{ x : \sup_{1\le n\le N} \left| \int_{r_n \le |s| \le 1/r_n} \frac{T_{t+v_n} f_s(x)}{s} ds \right| > \lambda \right\} \\ &\subset \left\{ x : \sup_{1\le n\le N} \int_{r_n \le |s| \le 1/r_n} \frac{a\,(t+v_n+s)}{s} ds > \lambda \right\} \\ &\cup \left\{ x : \inf_{1\le n\le N} \int_{r_n \le |s| \le 1/r_n} \frac{b\,(t+v_n+s)}{s} ds < -\lambda \right\} \\ &= \left\{ x : \sup_{1\le n\le N} \int_{r_n \le |s| \le 1/r_n} \frac{a\,(t+v_n+s)}{s} ds > \lambda \right\} \\ &\cup \left\{ x : \sup_{1\le n\le N} \int_{r_n \le |s| \le 1/r_n} \frac{b\,(t+v_n-s)}{s} ds > \lambda \right\} \end{aligned}$$

Therefore, we have

$$\begin{aligned} \mu\,(E_N) &= \frac{1}{2M} \int_{|t|<M} \mu \left\{ x : \sup_{1\le n\le N} \left| \int_{r_n \le |s| \le 1/r_n} \frac{T_{t+v_n} f_s(x)}{s} ds \right| > \lambda \right\} dt \\ &= \frac{1}{2M} \int_{|t|<M} \left[ \mu \left\{ x : \sup_{1\le n\le N} \int_{r_n \le |s| \le 1/r_n} \frac{a\,(t+v_n+s)}{s} ds > \lambda \right\} \right. \\ &\quad \left. + \mu \left\{ x : \sup_{1\le n\le N} \int_{r_n \le |s| \le 1/r_n} \frac{b\,(t+v_n-s)}{s} ds > \lambda \right\} \right] dt \end{aligned}$$

Now, by Fubini's theorem,

$$\begin{aligned}
&\int_{|t|<M} \mu\left\{x : \sup_{1\le n\le N} \int_{r_n\le|s|\le 1/r_n} \frac{a\left(t+v_n+s\right)}{s} ds > \lambda\right\} dt \\
&\le (m\times\mu)\left\{(t,x) : \sup_{1\le n\le N} \int_{r_n\le|s|\le 1/r_n} \frac{a\left(t+v_n+s\right)}{s} ds > \lambda\right\} \\
&\le \int \left|\left\{t : \sup_{1\le n\le N} \int_{r_n\le|s|\le 1/r_n} \frac{a\left(t+v_n+s\right)}{s} ds > \lambda\right\}\right| d\mu \\
&\le \int \left[\frac{C_1}{\lambda}\int_{\mathbb{R}} |a(t)dt|\right] d\mu \text{ by Proposition 2.1} \\
&\le \frac{C_1}{\lambda}\left[\int_{-3M}^{3M}\left(\int |a(t)|d\mu\right) dt \le \frac{C_1}{\lambda}\left[\int_{-3M}^{3M} \|\delta\|_1 dt\right.\right. \\
&= \frac{6C_1M}{\lambda}\|\delta\|_1
\end{aligned}$$

for some constant $C_1$, where $m$ stands for the Lebesgue measure on $\mathbb{R}$. In the same fashion, we also obtain that

$$\int_{|t|<M} \mu\left\{x : \sup_{1\le n\le N} \int_{r_n\le|s|\le 1/r_n} \frac{b\left(t+v_n-s\right)}{s} ds > \lambda\right\} dt \le \frac{6C_2M}{\lambda}\|\delta\|_1$$

Thus, $\mu\left(E_N\right) \le \dfrac{6\left(C_1+C_2\right)M}{2\lambda M}\|\delta\|_1 = \dfrac{3\left(C_1+C_2\right)}{\lambda}\|\delta\|_1$. Letting $N\to\infty$ and $C = 3\left(C_1+C_2\right)$, we obtain

$$\mu\left(\{x : \mathcal{H}^*_{\mathbf{w}}F(x) > \lambda\}\right) \le \frac{C}{\lambda}\|\delta\|_1. \qquad \square$$

## 4. Proofs of maximal inequalities for the discrete moving ergodic Hilbert transform

In this section we will prove the maximal inequalities for the discrete ergodic Hilbert transform along sequences satisfying CC. The the proof Theorem 2.4 follows the standard method of converting the local counterpart to discrete version; hence, we will sketch the proof only.

PROOF OF THEOREM 2.4. Let $\mathbf{a} = \{a(n)\}_{n\in\mathbb{Z}} \in l_p, 1\le p<\infty$, and define $f(t) = \sum_{k\in\mathbb{Z}} a(k)\chi_{I_k}(t)$, where $I_k = \left[k-\frac{1}{4}, k+\frac{1}{4}\right]$. Then $\|f\|_{L_p} = \frac{1}{2}\|\mathbf{a}\|_{l_p}$. Fix a positive integer $m$, and for the given sequence $\mathbf{w} = \{(v_n, r_n)\}$ with CC, consider

$\sum_{|k-j|>r_n} \frac{a(k+v_n)}{j-k}$, where $r_n$ is closest to $m$. Then

$$
\begin{aligned}
&\sum_{|k-j|>r_n} \frac{a(k+v_n)}{j-k} - 2\int_{|t-y|>r_n+\frac{1}{2}} \frac{f(y+v_n)}{t-y}dy \\
&\quad = 2\left[\sum_{|k-j|>r_n} \frac{a(k+v_n)}{j-k}\int_{k+v_n-\frac{1}{4}}^{k+v_n+\frac{1}{4}} 1dy - \int_{|t-y|>r_n+\frac{1}{2}} \frac{f(y+v_n)}{t-y}dy\right] \\
&\quad = 2\left[\sum_{|k-j|>r_n} \int \frac{a(k+v_n)}{j-k}\chi_{I_k+v_n}(t)dt - \sum_{|k-j|>r_n}\int_{I_k} \frac{f(y+v_n)}{t-y}dy\right]
\end{aligned}
$$

where $k$ is the integer closest to $y$. Consequently, for such $k$ and $y$,

$$
\begin{aligned}
&\left|\sum_{|k-j|>r_n} \frac{a(k+v_n)}{j-k} - 2\int_{|t-y|>r_n+\frac{1}{2}} \frac{f(y+v_n)}{t-y}dy\right| \\
&\qquad \leq 2\sum_{|k-j|>r_n}\int_{I_k}\left|\left(\frac{1}{j-k}-\frac{1}{t-y}\right)f(y+v_n)\right|dy.
\end{aligned}
$$

If $t \in I_j$ and $y \in I_k$ with $j \neq k$, then we have $|(t-y)-(k-j)| \leq \frac{1}{2} \leq \frac{1}{2}|k-j|$. Hence, we have $\left|\frac{1}{j-k} - \frac{1}{t-y}\right| \leq \frac{C}{(t-y)^2}$, for some constant $C$, and hence,

$$
\begin{aligned}
\left|\sum_{|k-j|>r_n} \frac{a(k+v_n)}{j-k} - 2\int_{|t-y|>r_n+\frac{1}{2}} \frac{f(y+v_n)}{t-y}dy\right| &\leq C\sum_{|k-j|>r_n}\int_{I_k}\frac{f(y+v_n)}{(t-y)^2}dy \\
&\leq C\int_{|t-y|>\frac{1}{2}}\frac{|f(y+v_n)|}{(t-y)^2}dy.
\end{aligned}
$$

This implies that,

$$
\left|\sum_{|k-j|>r_n} \frac{a(k+v_n)}{j-k}\right| \leq C\left[\sup_n\left|\int_{|t-y|>r_n}\frac{f(y+v_n)}{t-y}dy\right| + \int_{|t-y|>\frac{1}{2}}\frac{|f(y+v_n)|}{(t-y)^2}dy\right].
$$

In particular, if $m = 0$, we have

$$
\begin{aligned}
\left|\sum_{|k|>0}\frac{a(k+v_n+j)}{k}\right| &= \left|\sum_{|k-j|>0}\frac{a(k+v_n)}{j-k}\right| \\
&\leq C\left[\sup_n\left|\int_{|t-y|>r_n}\frac{f(y+v_n)}{t-y}dy\right| + \int_{|t-y|>\frac{1}{2}}\frac{|f(y+v_n)|}{(t-y)^2}dy\right].
\end{aligned}
$$

Therefore,

$$\sup_n \left| \sum_{|k|\le r_n} \frac{a(k+v_n+j)}{k} \right| = \sup_n \left| \sum_{|k-j|\le r_n} \frac{a(k+v_n)}{j-k} \right|$$
$$\le C \left[ \sup_n \left| \int_{|y|>r_n} \frac{f(y+v_n-t)}{y} dy \right| + \int_{|y|>\frac{1}{2}} \frac{|f(y+v_n-t)|}{y^2} dy \right].$$

Now, observe that the operator $Sf(x) := \int_{|y|>\frac{1}{2}} \frac{|f(y+v_n+t)|}{y^2} dy$ is weak (1,1) and strong (p,p), and the same is also the case for the operator

$$\sup_n \left| \int_{|y|>r_n} \frac{f(y+v_n+t)}{y} dy \right|$$

by Proposition 2.1. This proves the theorem. □

Now, the version of this maximal estimate (Theorem 2.5) in the ergodic setting follows from Calderón's transfer principle.

Proof of Theorem 2.5. Let $N > 0$ be a large integer, and $M' = 1 + \max_{n\le N}\{r_n, |v_n|\}$. Choose $M > M'$ and, for $f \in L_p$, define $\mathbf{a}_x = \{a_x(k)\}$ by

$$a_x(k) = \begin{cases} f(T^k x) & \text{if } |k| \le 3M \\ 0 \text{ if otherwise.} \end{cases}$$

Then, for almost every $x$ we have $\mathbf{a}_x \in l_p$, since

$$\|\mathbf{a}\|_p^p = \int_X \sum_k |a_x(k)|^p \, dx = \sum_{|k|\le r_N} \int_X \left| f\left(T^k x\right)\right|^p = 6M\|f\|_p^p.$$

Since $T$ is measure preserving, by Theorem 2.2,

$$\int_X \left| \left\{ |j| \le M : \sup_{n\le N} \left| \sum_{k,0<|k-j|\le r_n} \frac{a_x(v_n+j)}{j-k} \right| > \lambda \right\} \right| dx$$
$$\le \frac{C}{\lambda^p} \int_X \|\mathbf{a}_x\|_p^p \, dx \le \frac{2NC}{\lambda^p} \|f\|_p^p.$$

Now, let

$$U = \left\{ (x,j) \in X \times \mathbb{Z} : \sup_{n\le N} \left| \sum_{k,0<|k-j|\le r_n} \frac{a_x(v_n+j)}{j-k} \right| > \lambda \right\}.$$

Then, if $|j| \le M$, and $|k| < r_n$, since $r_n < M'$, $|v_n| < M'$, we have

$$-3M < -M - 2M' < v_n + j + k < 2M' + M < 3M.$$

In that range of values, we have, $a_x(v_n+j+k) = f\left(T^{v_n+j+k}x\right)$. Therefore,

$$\int_X \left|\left\{|j| \le M : \sup_{n\le N}\left|\sum_{k,0<|k-t|\le r_n}\frac{a_x(v_n+j)}{j-k}\right| > \lambda\right\}\right| dx$$
$$\ge \sum_{j\in\mathbb{Z}}\int_X \left[\chi_U(x,j)\chi_{\{|i|\le M\}}(j)\right] dx$$
$$\ge \sum_{|j|\le M}\mu\left(\left\{x : \mathcal{H}^*_{\mathbf{w},N}f(T^jx) > \lambda\right\}\right) = (2M)\mu\left(\left\{x : \mathcal{H}^*_{\mathbf{w},N}f(x) > \lambda\right\}\right).$$

Therefore,

$$\mu\left(\left\{x : \mathcal{H}^*_{\mathbf{w},N}f(x) > \lambda\right\}\right) \le \frac{1}{2M}\frac{6MC}{\lambda^p}\|f\|_p^p < \frac{3C}{\lambda^p}\|f\|_p^p.$$

Now, letting $N \to \infty$, we obtain that $\mathcal{H}^*_{\mathrm{w}}f(x)$ is weak (p,p) for $1 \le p < \infty$. The strong (p,p) property of the maximal operator $\mathcal{H}^*_{\mathbf{w}}f$ follows from the Marcinkiewicz interpolation theorem (see Proposition 4.2, [**T**]). □

Next, we will prove Theorem 2.6 which extends the maximal inequality in Theorem 2.5 to the admissible case. We need some preparation. The proposition below is the discrete analogue of Proposition 3.1.

PROPOSITION 4.1 ([**Ç1**]). *Let $T$ be an invertible measure preserving transformation and $F = \{f_n\}$ be a positive symmetric strongly bounded $T$-admissible process. Then there exists a sequence of non-negative functions $\{u_k\} \in L_p$ such that $f_n = T^n u_{|n|}$ for all $n \in \mathbb{Z}$. Furthermore, there exists $\delta \in L_p$ such that $f_n \le T^n\delta$ for all $n \in \mathbb{Z}$ and $\|\delta\|_p = \sup_{n\in\mathbb{Z}}\|f_n\|_p$.*

REMARKS.

1. The function $\delta \in L_p^+$is an exact dominant for $F$ if $p = 1$. When $p > 1$, it is called a dominant for $F$.
2. Proposition 4.1 also implies that if $F$ is a strongly bounded admissible process relative to an invertible measure preserving transformation, and if $R = \left\{\sum_{i=0}^{n-1} h_i\right\}$ is the purely subadditive part of $F$, then each $h_i \ge 0$. Indeed, $h_i = T^i\delta - f_i = T^i(\delta - u_i)$.
3. If $\sup_i \|f_i\|_1 < \infty$, obviously $F$ is bounded. Conversely, if $F \subset L_1$ is bounded, since $T$ is measure preserving and $F$ is $T$-admissible,
$$\|f_i\|_1 = \frac{1}{m}\sum_{k=0}^{m-1}\int T^k f_i \le \frac{1}{m}\sum_{i=k}^{m+k-1}\int f_i$$
$$\le \frac{1}{m}\int [F_{m+k} - F_k] \le \frac{1}{m}\int F_{m+k} \le \frac{m+k}{m}\gamma_F$$
and hence, letting $m \to \infty$ we have $\sup_{i\ge 0}\|f_i\|_1 \le \gamma_F$. Thus A positive $T$-admissible process $F \subset L_1$ is bounded if and only if it is strongly bounded. This property is not valid if $p \ne 1$. Indeed, there are admissible processes in $L_2$ satisfying $\sup_n \frac{1}{|n|}\|F_n\|_2 < \infty$, which are not strongly bounded [**Ç1**].

When $F$ is a $T$-admissible process on $\mathbb{Z}$, we have $f_i - T^i f_0 \geq 0$, and consequently, if a result holds for $\mathcal{H}_n f_0$, then the same is also valid for $\mathcal{H}_n F$ if and only if it is valid for $\mathcal{H}_n G$, where $g_i = f_i - T^i f_0$. Therefore, in such a case, we can assume that $f_i \geq 0$, for each $i \in \mathbb{Z}$. In particular, since we will consider sums without 0th-term, for convenience, we will always assume that $f_0 = 0$.

PROOF OF THEOREM 2.6. Let $N > 0$ be a large integer, and define

$$H_N^* F(x) = \sup_{1 \leq n \leq N} \left| \sum_{-r_n \leq |k| \leq r_n} \frac{f_k\left(T^{v_n} x\right)}{k} \right|, \text{ and } E_N = \left\{x \in X : H_N^* F(x) > \lambda\right\}.$$

Without loss of generality we can assume that each $f_i \geq 0$. Now, if $M' = 1 + \max\limits_{1 \leq n \leq N} \{r_n, |v_n|\}$, let $M = 2M'$.

Observe that $E_N = E_N^1 \cup E_N^2$, where

$$E_N^1 = \left\{ x \in X : \sup_{1 \leq n \leq N} \sum_{-r_n \leq |k| \leq r_n} \frac{f_k\left(T^{v_n} x\right)}{k} > \lambda \right\} \text{ and}$$

$$E_N^2 = \left\{ x \in X : \inf_{1 \leq n \leq N} \sum_{-r_n \leq |k| \leq r_n} \frac{f_k\left(T^{v_n} x\right)}{k} < -\lambda \right\}.$$

Now, consider $E_N^1$. By Proposition 4.1, $f_k = T^k u_{|k|}$ for all $k \in \mathbb{Z}$, where $u_k \uparrow \delta, u_k \leq u_{k+1}$ for all $k \geq 1$. Hence, we have

$$T^k u_0 \leq f_k \leq T^k \delta, \text{ for all } k \in \mathbb{Z}. \tag{2}$$

By the measure preserving property of $T$, for any $m \in \mathbb{Z}$,

$$\begin{aligned}
\mu\left(E_N^1\right) = \mu\left(T^m E_N^1\right) &= \mu\left\{ x : \sup_n \sum_{-r_n \leq |i| \leq r_n} \frac{T^{m+v_n+i} f_i(x)}{i} > \lambda \right\} \\
&= \mu\left\{ x : \sup_n \sum_{-r_n \leq |i| \leq r_n} \frac{T^{m+v_n+i} u_{|i|}(x)}{i} > \lambda \right\}.
\end{aligned}$$

By (2),

$$\begin{aligned}
\lambda < \sum_{-r_n \leq |i| \leq r_n} \frac{T^{m+v_n+i} u_{|i|}}{i} &= \sum_{i=-r_n}^{-1} \frac{T^{m+v_n+i} u_{|i|}}{i} + \sum_{i=1}^{r_n} \frac{T^{m+v_n+i} u_{|i|}}{i} \\
&\leq \sum_{i=-r_n}^{-1} \frac{T^{m+v_n+i} v_0}{i} + \sum_{i=1}^{r_n} \frac{T^{m+v_n+i} \delta}{i}.
\end{aligned}$$

Hence,

$$\mu\left(E_N^1\right) \leq \left\{ x : \sup_n \left[ \sum_{i=-r_n}^{-1} \frac{T^{m+v_n+i} u_0(x)}{i} + \sum_{i=1}^{r_n} \frac{T^{m+v_n+i} \delta(x)}{i} \right] > \lambda \right\}.$$

For $x \in X$, not belonging a set of measure zero, and for any $|m| \leq M$, define a sequence $\mathbf{a} = \{a_k\}$ by

$$a_k = \begin{cases} \delta(T^k x) & \text{if } 0 \leq k \leq 3M, \\ u_0(T^k x) & \text{if } -3M \leq |k| < 0, \\ 0 & \text{otherwise.} \end{cases}$$

Therefore, since $|m| + |v_n| + |i| < 3M$, we have $\{a_k\} \in l_1$. Thus,

$$\lambda < \sum_{-r_n \leq |i| \leq r_n} \frac{T^{m+v_n+i} u_{|k|}(x)}{i} \leq \sum_{i=-r_n}^{-1} \frac{a_{i+v_n+m}}{i} + \sum_{i=1}^{r_n} \frac{a_{i+v_n+m}}{i},$$

which implies that $E_N^1 \subset \left\{ x : \sup_n \sum_{-r_n \leq |i| \leq r_n} \frac{a_{i+v_n+m}}{i} > \lambda \right\}$.

For $E_N^2$, similarly, by (2), we have

$$\sum_{-r_n \leq |i| \leq r_n} \frac{T^{m+v_n+i} u_{|i|}}{i} \geq \sum_{i=-r_n}^{-1} \frac{T^{m+v_n+i}\delta}{i} + \sum_{i=1}^{r_n} \frac{T^{m+v_n+i} u_0}{i}.$$

Hence, for $x \in X$, not belonging a set of measure zero, and for any $|m| \leq M$, if we define a sequence $\mathbf{b} = \{b_k\}$ by

$$\mathbf{b}_k = \begin{cases} \delta(T^k x) & \text{if } -3M \leq |k| < 0, \\ v_0(T^k x) & \text{if } 0 \leq |k| \leq 3M, \\ 0 & \text{otherwise,} \end{cases}$$

it follows that $\{b_k\} \in l_1$. Therefore, as before,

$$E_N^2 \subset \left\{ x : \inf_n \sum_{-r_n \leq |i| \leq r_n} \frac{b_{i+v_n+m}}{i} < -\lambda \right\} = \left\{ x : \sup_n \sum_{-r_n \leq |i| \leq r_n} \frac{b_{-i+v_n+m}}{i} > \lambda \right\}.$$

Consequently, we observe that

$$\begin{aligned} &\left\{ x : \sup_{1 \leq n \leq N} \sum_{-r_n \leq |i| \leq r_n} \left| \frac{T^{m+v_n+i} f_i(x)}{i} \right| > \lambda \right\} \\ &\subset \left\{ x : \sup_{1 \leq n \leq N} \sum_{-r_n \leq |i| \leq r_n} \frac{a_{i+v_n+m}}{i} > \lambda \right\} \\ &\cup \left\{ x : \inf_{1 \leq n \leq N} \sum_{-r_n \leq |i| \leq r_n} \frac{b_{i+v_n+m}}{i} < -\lambda \right\} \\ &= \left\{ x : \sup_{1 \leq n \leq N} \sum_{-r_n \leq |i| \leq r_n} \frac{a_{i+v_n+m}}{i} > \lambda \right\} \\ &\cup \left\{ x : \sup_{1 \leq n \leq N} \sum_{-r_n \leq |i| \leq r_n} \frac{b_{-i+v_n+m}}{i} > \lambda \right\}. \end{aligned}$$

Therefore,

$$\begin{aligned}\mu(E_N) =& \frac{1}{2M+1} \sum_{m=-M}^{M} \mu\left\{x : \sup_{1\le n\le N}\left|\sum_{r_n\le|i|\le r_n} \frac{T^{i+v_n+m} f_i(x)}{i}\right| > \lambda\right\} \\ \le& \frac{1}{2M+1} \sum_{m=-M}^{M} \left[\mu\left\{x : \sup_{1\le n\le N} \sum_{-r_n\le|i|\le r_n} \frac{a_{i+v_n+m}}{i} > \lambda\right\}\right. \\ &\left. +\mu\left\{x : \sup_{1\le n\le N} \sum_{-r_n\le|i|\le r_n} \frac{b_{-i+v_n+m}}{i} > \lambda\right\}\right].\end{aligned}$$

Now, by Fubini's theorem,

$$\begin{aligned}&\sum_{m=-M}^{M} \mu\left\{x : \sup_{1\le n\le N} \sum_{-r_n\le|i|\le r_n} \frac{a_{i+v_n+m}}{i} > \lambda\right\} \\ &\le (\text{card} \times\mu)\left\{(m,x) : \sup_{1\le n\le N} \sum_{-r_n\le|i|\le r_n} \frac{a_{i+v_n+m}}{i} > \lambda\right\} \\ &\le \int\left|\left\{m : \sup_{1\le n\le N} \sum_{-r_n\le|i|\le r_n} \frac{a_{i+v_n+m}}{i} > \lambda\right\}\right| d\mu \\ &\le \int\left(\frac{C_1}{\lambda}\sum_{j\in\mathbb{Z}}|a_j|\right) d\mu \text{ by Theorem 2.4 (a)} \\ &\le \frac{C_1}{\lambda}\sum_{-N-M+v_n}^{N+M+v_n} \int |a_j|\, d\mu \le \frac{C_1}{\lambda}\sum_{-N-M+v_n}^{N+M+v_n} \|\delta\|_1 = \frac{2C_1(M+N)}{\lambda}\|\delta\|_1\end{aligned}$$

for some constant $C_1$, where *card* stands for the counting measure on $\mathbb{Z}$. In the same fashion, we also obtain that

$$\sum_{m=-M}^{M} \mu\left\{x : \sup_{1\le n\le N} \sum_{-r_n\le|i|\le r_n} \frac{b_{-i+v_n+m}}{i} > \lambda\right\} \le \frac{2C_2(M+N)}{\lambda}\|\delta\|_1.$$

Therefore, $\mu(E_N) \le \dfrac{2(C_1+C_2)(M+N)}{\lambda(2M+1)}\|\delta\|_1$. Denoting $C = C_1+C_2$, and letting $M \to \infty$, we deduce that $\mu(E_N) \le \dfrac{C}{\lambda}\|\delta\|_1$. The assertion follows by letting $N \to \infty$. □

## 5. Almost everywhere existence of the moving ergodic Hilbert transform

In this section, having the necessary maximal inequality and the a.e. existence of local ergodic Hilbert transform proved in [**FJR**], we will extend the a.e. existence of local ergodic Hilbert transform to the setting of strongly bounded symmetric admissible processes relative to measure preserving flows. The proof is similar to the proof of Theorem 3.2 in [**Ç2**] (and uses Theorem 2.3 above instead of Theorem 3.1 in [**Ç2**]).

THEOREM 5.1. *Let $\tau$ be a measure preserving flow and $F = \{f_t\} \subset L_1$ be a strongly bounded symmetric $\tau$-admissible process with exact dominant $\delta$. If $\mathbf{w} = \{(v_q, r_q)\}$ is a sequence satisfying CC such that $v_q, r_q \to 0$ as $q \to 0$. Then*

$$\mathcal{H}_{\mathrm{w}}F(x) = \lim_{q\to 0} \int_{r_n\le|s|\le 1/r_n} \frac{f_s\left(T_{v_q}x\right)}{s} ds \text{ exists a.e.}$$

PROOF. Since by Theorem 2.2 the local ergodic Hilbert transform exists for additive processes relative to measure preserving flows, we can assume that $f_t \geq 0$ for each $t \in \mathbb{R}$. Fix $\epsilon > 0$, and consider the additive process $G^\epsilon = \{T_t g_\epsilon\}$; hence, $G_t^\epsilon \leq F_t$ for all $t \in \mathbb{R}$ such that $\gamma_G = \|g_\epsilon\|_1 > \gamma_F - \epsilon$. Define a new process $Z = \{Z_t\}$ where, for each $t > 0$, $z_t(x) = f_t(x) - T_t g_\epsilon(x)$. Then $0 < z_t(x) \leq T^i\left(\delta - g_\epsilon\right)(x)$, and $\|\delta - g_\epsilon\|_1 < \epsilon$. Since

$$T_s z_t = T_s\left(f_t - T_t g_\epsilon\right) \leq f_{t+s} - T_{t+s} g_\epsilon = z_{t+s}$$

and

$$T_{-2t} z_t = T_{-2t} f_t - T_{-2t} T_t g_\epsilon = f_{-t} - T_{-t} g_\epsilon = z_{-t}$$

$Z = \{z_i\}$ is a strongly bounded, symmetric $T$-admissible process with exact dominant $\gamma_Z = \|\delta - g_\epsilon\| < \epsilon$. Now,

$$\int_{r_n\le|s|\le 1/r_n} \frac{f_s\left(T_{v_q}x\right)}{s} ds - \int_{r_n\le|s|\le 1/r_n} \frac{g_\epsilon\left(T_{v_q+s}x\right)}{s} ds = \int_{r_n\le|s|\le 1/r_n} \frac{z_s\left(T_{v_q}x\right)}{s} ds$$

and since $\lim_n \int_{r_n\le|s|\le 1/r_n} \frac{g_\epsilon\left(T_{v_q+s}x\right)}{s} ds$ exists a.e., by Theorem 2.2, it's enough to prove that

$$\lim_n \int_{r_n\le|s|\le 1/r_n} \frac{z_s\left(T_{v_q}x\right)}{s} ds$$

exists a.e. For, if $E = \{x : f^*(x) - f_*(x) > \lambda\}$, where

$$f^* = \limsup_n \int_{r_n\le|s|\le 1/r_n} \frac{f_s\left(T_{v_q}x\right)}{s} ds \text{ and } f_* = \liminf_n \int_{r_n\le|s|\le 1/r_n} \frac{f_s\left(T_{v_q}x\right)}{s} ds,$$

then

$$E \subset \left\{ x : \limsup \left| \int_{r_n\le|s|\le 1/r_n} \frac{z_s\left(T_{v_q}x\right)}{s} \right| ds > \frac{\lambda}{2} \right\}.$$

From Theorem 2.3 it follows that, for some constant $C > 0$, $\mu(E) \leq \dfrac{C}{\lambda}\|\delta - g_\epsilon\|_1 = \gamma_Z < \epsilon$. Since $\epsilon > 0$ is arbitrary, we obtain that $\mu(E) = 0$. Thus $\lim_{q\to 0} H_{(q,\mathbf{w})}F(x)$ exists a.e. □

REMARKS.

1. Since $L_p \subset L_1$, the strong boundedness in $L_p$ implies strong boundedness in $L_1$ (we deal with a probability space). Consequently, the assertion of Theorem 5.1 is also valid for strongly bounded symmetric admissible processes $F \subset L_p, 1 < p < \infty$.
2. It should be stressed here that without the symmetry condition, which is needed for cancellation, Theorem 5.1 is false. In fact, there are functions $f \in L_\infty$ with integral zero for which the one-sided ergodic Hilbert

transform (i.e., $\sum_{k=1}^{\infty} \frac{T^k f}{k}$ ) need not exist (see [**P2**], pp: 94-99). Let $\mathbf{w} = \{(0, n)\}$, take the above $f$ and define $\{f_k\}$ by $f_k = T^k f$ for $k > 0$ and $f_k = 0$ for $k \leq 0$. Then $\{f_k\}$ is $T$-admissible, and $\lim_n \mathcal{H}_{(n,\mathbf{w})} F(x)$ does not exist. Also, it should be observed that, given any invertible ergodic measure preserving transformation $T$, the one-sided ergodic Hilbert transform diverges a.e. for any non-negative $f$ (not identically zero), since the convergence of one-sided ergodic Hilbert transform implies convergence of the ergodic averages to zero (Kronecker's Lemma), so $\int f = 0$.

As in the local setting, since we have the maximal inequality for the discrete moving ergodic Hilbert transform, one naturally proceeds to proving the existence of the discrete moving ergodic Hilbert transform along sequences with CC. For, one needs to show the existence for a dense class in $L_p$. However, the following example, due to R. Jones [**J**], clearly indicates that in general there is no such dense class [**Ç3**]. We provide the example here for completeness.

EXAMPLE 5.2. Consider the dynamical system $(X, \mathcal{F}, \mu, T)$, where $X = \{0, 1, 2\}$, $\mathcal{F} = \mathcal{P}(X)$, $\mu(\{i\}) = 1/3, 0 \leq i \leq 2$, and $T(0) = 1, T(1) = 2, T(2) = 0$. Since $T$ is cyclic, it is ergodic. Consider the moving averages sequence $\{(n, n)\}$. Hence, there are infinitely many $v_n = n$ such that $n = 0(\text{mod}3), n = 1(\text{mod}3)$, and $r_n = n \to \infty$. Let $\mathcal{A} = \{n : n = 0(\text{mod}3)\}$ and $\mathcal{B} = \{n : n = 1(\text{mod}3)\}; \mathbf{w_1}$ and $\mathbf{w_2}$ be subsequences along $\mathcal{A}$ and $\mathcal{B}$, respectively. Let $f : X \to \mathbb{R}$ be such that $f(0) = f(2) = 0$ and $f(1) = 1$. Then, we have

$$\mathcal{H}^T f(1) = \lim_{n\to\infty} \sum_{0<|k|\leq n} \frac{f\left(T^k 1\right)}{k} = 0,$$

$$\mathcal{H}^T f(0) = \lim_{n\to\infty} \sum_{0<|k|\leq n} \frac{f\left(T^k 0\right)}{k} = \left(1 - \frac{1}{2}\right) + \left(\frac{1}{4} - \frac{1}{5}\right) + \left(\frac{1}{7} - \frac{1}{8}\right) + \ldots \neq 0.$$

On the other hand, for $n \in \mathcal{A}$, i.e., along $\mathbf{w_1}$, we have

$$\mathcal{H}^T_{(v_n, r_n)} f(0) = \sum_{0<|k|\leq r_n} \frac{f\left(T^{v_n + k} 0\right)}{k} = \sum_{0<|k|\leq r_n} \frac{f\left(T^k 0\right)}{k} \to \mathcal{H}^T f(0) \neq 0$$

whereas, for $n \in \mathcal{B}$ (along $\mathbf{w_2}$ ) we have

$$\begin{aligned}\mathcal{H}^T_{(v_n, r_n)} f(0) &= \sum_{0<|k|\leq r_n} \frac{f\left(T^{v_n + k} 0\right)}{k} = \sum_{|k|\leq r_n} \frac{f\left(T^{1+k} 0\right)}{k} \\ &= \sum_{0<|k|\leq r_n} \frac{f\left(T^k 1\right)}{k} \to \mathcal{H}^T f(1) = 0,\end{aligned}$$

which implies that $\lim_n \mathcal{H}^T_{(v_n, r_n)} f(0)$ does not exist.

The example above clearly indicates that, if one wants to obtain the a.e. existence of the ergodic Hilbert transform along moving averages sequences, then some restrictions need to be imposed. Observe that the measure preserving transformation in the example (which can be replicated in non-atomic cases by use of Rohlin tower constructions) does not have Lebesgue spectrum. We conjecture that, one has an affirmative result if the dynamical system $(X, \mathcal{F}, \mu, T)$ has Lebesgue spectrum.

## Acknowledgment

The author would like to express his gratitude to the anonymous referee for indicating some issues on the support of functions in Theorems 2.3 and 2.6, and for the suggestion that improved the presentation of Theorem 2.5.

## References

[AD] M. A. Akcoglu and Y. Déniel, *Moving weighted averages*, Canad. J. Math. **45** (1993), no. 3, 449–469, DOI 10.4153/CJM-1993-023-x. MR1222510

[AK] M. A. Akcoglu and U. Krengel, *Ergodic theorems for superadditive processes*, J. Reine Angew. Math. **323** (1981), 53–67, DOI 10.1515/crll.1981.323.53. MR611442

[BJR] Alexandra Bellow, Roger Jones, and Joseph Rosenblatt, *Convergence for moving averages*, Ergodic Theory Dynam. Systems **10** (1990), no. 1, 43–62, DOI 10.1017/S0143385700005381. MR1053798

[C] Mischa Cotlar, *A unified theory of Hilbert transforms and ergodic theorems* (English, with Spanish summary), Rev. Mat. Cuyana **1** (1955), 105–167 (1956). MR84632

[CaJRW] James T. Campbell, Roger L. Jones, Karin Reinhold, and Máté Wierdl, *Oscillation and variation for the Hilbert transform*, Duke Math. J. **105** (2000), no. 1, 59–83, DOI 10.1215/S0012-7094-00-10513-3. MR1788042

[CaP] James Campbell and Karl Petersen, *The spectral measure and Hilbert transform of a measure-preserving transformation*, Trans. Amer. Math. Soc. **313** (1989), no. 1, 121–129, DOI 10.2307/2001068. MR958884

[Ç1] Doğan Çömez, *Moving averages of superadditive processes with respect to $L_p$-contractions*, $1 < p < \infty$, Rocky Mountain J. Math. **28** (1998), no. 3, 875–883, DOI 10.1216/rmjm/1181071744. MR1657055

[Ç2] Doğan Çömez, *The ergodic Hilbert transform for admissible processes*, Canad. Math. Bull. **49** (2006), no. 2, 203–212, DOI 10.4153/CMB-2006-021-2. MR2226243

[Ç3] Doğan Çömez, *Existence of discrete ergodic singular transforms for admissible processes*, Colloq. Math. **112** (2008), no. 2, 335–343, DOI 10.4064/cm112-2-8. MR2383337

[F] S. E. Ferrando, *Moving ergodic theorems for superadditive processes*, Canad. J. Math. **47** (1995), no. 4, 728–743, DOI 10.4153/CJM-1995-038-3. MR1346161

[FJR] S. Ferrando, R. L. Jones, and K. Reinhold, *On approach regions for the conjugate Poisson integral and singular integrals*, Studia Math. **120** (1996), no. 2, 169–182, DOI 10.4064/sm220-2-5. MR1406535

[J] R. Jones, Personal communication.

[K] J. F. C. Kingman, *The ergodic theory of subadditive stochastic processes*, J. Roy. Statist. Soc. Ser. B **30** (1968), 499–510. MR254907

[NS] Alexander Nagel and Elias M. Stein, *On certain maximal functions and approach regions*, Adv. in Math. **54** (1984), no. 1, 83–106, DOI 10.1016/0001-8708(84)90038-0. MR761764

[P1] Karl Petersen, *Another proof of the existence of the ergodic Hilbert transform*, Proc. Amer. Math. Soc. **88** (1983), no. 1, 39–43, DOI 10.2307/2045106. MR691275

[P2] Karl Petersen, *Ergodic theory*, Cambridge Studies in Advanced Mathematics, vol. 2, Cambridge University Press, Cambridge, 1983, DOI 10.1017/CBO9780511608728. MR833286

[T] Alberto Torchinsky, *Real-variable methods in harmonic analysis*, Pure and Applied Mathematics, vol. 123, Academic Press, Inc., Orlando, FL, 1986. MR869816

*Email address*: `mrinal.roychowdhury@utrgv.edu`

Contemporary Mathematics
Volume **839**, 2026
https://doi.org/10.1090/conm/839/16727

# Orders of oscillation motivated by Sarnak's conjecture—Part II

Yunping Jiang

ABSTRACT. This work is a continuation of the author's work in [Proc. Amer. Math. Soc., 147 (2019), 3075–3085]. We study the linear disjointness between higher-order oscillating sequences and nonlinear dynamical systems. Specifically, we prove that any oscillating sequence of order $m = d + k - 1$ and any simple polynomial skew product of degree $k$ on the $d$-Euclidean space are linearly disjoint. Additionally, we demonstrate that any oscillating sequence of order $d$ and any minimal mean attractable and minimal quasi-discrete spectrum dynamical system of order $d$ are linearly disjoint. Finally, we introduce multi-linearly disjoint sequences and construct examples of such sequences.

## 1. Introduction

Let $\mathbb{N} = \{1, 2, \cdots, n \cdots\}$ be the set of all natural numbers. Let $X$ be a compact metric space and let $C(X)$ be the space of all continuous functions $\phi$ on $X$ equipped with the maximum norm $\|\phi\| = \max_{x\in X} |\phi(x)|$.

Let $f : X \to X$ be a continuous map. We use $f$ to represent a (discrete) dynamical system $\{f^n\}_{n\in\{0\}\cup\mathbb{N}}$, where $f^n$ means the $n$-time composition of $f$ and $f^0$ means the identity.

Let $\mathbb{C}$ be the complex plane. Let $\mathbf{c} = (c_n)_{n\in\mathbb{N}}$ be a sequence of complex numbers in $\mathbb{C}$. Without loss of generality, we assume that the asymptotic mean of $\mathbf{c}$ is zero, that is,

$$\mathbb{E}(\mathbf{c}) = \lim_{N\to\infty} \frac{1}{N} \sum_{n=1}^{N} c_n = 0.$$

We also assume that $\mathbf{c}$ satisfies the following technical condition: there are constants $C > 0$ and $\lambda > 1$ such that

$$\frac{1}{N} \sum_{n=1}^{N} |c_n|^\lambda \leq C, \quad \forall\, N \geq 1. \tag{1}$$

An example is the Liouville sequence $\mathbf{l} = (\lambda(n))_{n\in\mathbb{N}}$ where $\lambda(n) = (-1)^{\Omega(n)}$ and $\Omega(n)$ is the number of prime factors of $n$ (counted by multiplicity). Another example

2020 *Mathematics Subject Classification.* Primary 37A35, 11K65; Secondary 37A25, 11N05.
*Key words and phrases.* oscillating sequence of order $d$, completely oscillating sequence, simple polynomial skew products, MMA and MQDS, linearly disjoint, multi-linearly disjoint.
This work was partially supported by grants from the Simons Foundation (#523341 and #942077), a PSC-CUNY Enhanced Award (#66680-54).

is the Möbius sequence $\mathbf{u} = (\mu(n))_{n\in\mathbb{N}}$, where $\mu(n) = 0$ if $p^2|n$ for a prime number $p$ and $\mu(n) = \lambda(n)$, otherwise. Note that $\mathbb{E}(\mathbf{l}) = \mathbb{E}(\mathbf{u}) = 0$ (the prime number theorem).

DEFINITION 1. *We say that* $\mathbf{c}$ *and* $f$ *are (asymptotically) linearly disjoint if for any* $\phi \in C(X)$,

$$\lim_{N\to\infty} \frac{1}{N} \sum_{n=1}^{N} c_n \phi(f^n x) = 0, \quad \forall\, x \in X. \tag{2}$$

The Sarnak conjecture in number theory [**17, 18**] states

CONJECTURE 1. *The sequence* $\mathbf{u}$ *and any continuous dynamical system* $f : X \to X$ *having zero topological entropy are linearly disjoint.*

Motivated by this conjecture, we have defined and studied oscillating sequences in [**9**].

DEFINITION 2. *We say that* $\mathbf{c}$ *is an oscillating sequence if for every* $0 \leq \theta < 1$,

$$\lim_{N\to\infty} \frac{1}{N} \sum_{n=1}^{N} c_n e^{2\pi i n\theta} = 0. \tag{3}$$

DEFINITION 3. *We say that* $\mathbf{c}$ *is an oscillating sequence in the arithmetic sense if for every* $0 \leq \theta < 1$ *and any pair of integers* $0 \leq l < k$,

$$\lim_{N\to\infty} \frac{1}{N} \sum_{1\leq n\leq N,\ n\equiv l \pmod k} c_n e^{2\pi i n\theta} = 0. \tag{4}$$

In addition, we define the orders of oscillation of $\mathbf{c}$ in [**13**]. Let $d \geq 2$ be an integer.

DEFINITION 4. *We say that* $\mathbf{c}$ *is an oscillating sequence of order* $d$ *if for every real coefficient polynomial* $P$ *whose degree is less than or equal to* $d$,

$$\lim_{N\to\infty} \frac{1}{N} \sum_{n=1}^{N} c_n e^{2\pi i P(n)} = 0. \tag{5}$$

*If* (5) *holds for all real coefficient polynomials* $P$, *then we call* $\mathbf{c}$ *a completely oscillating sequence.*

DEFINITION 5. *We say that* $\mathbf{c}$ *is an oscillating sequence of order* $d$ *in the arithmetic sense if for every real coefficient polynomial* $P$ *whose degree is less than or equal* $d$ *and any pair of integers* $0 \leq l < k$,

$$\lim_{N\to\infty} \frac{1}{N} \sum_{1\leq n\leq N,\ n\equiv l \pmod k} c_n e^{2\pi i P(n)} = 0. \tag{6}$$

*If* (6) *holds for all real coefficient polynomials* $P$ *and all pairs of integers* $0 \leq l < k$, *then we call* $\mathbf{c}$ *a completely oscillating sequence in the arithmetic sense.*

PROPOSITION 1. *Definition* 2 *and Definition* 3 *are equivalent. And Definition* 4 *and Definition* 5 *are equivalent.*

PROOF. It is clear that Definition 3 and Definition 5 imply Definition 2 and Definition 4, respectively, by taking $l = 0$ and $k = 1$.

Now, suppose that we have Definition 4 (or Definition 2). Then for any real coefficient polynomial $Q(x)$, denote

$$S_N(Q) = \frac{1}{N} \sum_{n=1}^{N} c_n e^{2\pi i Q(n)},$$

we have that

$$\lim_{N\to\infty} S_N(Q) = 0.$$

For any pair of integers $0 \leq l < k$, let

$$\omega = e^{2\pi i \frac{1}{k}}.$$

We use the facts:

$$\sum_{j=1}^{k} \omega^{j(n-l)} = k \quad \text{if } n - l \equiv 0 \pmod{k}$$

and

$$\sum_{j=1}^{k} \omega^{j(n-l)} = 0 \quad \text{if } n - l \not\equiv 0 \pmod{k}.$$

For any real coefficient polynomial $P(x)$,

$$\frac{1}{N} \sum_{1\leq n\leq N, n-l\equiv 0 \pmod{k}} c_n e^{2\pi i P(n)} = \frac{1}{k}\frac{1}{N} \sum_{n=1}^{N} c_n \Big( \sum_{j=1}^{k} \omega^{j(n-l)} \Big) e^{2\pi i P(n)}$$

$$= \frac{1}{k} \sum_{j=1}^{k} \frac{1}{N} \sum_{n=1}^{N} c_n \Big( \omega^{j(n-l)} \Big) e^{2\pi i P(n)} = \frac{1}{k} \sum_{j=1}^{k} \Big( \omega^{-jl} \Big) \frac{1}{N} \sum_{n=1}^{N} c_n e^{2\pi i (P(n) + \frac{jn}{k})}$$

$$= \frac{1}{k} \sum_{j=1}^{k} \Big( \omega^{-jl} \Big) S_N(Q_j),$$

where

$$Q_j(x) = P(x) + \frac{xj}{k}, \quad 1 \leq j \leq k.$$

Since

$$\lim_{N\to\infty} S_N(Q_j) = 0, \quad \forall\, 1 \leq j \leq k,$$

we have that

$$\lim_{N\to\infty} \frac{1}{N} \sum_{1\leq n\leq N, n\equiv l \pmod{k}} c_n e^{2\pi i P(n)} = \frac{1}{k} \sum_{j=1}^{k} \Big( \omega^{-jl} \Big) \lim_{N\to\infty} S_N(Q_j) = 0.$$

This completes the proof. □

REMARK 1. *From Davenport* [**7**], *we know that the Möbius sequence* **u** *is an oscillating sequence, and from Hua* [**12**], *we know that the Möbius sequence* **u** *is a completely oscillating sequence.*

In [**9**], we studied the linear disjointness of oscillating sequences and mean Lyapunov stable (abbreviated MLS) dynamical systems. All MLS dynamical systems have zero topological entropy (see [**16**]). In [**13**], we studied the linear disjointness of oscillating sequences of order $d$ and affine torus maps on the $d$-torus of zero topological entropy for $d \geq 2$ as follows.

Let $\mathbb{Z}$ and $\mathbb{R}$ mean the sets of integers and real numbers, respectively. Let $d \geq 2$ be an integer. Let $\mathbb{R}^d$ be the $d$-Euclidean real vector space and let $\mathbb{Z}^d$ be the integer lattice in $\mathbb{R}^d$. Then $\mathbb{T}^d = \mathbb{R}^d/\mathbb{Z}^d$ is the $d$-torus.

For $\mathbf{x} \in \mathbb{T}^d$, let $\mathbf{x}^t = (x_1, \cdots, x_d)$ mean the transpose of $\mathbf{x}$. Let $\mathrm{GL}(d, \mathbb{Z})$ be the space of all $d \times d-$matrices $A$ of integer entries with determinants $\det(A) = \pm 1$. For any $A \in \mathrm{GL}(d, \mathbb{Z})$, the map $A\mathbf{x}$ is an automorphism of $\mathbb{T}^d$. For $\mathbf{v} \in \mathbb{T}^d$ with $\mathbf{v}^t = (v_1, \cdots, v_d)$, we have an affine dynamical system on $\mathbb{T}^d$,

$$T_{d,A,\mathbf{v}}(\mathbf{x}) = A\mathbf{x} + \mathbf{v} \tag{7}$$

A result of Sinai (see [**19**]) says that $T_{d,A,\mathbf{v}}$ has zero topological entropy if and only if the absolute values of all eigenvalues of $A$ are 1; furthermore, a result of Kronecker (see [**14**]) implies that in this case, all eigenvalues of $A$ are roots of unity. Therefore, for some positive integer $m$, all the eigenvalues of $A^m$ are 1. Without loss of generality, we only need to consider $A \in \mathrm{GL}(d, \mathbb{Z})$ such that all its eigenvalues are 1. We have

LEMMA 1. *Suppose $A \in \mathrm{GL}(d, \mathbb{Z})$ such that all its eigenvalues are 1. Then we have a matrix $P \in \mathrm{GL}(d, \mathbb{Z})$ whose determinant is 1 such that $L = P^{-1}AP \in \mathrm{GL}(d, \mathbb{Z})$ is a lower triangular matrix.*

See [**13**, Lemma 1] for a proof. Thus, without loss of generality, we only need to consider a lower triangular matrix $A \in \mathrm{GL}(d, \mathbb{Z})$ whose entries on the main diagonal are 1. Under this assumption, the map in (7) is an affine skew product,

$$T_{d,A,\mathbf{v}}(\mathbf{x}) = \begin{pmatrix} x_1 + v_1 \\ x_2 + b_{21}x_1 + v_2 \\ \vdots \\ x_d + b_{d(d-1)}x_{d-1} \cdots + b_{d1}x_1 + v_d \end{pmatrix} \tag{8}$$

where $b_{ij}$, $2 \leq i \leq d$, $1 \leq j \leq i-1$, are integers. In [**13**], I proved the following theorem.

THEOREM A. *Any oscillating sequence $\mathbf{c}$ of order $d \geq 2$ and any affine skew product $T_{d,A,\mathbf{v}}$ in (8) are linearly disjoint.*

From the proof of Theorem A, we can obtain directly that

COROLLARY A. *Any oscillating sequence $\mathbf{c}$ of order $d \geq 2$ in the arithmetic sense (equivalently, any oscillating sequence $\mathbf{c}$ of order $d \geq 2$ (see Proposition 1)) and any affine torus map $T_{d,A,\mathbf{v}}$ in (7) having zero topological entropy are linearly disjoint.*

We are interested in the following problem.

PROBLEM 1. *Are any completely oscillating sequence $\mathbf{c}$ (having positive entropy) and any continuous torus map $f : \mathbb{T}^d \to \mathbb{T}^d$ having zero topological entropy linearly disjoint?*

## 2. A generalized notation of linear disjointness

To demonstrate that all orders of oscillation needed in the study of Problem 1, we consider a torus continuous map $f : \mathbb{T}^d \to \mathbb{T}^d$ having zero topological entropy whose lift $\widetilde{T}$ has the form

$$\widetilde{T}(\mathbf{x}) = A\mathbf{x} + C(\mathbf{x}) : \mathbb{R}^d \to \mathbb{R}^d,$$

where $A \in \mathrm{GL}(d, \mathbb{Z})$ with all absolute values of eigenvalues 1 and $C(\mathbf{x})$ is a continuous periodic map, that is, $C(\mathbf{x} + \mathbf{1}) = C(\mathbf{x})$ and $\mathbf{1}^t = (1, \cdots, 1)$. If we consider a polynomial approximation $P(\mathbf{x})$ of $C(\mathbf{x})$, then we have a continuous map

$$F(\mathbf{x}) = A\mathbf{x} + P(\mathbf{x}) : \mathbb{R}^d \to \mathbb{R}^d.$$

However, $F(\mathbf{x})$ may not induce a continuous torus map. So, we would like to generalize the definition of linear disjointness.

Let $Y$ be a Polish space, not necessarily compact. Suppose $\sim$ is an equivalence relation on $Y$ such that the quotient space $X = Y/\sim$ is a compact metric space. Let $\pi : Y \to X$ be the natural projection, that is, $\pi(y) = [y]$, where $[y]$ means the equivalence class of $y$.

For any continuous map $f : Y \to Y$, we have a (discrete) continuous dynamical system $\{f^n\}_{n\in\{0\}\cup\mathbb{N}}$ on $Y$. We just use $f$ to mean this dynamical system. Let $C(X)$ be the space of all continuous functions on $X$ with the maximum norm $\|\phi\| = \max_{x\in X} |\phi(x)|$. We can have a generalized notation of linear disjointness.

DEFINITION 6. *We say that* $\mathbf{c}$ *and a continuous dynamical system* $f : Y \to Y$ *are (asymptotically) linearly disjoint if for any* $\phi \in C(X)$,

$$\lim_{N\to\infty} \frac{1}{N} \sum_{n=1}^{N} c_n \phi(\pi(f^n y)) = 0, \quad \forall\, y \in Y. \tag{9}$$

We have a generalized conjecture (refer to Conjecture 1).

CONJECTURE 2. *The Möbius sequence* $\mathbf{u}$ *and any continuous dynamical system* $f : Y \to Y$ *having zero entropy are linearly disjoint.*

REMARK 2. *In the above conjecture, we use the definition of entropy given in* [**6**] *for a continuous dynamical system* $f$ *from a Polish space* $Y$ *into itself. When the space* $Y = X$ *itself is compact, the entropy defined in* [**6**] *is the same as the topological entropy.*

## 3. Polynomial skew products

Consider $Y = \mathbb{R}^d$, $X = \mathbb{T}^d$. Then $\pi : \mathbb{R}^d \to \mathbb{T}^d$ is the universal cover. We call a map $f$ a polynomial skew product of degree $k$ if

$$f(\mathbf{x}) = \begin{pmatrix} x_1 + a \\ x_2 + h_2(x_1) \\ x_3 + h_3(x_1, x_2) \\ \vdots \\ x_d + h_d(x_1, x_2, \cdots, x_{d-1}) \end{pmatrix} \tag{10}$$

where $a$ is a constant and $h_2(x_1)$ is a polynomial of $x_1$ of degree $\leq k$, $h_3(x_1, x_2)$ is a polynomial of $x_1$ and $x_2$ of degree $\leq k$, $\cdots$, $h_d(x_1, \cdots, x_{d-1})$ is a polynomial of

$x_1, \cdots, x_{d-1}$ of degree $\leq k$. We call $f$ a simple polynomial skew product of degree $k$ if it is in the form

$$f(\mathbf{x}) = \begin{pmatrix} x_1 + a \\ x_2 + h_2(x_1) \\ x_3 + b_{32}x_2 + h_3(x_1) \\ \vdots \\ x_d + b_{d(d-1)}x_{d-1} \cdots + b_{d2}x_2 + h_d(x_1) \end{pmatrix} \tag{11}$$

where $a$ is a constant and $b_{ij}$, $3 \leq i \leq d$, $2 \leq j \leq i-1$, are integers and $h_i(x_1)$, $2 \leq i \leq d$, are polynomials of $x_1$ of degree $\leq k$. One of the results in this paper, which generalizes Theorem A, is the following theorem.

THEOREM 1. *Suppose $d \geq 2$ and $k \geq 1$. Any oscillating sequence $\mathbf{c}$ of order $m = d + k - 1$ and any simple polynomial skew products $f(\mathbf{x})$ of degree $k$ in (11) are linearly disjoint.*

COROLLARY 1. *Suppose $A \in \mathrm{GL}(d, \mathbb{Z})$ with all absolute values of the eigenvalues 1. Let $q \geq 1$ be the smallest integer such that all the eigenvalues of $A^q$ are 1. Suppose*

$$f^q(\mathbf{x}) = A^q\mathbf{x} + \mathbf{h}(x_1)$$

*where $\mathbf{h}^t(x_1) = (a, h_2(x_1), \cdots, h_d(x_1))$ and $a$ is a constant and $h_i(x_1)$, $2 \leq i \leq d$, are polynomials of $x_1$ of degree $\leq k$ for $k \geq 1$. Then, any oscillating sequence $\mathbf{c}$ of order $m = d + k - 1$ and $f$ are linearly disjoint.*

PROOF OF THEOREM 1. Let $\mathbf{k} \in \mathbb{Z}^d$ with $\mathbf{k}^t = (k_1, \cdots, k_d)$, define

$$e(\mathbf{k} \cdot \mathbf{x}) = e^{2\pi i(k_1x_1 + \cdots + k_dx_d)}.$$

From the Stone-Weierstrass theorem, the set $S = \Big\{e(\mathbf{k} \cdot \mathbf{x})\Big\}_{\mathbf{k} \in \mathbb{Z}^d}$ forms a dense subset in $C(\mathbb{T}^d)$. A trigonometric polynomial $p$ is a linear combination of elements in $S$. We can write $p$ as

$$p(\mathbf{x}) = \sum_{m_1 \leq k_1 \leq s_1} \cdots \sum_{m_d \leq k_d \leq s_d} a_{\mathbf{k}} e^{2\pi i(k_1x_1 + \cdots + k_dx_d)}.$$

For any $\phi \in C(\mathbb{T}^d)$, we have a sequence of trigonometric polynomials

$$p_q(\mathbf{x}) = \sum_{m_{1q} \leq k_1 \leq s_{1q}} \cdots \sum_{m_{dq} \leq k_d \leq s_{dq}} a_{\mathbf{k},q} e^{2\pi i(k_1x_1 + \cdots + k_dx_d)}. \tag{12}$$

such that $\|\phi - p_q\| \to 0$ as $q \to \infty$. The sequence $\{p_q\}_{q \in \mathbb{N}}$ is called the trigonometric approximation of $\phi$. Consider

$$S_N\phi(\mathbf{x}) = \frac{1}{N}\sum_{n=1}^{N} c_n \phi(f^n\mathbf{x}).$$

For any $\epsilon > 0$, we have an integer $r > 0$ such that

$$\|\phi - p_r\| < \frac{\epsilon}{2C^{\frac{1}{\lambda}}}.$$

Then

$$S_N\phi(\mathbf{x}) = \Big(\frac{1}{N}\sum_{n=1}^{N} c_n\big(\phi(f^n\mathbf{x}) - p_r(f^n\mathbf{x})\big)\Big) + \Big(\frac{1}{N}\sum_{n=1}^{N} c_n p_r(f^n\mathbf{x})\Big) = I + II.$$

For the estimation of $I$, we apply the Hölder inequality. Let $\lambda$ and $C$ be the numbers in (1) and $\lambda' > 1$ be the dual number of $\lambda$, that is,

$$\frac{1}{\lambda} + \frac{1}{\lambda'} = 1.$$

Then we have that

$$|I| \leq \Big(\frac{1}{N}\sum_{l=1}^{N}|c_n|^{\lambda}\Big)^{\frac{1}{\lambda}}\Big(\frac{1}{N}\sum_{n=1}^{N}|\phi(f^n\mathbf{x}) - p_r(f^n\mathbf{x})|^{\lambda'}\Big)^{\frac{1}{\lambda'}} \leq C^{\frac{1}{\lambda}} \cdot \frac{\epsilon}{2C^{\frac{1}{\lambda}}} = \frac{\epsilon}{2}.$$

For the estimation of $II$, let $\mathbf{x}_n = f^n\mathbf{x}$ for $n = 1, 2, \cdots$. Denote

$$\mathbf{x}_n^t = (x_1^n, \cdots, x_d^n).$$

Then

$$x_1^n = x_1 + an$$

is a polynomial of $n$ of degree less than or equal to 1. Assume $x_1^0 = x_1$.

Suppose $h_2(x_1) = \sum_{i=0}^{k} r_i \cdot (x_1)^i$. Then

$$\begin{aligned} x_2^n = x_2 + \sum_{l=0}^{n-1} h_2(x_1^l) &= x_2 + \sum_{l=0}^{n-1}\sum_{i=0}^{k} r_i \cdot (x_1 + la)^i \\ &= \sum_{l=0}^{n-1}\sum_{i=0}^{k} s_i l^i = \sum_{i=0}^{k} s_i \sum_{l=0}^{n-1} l^i = \sum_{i=0}^{k} s_i Q_i(n) \end{aligned}$$

where all $s_i$ are numbers only depending on $r_i$, $k$, $x_1$, and $a$ but not on $n$, and

$$Q_i(n) = \sum_{l=0}^{n-1} l^i$$

are polynomials of $n$ of degrees $i+1$ for $0 \leq i \leq k$. Thus, $x_2^n$ is a polynomial of $n$ of degree less than or equal to $k+1$. Assume $x_2^0 = x_2$.

Now

$$x_3^n = x_3 + b_{32}\Big(\sum_{l=0}^{n-1} x_2^l\Big) + \sum_{l=0}^{n-1} h_3(x_1^l).$$

Similar to the calculation of $x_2^n$, we have that $\sum_{l=0}^{n-1} h_3(x_1^l)$ is a polynomial of $n$ of degree less than or equal to $k+1$. Since $x_2^n$ is a polynomial of $n$ of degree less than or equal to $k+1$, $\sum_{l=0}^{n-1} x_2^l$ is a polynomial of $n$ of degree less than or equal to $k+2$. Thus, $x_3^n$ is a polynomial of $n$ of degree less than or equal to $k+2$. Inductively, we have that $x_i^n$ is a polynomial of $n$ of degree of $k+i-1$ for $4 \leq i \leq d$. Furthermore,

$$P_{\mathbf{k}}(n) = k_1 x_1^n + \cdots + k_d x_d^n$$

is a polynomial of $n$ of degree less than or equal to $d+k-1$ and

$$p_r(f^n\mathbf{x}) = \sum_{m_{1r}\leq k_1 \leq s_{1r}} \cdots \sum_{m_{dr}\leq k_d \leq s_{dr}} a_{\mathbf{k},r} e^{2\pi i P_{\mathbf{k}}(n)}.$$

We can now continue to estimate

$$\begin{aligned} |II| &= \Big|\frac{1}{N}\sum_{n=1}^{N} c_n \sum_{m_{1r}\leq k_1 \leq s_{1r}} \cdots \sum_{m_{dr}\leq k_d \leq s_{dr}} a_{\mathbf{k},r} e^{2\pi i P_{\mathbf{k}}(n)}\Big| \\ &= \Big|\sum_{m_{1r}\leq k_1 \leq s_{1r}} \cdots \sum_{m_{dr}\leq k_d \leq s_{dr}} a_{\mathbf{k},r} \frac{1}{N}\sum_{n=1}^{N} c_n e^{2\pi i P_{\mathbf{k}}(n)}\Big|. \end{aligned}$$

Let

$$L = \max\{|m_{1r}|, \cdots, |m_{dr}|, |s_{1r}|, \cdots, |s_{dr}|, |a_{\mathbf{k},r}| \ ; \ m_{jr} \le k_j \le s_{jr}, 1 \le j \le d\}.$$

Since $\mathbf{c}$ is an oscillating sequence of order $d+k-1$, we can find an integer $M > r$ such that for $N > M$,

$$\Big| \frac{1}{N} \sum_{n=1}^{N} c_n e^{2\pi i P_{\mathbf{k}}(n)} \Big| < \frac{\epsilon}{2L^d}, \quad \forall\, m_{1r} \le k_1 \le s_{1r}, \cdots, m_{dr} \le k_n \le s_{dr}.$$

This implies that

$$|II| < \epsilon/2.$$

Combining the estimations of $I$ and $II$, we get that, for all $N > M$,

$$|S_N \phi(\mathbf{x})| < \epsilon.$$

This says that $\lim_{N\to\infty} S_N\phi(\mathbf{x}) = 0$. This completes the proof. □

Proof of Corollary 1. Using the same proof as that of Theorem 1, we can show that any oscillating sequence $\mathbf{c}$ of order $d \ge 2$ in the arithmetic sense and $f$ are linearly disjoint. Then, using Proposition 1, we get any oscillating sequence $\mathbf{c}$ of order $d \ge 2$ and $f$ are linearly disjoint. □

Theorem 1 can be generalized to the following theorem.

Theorem 2. *Suppose $f$ is a polynomial skew product of degree $k$ in the form (10). Then there is a positive integer $m = m(d, k, h_1, \cdots, h_d)$ such that any oscillating sequence $\mathbf{c}$ of order $m$ and $f$ are linearly disjoint.*

Proof. Most of the proof is similar to the proof of Theorem 1. Here, we outline some different steps. Let $\mathbf{x}_n = f^n\mathbf{x}$ for $n = 1, 2, \cdots$ and $\mathbf{x}_0$. Denote $\mathbf{x}_n^t = (x_1^n, \cdots, x_d^n)$. From the proof of Theorem 1, we know that $x_1^n$ and $x_2^n$ are polynomials of $n$ of degrees less than or equal to $k+2$.

Let $h_3(x_1, x_2) = \sum_{0\le i+j\le k} r_{ij}(x_1)^i(x_2)^j$. Then

$$\begin{aligned} x_3^n &= x_3 + \sum_{l=0}^{n-1} h_3(x_1^l, x_2^l) = x_3 + \sum_{l=0}^{n-1} \sum_{0\le i+j\le k} r_{ij}\cdot(x_1^l)^i\cdot(x_2^l)^j \\ &= x_3 + \sum_{0\le i+j\le k} r_{ij}\cdot\Big(\sum_{l=0}^{n-1}(x_1^l)^i(x_2^l)^j\Big). \end{aligned}$$

Here $\sum_{l=0}^{n-1}(x_1^l)^i(x_2^l)^j$ is a polynomial of $n$. Inductively, we have that

$$x_i^n = x_i + \sum_{l=0}^{n-1} h_i(x_1^l, \cdots, x_{i-1}^l)$$

is a polynomial of $n$ for $4 \le i \le d$.

Let $m$ be the maximum of all degrees of $x_i^n$ for $1 \le i \le d$. The rest of the proof is the same as in the proof of Theorem 1 □

Corollary 2. *Suppose $A \in \mathrm{GL}(d, \mathbb{Z})$ with all absolute values of the eigenvalues 1. Let $q \ge 1$ be the smallest integer such that all the eigenvalues of $A^q$ are 1. Suppose*

$$f^q(\mathbf{x}) = A^q\mathbf{x} + \mathbf{h}$$

*where* $\mathbf{h}^t = (a, h_2(x_1), h_3(x_1, x_2), \cdots, h_d(x_1, \cdots, x_{d-1}))$ *and* $h_i(x_1, \cdots, x_{i-1})$, $2 \leq i \leq d$, *are polynomials of* $x_1, \cdots, x_{i-1}$ *degree* $\leq k$. *Then there is a positive integer* $m = m(d, k, h_1, \cdots, h_d)$ *such that any oscillating sequence* $\mathbf{c}$ *of order* $m$ *and* $f$ *are linearly disjoint.*

## 4. Minimal mean attractable and minimal quasi-discrete spectrum dynamical systems

In [**9**], we extended the concept of MLS dynamical systems of Fomin [**8**] to the concept of minimal MLS (abbreviated MMLS) dynamical systems. We also define a new concept called minimal mean attractable (abbreviated MMA) dynamical systems in [**9**]. We studied the linear disjointness of any oscillating sequence and any MMA and MMLS dynamical system by using these two concepts. In this paper, I will extend the concept of the quasi-discrete spectrum (abbreviated QDS) dynamical systems of Abramov, Hahn, and Parry [**2, 11**] to the concept of minimal QDS (abbreviated MQDS) dynamical systems of order $d$ and study the linear disjointness of any oscillating sequence of order $d$ and any MMA and MQDS (d) dynamical system.

Suppose $X$ is a compact metric space. Let $f : X \to X$ be a continuous dynamical system. A subset $K$ of $X$ is called minimal (respective to $f$) if for any $x \in K$, the closure of the forward orbit of $x$ is $K$, that is,

$$\overline{\{f^n x\}_{n=0}^{\infty}} = K.$$

DEFINITION 7 (MMA). *Suppose* $K \subseteq X$ *is minimal. We say* $x \in X$ *is* mean attracted *to* $K$ *if for any* $\epsilon > 0$, *there is a point* $z = z_{\epsilon,x} \in K$ *such that*

$$\limsup_{N\to\infty} \frac{1}{N} \sum_{n=1}^{N} d(f^n x, f^n z) < \epsilon. \tag{13}$$

*The* basin of attraction *of* $K$, *denoted as* Basin$(K)$, *is defined to be the set of all points* $x$ *that are mean attracted to* $K$. *We call* $f$ minimal mean attractable *(abbreviated MMA) if*

$$X = \bigcup_K \mathrm{Basin}(K) \tag{14}$$

*where* $K$ *varies among all minimal subsets of* $X$.

REMARK 3. *It is clear that* $K \subseteq$ Basin$(K)$. *The union in the above definition could be uncountable.*

We use [**10**] as a general reference in reviewing a QDS dynamical system. Suppose $G$ is an Abelian group in multiplication and $\Lambda : G \to G$ is a homomorphism. Consider another homomorphism $\Phi : G \to G$ defined as

$$\Phi g := g \cdot \Lambda g, \quad g \in G.$$

Then we have that

$$\Phi^n g = \prod_{j=0}^{n} (\Lambda^j g)^{\binom{n}{j}}, \quad g \in G.$$

Let $G_n = \ker(\Lambda^n)$ be the kernel of $\Lambda^n$, which is a subgroup of $G$. Then, we have a filter

$$\{1\} = G_0 \subseteq G_1 \subseteq \cdots \subseteq G_n \subseteq G_{n+1} \subseteq \cdots \subseteq G$$

and $\Lambda: G_n \to G_{n-1}$ for all $n \geq 1$. Note that $\Phi: G_n \to G_n$ is an automorphism for every $n \geq 1$ (one can check it by using induction). The homomorphism $\Lambda: G \to G$ is called nilpotent if $G = G_n$ for some $n \geq 0$. It is called quasi-nilpotent if $G = \cup_{n=0}^{\infty} G_n$.

A triple $(G, \Lambda, \iota)$ is called a signature if $G$ is an Abelian group, $\Lambda: G \to G$ is a quasi-nilpotent homomorphism and

$$\iota: G_1 \to \mathbb{T}$$

is an injective homomorphism (i.e., monomorphism), where $G_1 = \ker(\Lambda)$. The order of the signature $(G, \Lambda, \iota)$ is

$$\operatorname{ord}(G, \Lambda, \iota) := \inf\{n \in \mathbb{N} \mid G = G_n\} \in \mathbb{N} \cup \{\infty\}.$$

Note that the order is infinite if $G \neq G_n$ for all $n \in \mathbb{N}$, that is, $G$ is not nilpotent.

Now let us return to $f: X \to X$. Suppose $K \subseteq X$ is minimal. Consider the Koopman operator $\Phi: C(K) \to C(K)$ defined as

$$\Phi\phi = \phi \circ f, \quad \phi \in C(K).$$

Let

$$G = \{g \in C(K) \mid |g| = 1\}.$$

Then, it is an Abelian group under multiplication. Consider the homomorphism $\Lambda: G \to G$ defined as

$$\Lambda g = \Phi g \cdot \overline{g}, \quad g \in G.$$

Then

$$\Phi g = g \cdot \Lambda g, \quad g \in G.$$

One can see that $G_1 = Fix(\Phi) \cap G$ where $Fix(\Phi)$ is the set of all fixed points of $\Phi$. Since $K$ is minimal, we have that $G_1 = \mathbb{T}$ and

$$Fix(\Phi) = \operatorname{span}_{\mathbb{C}}(G_1) = \mathbb{C}.$$

Let us consider $G_2 = \ker(\Lambda^2)$. For any $g \in G_2$, we have $\Lambda^2 g = 1$. This implies that

$$\Phi^2 g = g \cdot (\Lambda g)^2 \cdot \Lambda^2 g = \Lambda g \cdot \Phi g$$

Since $\Lambda g \in G_1$, $\lambda = \Lambda g$ with $|\lambda| = 1$ is a unimodular eigenvalue of $\Phi$ with an unimodular eigenvector $g' = \Phi g \in G_2$. Conversely, suppose $\lambda$ is a unimodular eigenvalue of $\Phi$ with a unimodular eigenvector $g$, that is, $\Phi g = \lambda g$ and $|\lambda| = 1$. Then $|g| = 1 \in G_1$ and $\Lambda g = \Phi g \cdot \overline{g} = \lambda g \cdot \overline{g} = \lambda \in G_1$. This says $g \in G_2$. Thus, $G_2$ is the set of all unimodular eigenvectors of $\Phi$. In general, elements in $G_n$ are called unimodular quasi-eigenvectors of order $n-1$. We say that $f: K \to K$ is quasi-discrete spectrum (abbreviated QDS) if the linear hull of all unimodular quasi-eigenvectors is dense in $C(K)$, that is, $G = \cup_{n=0}^{\infty} G_n$ and

$$\overline{\operatorname{span}_{\mathbb{C}}(G)} = C(K).$$

We say that $f: K \to K$ is quasi-discrete spectrum of order $d$ (abbreviated QDS(d)) if it is QDS and $\operatorname{ord}(G, \Lambda, \iota) = d + 1$.

Definition 8 (MMA and MQDS). *We call $f: X \to X$ MMA and MQDS (d) if $f$ is MMA and for every minimal set $K \subseteq X$, $f: K \to K$ is QDS (k) for some $k \leq d$.*

Theorem 3. *Any oscillating sequence $\mathbf{c}$ of order $d \geq 1$ and any MMA and MQDS (d) $f$ are linearly disjoint.*

PROOF. Let $\lambda > 1$ and $C$ be the numbers in (1). Let $\lambda' > 1$ be the dual number of $\lambda$, that is,

$$\frac{1}{\lambda} + \frac{1}{\lambda'} = 1.$$

For any $x \in X$ and any $\phi \in C(X)$, we need to prove that (2) holds. Since $f$ is MMA, $x \in Basin(K)$ for some minimal subset $K \subseteq X$. Assume that we already know that (2) holds for $z \in K$. For an arbitrarily small number $\epsilon > 0$, the uniform continuity of $\phi$ implies that there is a $\delta > 0$ such that

$$|\phi(u) - \phi(v)| < \frac{\epsilon}{2^{\frac{1+\lambda'}{\lambda'}} C^{\frac{1}{\lambda}}}$$

whenever $d(u, v) < \delta$. Since $f$ is MMA, there exists $z = z_{\delta,x} \in K$ such that

$$\limsup_{N\to\infty} \frac{1}{N} \sum_{n=1}^{N} d(f^n x, f^n z) < \delta^2.$$

Let $E = \{n \geq 1 \mid d(f^n x, f^n z) \geq \delta\}$. Then the above inequality implies an estimation of the upper density $\overline{D}(E)$ of $E$ in $\mathbb{N}$:

$$\overline{D}(E) = \limsup_{n\to\infty} \frac{\#(E \cap [1, N])}{N} \leq \delta$$

because

$$\delta\sharp(E \cap [1, N]) \leq \sum_{n=1}^{N} d(f^n x, f^n z).$$

Consider

$$S_N\phi(x) = \frac{1}{N} \sum_{n=1}^{N} c_n \phi(f^n x).$$

Write

$$S_N\phi(x) = \Big(S_N\phi(x) - S_N\phi(z)\Big) + S_N\phi(z) = I + II$$

Since we assume that (2) holds for $z \in K$, we have a $N_0 > 0$ such that

$$|II| = |S_N\phi(z)| < \epsilon/2$$

for all $N > N_0$. From the Hölder inequality,

$$|I| = |S_N\phi(x) - S_N\phi(z)| \leq \Big(\frac{1}{N}\sum_{n=1}^{N} |c_n|^\lambda\Big)^{\frac{1}{\lambda}} \Big(\frac{1}{N}\sum_{n=1}^{N} |\phi(f^n x) - \phi(f^n z)|^{\lambda'}\Big)^{1/\lambda'}$$

$$\leq C^{\frac{1}{\lambda}} \Big(\frac{1}{N}\sum_{n=1}^{N} |\phi(f^n x) - \phi(f^n z)|^{\lambda'}\Big)^{1/\lambda'}.$$

This implies that

$$|I| \leq C^{\frac{1}{\lambda}} \Big(\frac{1}{N} \sum_{n\in[1,N]\setminus E} |\phi(f^n x) - \phi(f^n z)|^{\lambda'} + \frac{1}{N}\sum_{n\in E} |\phi(f^n x) - \phi(f^n z)|^{\lambda'}\Big)^{1/\lambda'}$$

$$\leq C^{\frac{1}{\lambda}} \Big(\frac{\epsilon^{\lambda'}}{2^{1+\lambda'} C^{\frac{\lambda'}{\lambda}}} + (2\|\phi\|_\infty)^{\lambda'} \delta\Big)^{\frac{1}{\lambda'}}.$$

Here we can take $\delta$ small enough such that

$$(2\|\phi\|_\infty)^{\lambda'} \delta < \frac{\epsilon^{\lambda'}}{2^{1+\lambda'} C^{\frac{\lambda'}{\lambda}}}.$$

So we have that

$$|S_N\phi(x)| \le |I| + |II| \le \epsilon$$

for all $N > N_0$. We proved (2) under the assumption.

Now we prove the assumption that for any $z \in K$, (2) holds. Since $f : K \to K$ is a QDS (d) flow, we have that

$$C(K) = \overline{\mathrm{span}_{\mathbb{C}}(G_{d+1})}.$$

Similar to the proof of Theorem 1, we need only to prove (2) for any $g \in G_{d+1}$ and any $z \in K$. Since $\Lambda^{d+1}g = 1$, for any $n \ge d+1$, we have

$$\Phi^n g(z) = g \circ f^n(z) = \prod_{j=0}^{n} (\Lambda^j g(z))^{\binom{n}{j}} = \prod_{j=0}^{d} (\Lambda^j g(z))^{\binom{n}{j}}.$$

Since $\Lambda^j g \in G_{d+1-j}$ for $0 \le j \le d$, we have $\Lambda^j g(z) = e^{2\pi i \theta_j}$ for some real number $\theta_j \in [0, 2\pi]$. This implies that, for $n > d$,

$$\Phi^n g(z) = e^{2\pi i \sum_{j=0}^{d} \theta_j \binom{n}{j}} = e^{2\pi i P(n)}$$

where

$$P(n) = \sum_{j=0}^{d} a_j n^j$$

is a real-coefficient polynomial of $n$ whose degree is less than or equal to $d$. Now

$$S_N g(z) = \frac{1}{N}\sum_{n=1}^{N} c_n g \circ f^n = \frac{1}{N}\sum_{n=1}^{d} c_n \Phi^n g(z) + \frac{1}{N}\sum_{n=d+1}^{N} c_n \Phi^n g(z)$$
$$= \frac{1}{N}\sum_{n=1}^{d} c_n \Phi^n g(z) + \frac{1}{N}\sum_{n=d+1}^{N} c_n e^{2\pi i P(n)}.$$

Since $\mathbf{c}$ is an oscillating sequence of order $d$,

$$\lim_{n\to\infty} \frac{1}{N}\sum_{n=d+1}^{N} c_n e^{2\pi P(n)} = 0.$$

Therefore, we have that

$$\lim_{N\to\infty} S_N g(z) = 0.$$

This completes the proof of the assumption and thus, the proof of the theorem. □

REMARK 4. *From Theorem 1, Theorem 2, and Theorem 3, we can see that a general nonlinear continuous torus map on $\mathbb{T}^d$ having zero topological entropy will not be MQDS(d).*

## 5. Multi-linearly disjoint sequences

It has been asked whether it is enough to study Sarnak's conjecture just by studying the orders of oscillation sequences. We can construct examples of completely oscillating sequences from automatic sequences such as the Thue-Morse and Rudin-Shapiro sequences and many other automatic sequences not correlated with periodic sequences (see [**1, 4**]). These examples have zero topological entropy, as they are viewed as dynamical systems. These sequences and themself viewed as dynamical systems are not linearly disjoint. However, they are not our main concern in the study of linear disjointness. We are more interested in completely oscillating sequences having positive entropy, such as the Möbius sequence **u**. In [**3**], we have constructed completely oscillating sequences other than the Möbius sequence (but still have positive entropy (refer to Remark 2)) as follows.

Let $\mathcal{C}_+^k((1,\infty))$ be the space of all positive real-valued $k$-time continuously differentiable functions on $(1,\infty)$, whose $i$-th derivative is nonnegative for $i \le k$. We proved the following theorem in [**3**].

THEOREM B. *Take $g \in \mathcal{C}_+^2((1,\infty))$. Then, for a fixed real number $\alpha \neq 0$ and almost all real numbers $\beta > 1$ (alternatively, for a fixed real number $\beta > 1$ and almost all real numbers $\alpha$),*

$$\mathbf{c} = \left(e^{2\pi i(\alpha\beta^n g(\beta))}\right)_{n\in\mathbb{N}}$$

*are completely oscillating sequences.*

We would like to know that

PROBLEM 2. *Are any completely oscillating sequence* $\mathbf{c}$ *in Theorem* B *and any dynamical system of zero topological entropy $f : X \to X$ linearly disjoint?*

Given the study of the above problem, we would like to study those multi-linearly disjoint sequences $\mathbf{c}$.

DEFINITION 9. *We call* $\mathbf{c}$ *multi-linearly disjoint if*

$$\lim_{N\to\infty} \frac{1}{N}\sum_{n=1}^{N} c_{n+l_1}^{k_1} c_{n+l_2}^{k_2}\cdots c_{n+l_r}^{k_r} = 0 \tag{15}$$

*for all $r \ge 1$, all choices of integers $0 \le l_1 < l_2 < \cdots < l_r$ and $k_j \in \mathbb{N}$, $j = 1, 2, \cdots, r$, such that not all $c_{n+l_j}^{k_j} = |c_{n+l_j}|^{k_j}$ hold.*

This definition is motivated by the Chowla conjecture in [**5**].

CONJECTURE 3. *For any $r$ integers $0 \le l_1 < l_2 < \cdots < l_r$ and $r$ indexes $i_1, i_2 \cdots, i_r$ with $i_k = 1$ or $2$, $1 \le k \le r$, but not all $2$,*

$$\lim_{N\to\infty} \frac{1}{N}\sum_{n=1}^{N} \mu(n+l_1)^{i_1}\mu(n+l_2)^{i_2}\cdots\mu(n+l_r)^{i_r} = 0.$$

Using the same proof as that of Theorem B, we have

THEOREM 4. *Take $g \in \mathcal{C}_+^2((1,\infty))$. Then, for a fixed real number $\alpha \neq 0$ and almost all real numbers $\beta > 1$ (alternatively, for a fixed real number $\beta > 1$ and almost all real numbers $\alpha$),*

$$\mathbf{c} = \left(e^{2\pi i(\alpha\beta^n g(\beta))}\right)_{n\in\mathbb{N}}$$

*are multi-linearly disjoint sequences.*

To have a self-contained paper, we provide a detailed proof of Theorem 4. Consider

$$c_{n+l_1}^{k_1} c_{n+l_2}^{k_2}\cdots c_{n+l_r}^{k_r} = e^{2\pi i\left(\sum_{i=1}^{r} k_i\left(\alpha\beta^{n+l_i} g(\beta)\right)\right)}.$$

This says that (15) relates with the uniformly distributed modulo 1 (abbreviated u.d.1) of the sequence

$$\left(\sum_{i=1}^{r} k_i\left(\alpha\beta^{n+l_i} g(\beta)\right)\right)_{n\in\mathbb{N}}$$

on the unit interval $[0,1]$. For a real number $x$, let $\{x\} = x \mod 1$ be the fractional part of $x$.

DEFINITION 10. *We say that a sequence* $\mathbf{x} = (x_n)_{n\in\mathbb{N}}$ *of real numbers is u.d.1 if for any* $0 \le a < b \le 1$, *we have*

$$\lim_{N\to\infty} \frac{\#(\{n \in [1,N] \mid \{x_n\} \in [a,b]\})}{N} = b - a.$$

The following Weyl criterion connects (15) and u.d.1.

THEOREM C. *The sequence* $\mathbf{x} = (x_n)_{n\in\mathbb{N}}$ *is u.d.1 if and only if*

$$\lim_{N\to\infty} \frac{1}{N} \sum_{n=1}^{N} e^{2\pi i h x_n} = 0 \quad \textit{for all integers } h \neq 0.$$

The following theorem is essentially [**3**, Theorem 1] (that is, take all $l_i = 0$, $1 \le i \le r$, in the proof in [**3**]).

THEOREM 5. *Let us take* $g \in \mathcal{C}_+^2((1,\infty))$. *Then, for a fixed real number* $\alpha \neq 0$ *and almost all real numbers* $\beta > 1$ *(alternatively, for a fixed real number* $\beta > 1$ *and almost all real numbers* $\alpha$*) and for all* $r \ge 1$, *all choices of integers* $0 \le l_1 < l_2 < \cdots < l_r$ *and* $k_i \in \mathbb{N}$, $i = 1, 2, \cdots, r$, *sequences*

$$\Big( \sum_{i=1}^{r} k_i \big(\alpha \beta^{n+l_i} g(\beta)\big) \Big)_{n\in\mathbb{N}} \tag{16}$$

*are u.d.1.*

Thus, Theorem 4 is just a corollary of Theorem 5 and Teorem C. The following theorem is credited as Koksma's Theorem in [**15**], which we will use in the proof of Theorem 5.

THEOREM D. *Let* $(y_n(x))_{n\in\mathbb{N}}$ *be a sequence of real-valued* $C^1$ *functions defined on an interval* $[a,b]$. *Suppose* $y'_m(x) - y'_n(x)$ *is monotone on* $[a,b]$ *for any two integers* $m \neq n$ *and suppose*

$$\inf_{m\neq n} \min_{x\in[a,b]} |y'_m(x) - y'_n(x)| > 0.$$

*Then for almost all* $x \in [a,b]$, *the sequence* $\mathbf{y} = (y_n(x))_{n\in\mathbb{N}}$ *is u.d.1.*

PROOF OF THEOREM 5. Given a choice of $0 \le l_1 < l_2 < \cdots < l_r$ and $k_i \in \mathbb{N}$, $i = 1, 2, \cdots, r$, consider the sequence

$$\sum_{i=1}^{r} k_i \beta^{n+l_i} g(\beta) = \beta^n g(\beta) \sum_{i=1}^{r} k_i \beta^{l_i} = G(\beta)\beta^n$$

where $G(\beta) = g(\beta)\sum_{i=1}^{r} k_i \beta^{l_i} \in C_+^2((0,\infty))$ since both $g(\beta)$ and $\sum_{i=1}^{r} k_i\beta^{l_i}$ are in $C_+^2((0,\infty))$. Define a function

$$y_n(x) = G(x)x^n, \quad x \in (1,\infty).$$

For $n > m$, we have

$$y'_n(x) - y'_m(x) = G(x)(nx^{n-1} - mx^{m-1}) + G'(x)(x^n - x^m)$$

Since $G', G'' \geq 0$ and since $nx^{n-m} - m \geq n - m \geq 1$ for $x > 1$, we see that every term in the expression of $y_n'(x) - y_m'(x)$ are in $\mathcal{C}_+^1([a, \eta])$ for any $1 < a < \eta < \infty$. By the closure property of $\mathcal{C}_+^1([a, \eta])$, we have that

$$y_n'(x) - y_m'(x) \in \mathcal{C}_+^1([a, \eta]), \quad \forall\, n > m. \tag{17}$$

In particular, this implies that $y_n' - y_m'$ is increasing for $n > m$. Furthermore, we see that there is a constant $L > 0$ such that

$$|y_n'(x) - y_m'(x)| \geq L, \quad \forall\, n > m \in \mathbb{N}, \ \forall\, a \leq x \leq \eta. \tag{18}$$

(17) and (18) say that the sequence $(y_n(x))_{n \in \mathbb{N}}$ of real-valued $C^1$ functions satisfies all the hypotheses of Theorem D.

Theorem D implies that for almost all $x$ in $[(2^k+1)/2^k, (2^{k-1}+1)/2^{k-1}]$ or $[k, k+1]$ for $k \geq 2$, the sequence $(y_n(x))_{n \in \mathbb{N}}$ is u.d.1. Further, this implies that for almost all

$$x \in (1, \infty) = \cup_{k=2}^{\infty} \Big[ \frac{2^k+1}{2^k}, \frac{2^{k-1}+1}{2^{k-1}} \Big] \cup \cup_{k=2}^{\infty} [k, k+1]$$

the sequence $(y_n(x))_{n \in \mathbb{N}}$ is u.d.1. For a fixed $\alpha > 0$ or $\alpha < 0$, using similar arguments, we have that for almost all $x \in (1, \infty)$, $(\alpha y_n(x))_{n \in \mathbb{N}}$ is u.d.1.

Let

$$A_{r,(l_1,\cdots,l_r),(k_1,\cdots,k_r)} = \{\beta > 1 \mid \alpha\beta^n g(\beta) \sum_{i=1}^{r} k_i \beta^{l_i} \text{ is not u.d.1 } \}.$$

Then the one dimensional Lebesgue measure of $A_{r,(l_1,\cdots,l_r),(k_1,\cdots,k_r)}$ is zero. Since the set

$$U = \{(r, (l_1, \ldots, l_r), (k_1, \cdots, k_r)) \mid r, k_1, \cdots, k_r, l_2 < \cdots < l_r \in \mathbb{N}, l_1 \in \mathbb{N} \cup \{0\}, l_1 < l_2\}$$

is countable, the one-dimensional Lebesgue measure of

$$\bigcup_{(r,(l_1,\ldots,l_r),(k_1,\cdots k_r)) \in U} A_{r,(l_1,\cdots,l_r),(k_1,\cdots,k_r)}$$

is zero too.

For a fixed real number $\alpha \neq 0$ in the theorem, take a real number

$$\beta \in (1, \infty) \setminus \bigcup_{(r,(l_1,\ldots,l_r),(k_1,\cdots k_r)) \in U} A_{r,(l_1,\cdots,l_r),(k_1,\cdots,k_r)}.$$

This says that the sequence

$$\Big( \sum_{i=1}^{r} \alpha\beta^{n+l_i} g(\beta) \Big)_{n \in \mathbb{N}}$$

is u.d.1 for all

$$(r, (l_1, \ldots, l_r), (k_1, \cdots k_r)) \in U.$$

This completes the proof. □

## Acknowledgments

I would like to thank Professor Andrew Török for his helpful comments following my talk at an AMS meeting. I am also grateful to the referees for their many valuable suggestions and comments that helped improve this paper.

## References


[1] el Abdalaoui, e. H., *Oscillating sequences, Gowers norms and Sarnak's conjecture*, arXiv 1704.07243v3, 2017.

[2] Abramov, L. M., *Metric automorphisms with quasi-discrete spectrum*, Izv. Akad. Nauk SSSR Ser. Mat. **26** (1962), 513–530.

[3] Akiyama, S. and Jiang, Y., *Higher order oscillation and uniform distribution. Uniform Distribution Theory*, Uniform Distribution Theory, Volume 14 (2019), no. 1, 1–10.

[4] Byszewski, J., Konieczny, J., and Mïllner, C., *Gowers Norms for Automatic Sequences* arXiv 2002.09509, 2020.

[5] Chowla, S., *The Riemann hypothesis and Hilbert's tenth problem. Mathematics and Its Applications, Vol 4*, Gordon and Breach Science Publishers, New York, 1965.

[6] Dai, X. and Jiang, Y., *Distance entropy of dynamical systems on noncompact phase spaces. Discrete and Continuous Dynamical Systems, Vol. 20, No. 2* (2008), 313–333.

[7] Davenport, H., *On some infinite series involving arithmetical functions (II).* Quart. J. Math. Oxford, **8** (1937), 313–320.

[8] Fomin, S., *On dynamical systems with a purely point spectrum*, Dokl. Akad. Nauk SSSR **77** (1951), 29–32 (in Russian).

[9] Fan, A. and Jiang, Y.. *Oscillating Sequences, MMA and MMLS Flows and Sarnak's Conjecture.* Ergodic Theory & Dynamical. Systems, August 2018, Vol. 38, no. 5, 1709–1744.

[10] Haase, M. and Moriakov, N., *On systems with quasi-discrete spectrum*, arXiv 1509.08961v3, 2015.

[11] Hahn, F. and Parry, W., *Minimal dynamical systems with quasi-discrete spectrum*, J. London Math. Soc. **40** (1965), 309–323.

[12] Hua, L., *Additive Theory of Prime Numbers* (Translations of Mathematical Monographs : Vol. **13**). American Mathematical Society. 1966.

[13] Jiang, Y., *Orders of oscillation motivated by Sarnak's conjecture*, Proceedings of AMS, Volume **147**, Number 7, July 2019, 3075–3085.

[14] Kronecker, L., *Zwei Sätze über Gleichungen mit ganzzahligen Coefficienten*, J. Reine Angew. Math., 53 (1857), 173–175 (in German).

[15] Kuipers, L. and Niederreiter, H., *Uniform Distribution of Sequences*, J. Wiley and Sons, New York, 1974.

[16] Li, J., Tu, S., and Ye, X., *Mean-equicontinuity and mean sensitivity*, Ergod. Th. & Dynam. Sys., **35** (2015), 2587–2612.

[17] Sarnak, P., *Three lectures on the Möbius function, randomness and dynamics*, IAS Lecture Notes, 2009; `http://publications.ias.edu/sites/default/files/MobiusFunctionsLectures(2).pdf`.

[18] Sarnak, P., *Möbius randomness and dynamics*, Not. S. Afr. Math. Soc. **43** (2012), 89–97.

[19] Sinai, Y., *On the concept of entropy for a dynamical system*, Dokl. Akad. Nauk SSSR, 124 (1959), 768–771.



Department of Mathematics, Queens College of the City University of New York, Flushing, New York 11367-1597; and The PhD Program in Mathematics, Graduate Center of the City University of New York; New York, New York 10016

*Email address*: `yunping.jiang@qc.cuny.edu`

Contemporary Mathematics
Volume **839**, 2026
https://doi.org/10.1090/conm/839/16728

# Reciprocal transformations and their discrete Maharam extensions

Chris Johnson

ABSTRACT. We introduce two abstract constructions for building new measurable dynamical systems from existing ones and study their ergodic properties. The first of these constructions, a "reciprocal transformation," produces a type of non-singular transformation where the measures of subsets are distorted in a simple way. We then introduce the "discrete Maharam extension" which associates an infinite measure-preserving transformation to each reciprocal transformation. We give some preliminary results about the ergodic theory of each of these constructions, mention ongoing work, as well as conjectures and questions for future research.

## 1. Introduction

In this paper we consider two simple constructions and explore ergodic theoretic properties of these constructions. We will also mention some possible applications of these constructions to the theory of interval exchanges and translation surfaces.

The first construction we introduce considers non-singular transformations that arise by composing a measure-preserving transformation with a special map we call a "scaling involution" to produce what we will refer to as a "reciprocal transformation." This composition gives us a map which no longer preserves measure, but where the measure is distorted in an easy-to-understand way. This gives us a family of non-singular transformations which we believe are similar enough to measure-preserving transformations that in some particular instances the dynamics can be easily understood.

The second construction gives a method for associating a measure-preserving transformation, albeit on a space of infinite measure, to each reciprocal transformation. As we will see, ergodic properties of these infinite transformations are reflected in the dynamics of the finite non-singular transformations.

While the definitions and theorems given in this article are expressed in terms of abstract ergodic theory, we will consider as running examples the special cases when the initial measure-preserving transformation is an interval exchange transformation, and the scaling involution is a piecewise affine involution of an interval. Indeed, the initial goal of these constructions was to provide a method of producing infinite "self-similar" interval exchanges which could easily be seen to be

2020 *Mathematics Subject Classification.* Primary 37A40; Secondary 37E05, 37E35.
*Key words and phrases.* Ergodic theory, nonsingular transformations, interval exchange transformations, translation surfaces.

ergodic, and then consider the (ergodic) vertical flow on translation surfaces of infinite area which arise as their suspensions. We note here that while the theory of compact translation surfaces has seen tremendous progress in the last two decades, non-compact translation surfaces are not as well understood, and so a method for producing infinite-area translation surfaces with an ergodic vertical flow seems to be an interesting problem.

Section 2 below contains a quick summary of the pertinent background. After very briefly mentioning some classical ergodic theory of a measure-preserving transformation on a probability space, we mention non-singular transformations and ergodic theory for a space of infinite measure. We then give a quick introduction to interval exchange transformations and translation surfaces. References to more thorough presentations are provided. In Sections 3 and 4 we introduce our two constructions and prove some basic theorems about their dynamics. Section 5 discusses ongoing work related to the Krieger type of the reciprocal transformations.

## 2. Background

For the convenience of the reader, in this section we give some very basic background from ergodic theory, as well as the theory of interval exchanges and translation surfaces. For more details on standard ergodic theory, see [**Pet89**] and [**Wal82**]. Information on infinite ergodic theory can be found in [**Aar97**], while [**DS23**] and the references therein contain information about non-singular ergodic theory. Finally, introductions to interval exchange transformations and translation surfaces can be found in [**Via06**], [**Vor96**], [**Wri15**], and [**Zor06**].

**2.1. Standard ergodic theory.** Let $(X, \mathcal{B}, \mu)$ be a probability space. We say that a map $F : X \to X$ is a *measure-preserving transformation (with respect to $\mu$)* if $F$ is a measurable bijection and for each measurable subset $E$ we have $\mu(E) = \mu(F(E))$. All definitions throughout this article will be up to a set of measure zero. That is, if $F$ was measurable and not a bijection from $X$ to itself, but if there existed a null set $N$ such that the restriction of $F$ to $X \setminus N$ is a bijection, we will still refer to $F$ as a "measurable bijection."

One of the most fundamental properties of measure-preserving transformations is given by the famous Poincaré recurrence theorem:

THEOREM (Poincaré recurrence theorem). *Let $F$ be a measure-preserving transformation of a probability space $(X, \mathcal{B}, \mu)$. For every set $E$ of positive measure and for almost every $x \in E$, there exists an $n$ such that $F^n(x) \in E$.*

In language to be described in the next subsection below, the Poincaré recurrence theorem states that every measure-preserving transformation of a probability space is "conservative."

We say that a measurable set $E$ is *$F$-invariant* if $F(E) = E$ (again, we allow this statement to hold up to a null set). The transformation $F$ is said to be *ergodic* if the only invariant sets have measure zero or measure one.

**2.2. Non-singular ergodic theory.** As above, continue to suppose that $(X, \mathcal{B}, \mu)$ is a probability space and that $F : X \to X$ is a measurable bijection. We say that $F$ is a *non-singular transformation (with respect to $\mu$)* if $F$ preserves the $\sigma$-ideal of $\mu$-null sets: i.e., $\mu \ll F_*\mu \ll \mu$. That is, given a measurable set $E$,

$\mu(E) = 0$ if and only if $\mu(F(E)) = 0$ as well; $F$ may distort the measure of a set, but it must respect whether or not a set has positive or zero measure.

Unlike measure-preserving transformations, non-singular transformations need not satisfy the conclusion of the Poincaré recurrence theorem. To be more precise, we must define "wandering sets," and the "conservative" and "dissipative" parts of the transformation.

A set $W$ is said to be *wandering* under $F$ if it is disjoint from all of its images: $W \cap F^n(W) = \emptyset$. The union of all wandering sets is called the *dissipative part* of $F$ and is denoted $\mathfrak{D}$. The complement of $\mathfrak{D}$ is the *conservative part* of $F$ and is denoted $\mathfrak{C}$. If $\mathfrak{C}$ has full measure, then $F$ is called *conservative*. If $\mathfrak{D}$ has full measure, then $F$ is called *totally dissipative*. Notice that both $\mathfrak{C}$ and $\mathfrak{D}$ are $F$-invariant sets. The decomposition of $X$ into the conservative and dissipative parts of $F$ is known as the *Hopf decomposition* of $F$.

The notion of ergodicity makes sense for non-singular transformations and is defined just as in the measure-preserving case: we say a non-singular transformation $F$ is *ergodic (with respect to $\mu$)* if each $F$-invariant set has full measure or zero measure. As shown in [**Aar97**, Proposition 1.2.1], if $F$ is a non-singular bijection and $\mu$ is non-atomic, then $F$ must be conservative to be ergodic. Throughout this article, all measures considered will be assumed to be non-atomic, and so conservativity will be a prerequisite for ergodicity.

**2.3. Infinite ergodic theory.** The notion of a measure-preserving transformation easily extends to measure spaces of infinite measure, but the definition of ergodicity must be rephrased slightly. Instead of simply requiring that an invariant set have full (infinite) measure, we should require that the measure of an invariant set or its complement is zero. We note too that measure-preserving transformations of infinite measure spaces need not be conservative: the conclusion of the Poincaré recurrence theorem does not necessarily hold. Consider, for example, the map $Fx = x + 1$ defined on the real line which preserves Lebesgue measure, but is totally dissipative.

**2.4. Interval exchanges and translation surfaces.** An *interval exchange transformation* is a bijection from an interval to itself which is a piecewise translation with finitely-many intervals of continuity. Each interval exchange is determined by a permutation describing how the intervals of continuity are exchanged by the transformation, together with a vector of length data giving the length of each interval of continuity. These transformations are often described pictorially by giving an image where each interval of continuity appears as a subsegment of the interval, and the image of each interval appears on an interval below. For example, Figure 1 describes an interval exchange on the interval $X = [0, 1)$ where the intervals of continuity are $[0, 0.3)$, $[0.3, 0.8)$, and $[0.8, 1)$ which are denoted $A$, $B$, and $C$. The transformation $T : X \to X$ indicated in Figure 1 reverses the order of the intervals so that $T(A) = [0.7, 1)$, $T(B) = [0.2, 0.7)$, and $T(C) = [0, 0.2)$.

By considering certain suspensions of an interval exchange, we can obtain a *translation surface*, which is a surface endowed with singular flat metric with trivial holonomy. These surfaces have a well-defined notion of direction at each point, aside from a finite amount of ambiguity at the singularities. That is, the notion of "vertical or "horizontal" or "with slope $m$" makes sense at all but finitely-many points on the surface, and this allows us to have a well-defined directional flow

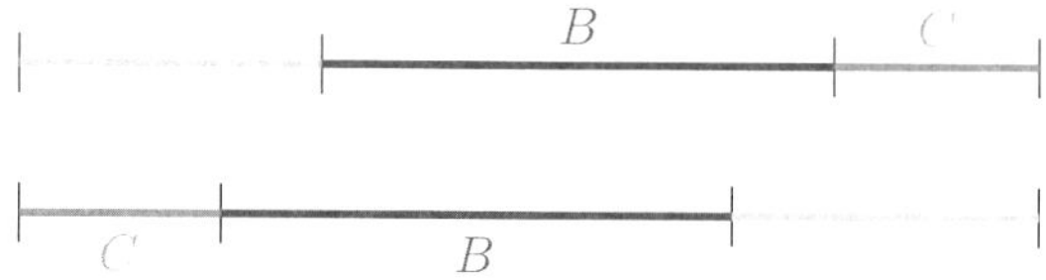


FIGURE 1. An example of an interval exchange with three intervals of continuity.

on the surface (minus the measure zero subset consisting of points where the flow would hit a singularity). The interval on which the interval exchange is defined then appears as a geodesic segment on the surface, and the first-return map of the vertical flow on the surface to that geodesic segment recovers the original interval exchange.

To be precise about the construction of a translation surface as a suspension over an interval exchange, we consider a complex vector $\zeta$ with dimension equal to the number of intervals of continuity. Suppose the $n$-th component of $\zeta$ has real part equal to the width of the $n$-th interval of continuity of the interval exchange. Now consider a broken line emanating from the left-hand endpoint of the interval consisting of a concatenation of segments described by $\zeta$. We then consider the corresponding broken line given by permuting the entries of $\zeta$ as described by the interval exchange. Identifying the segments corresponding to the same entry of $\zeta$ gives the resulting interval exchange. In Figure 2 we see the surface obtained by the construction just described using the interval exchange denoted in Figure 1 and the vector $\zeta = (0.3 + i, 0.5 + 0.2i, 0.2 - i)$.

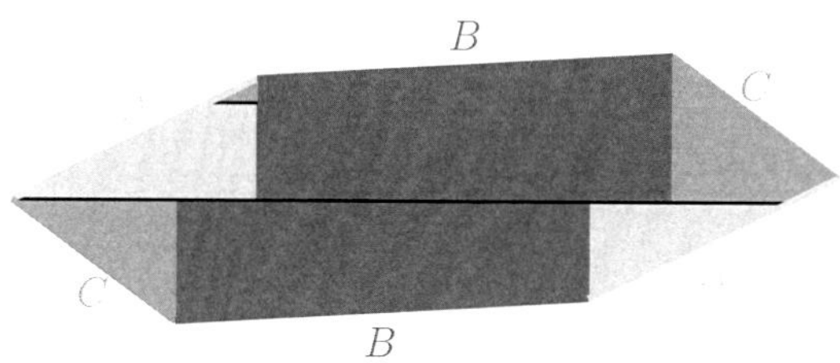


FIGURE 2. A translation surface obtained as a suspension over the interval exchange described in Figure 1.

In addition to interval exchanges which preserve the Lebesgue measure of an interval, we will also be concerned with *affine interval exchanges* which are non-singular transformations of an interval given by a piecewise affine bijection of the interval to itself. Just as interval exchanges can be denoted pictorially, affine interval exchanges can be denoted in a similar way, such as in Figure 3.

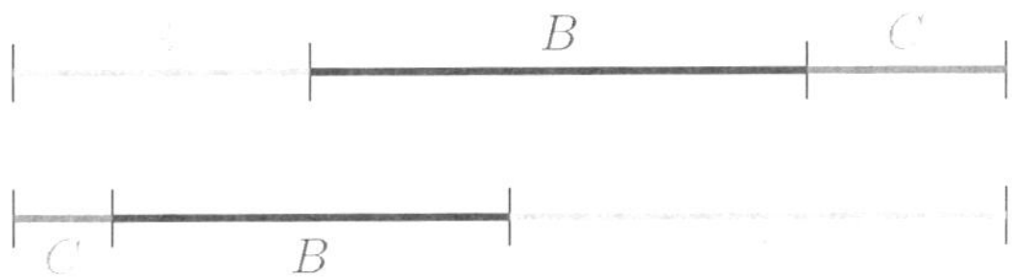


FIGURE 3. An example of an affine interval exchange with three intervals of continuity. The intervals are permuted just as in the interval exchange in Figure 1, but the lengths of the intervals can be distorted.

## 3. Reciprocal transformations

**3.1. Introduction and an example.** Let $(X, \mathcal{B}, \mu)$ be a probability space. We call a measurable bijection $\Phi : X \to X$ a *scaling involution* if $\Phi^2 = \text{id}$ and there exists a set $S$ of positive measure such that the following conditions are satisfied:

- $X = S \sqcup \Phi(S)$
- $\mu(S) < \mu(\Phi(S))$ ($S$ is the smaller set)
- For each measurable $E \subseteq S$, $\mu(\Phi(E)) = \frac{\mu(\Phi(S))}{\mu(S)}\mu(E)$.

That is, $\Phi$ simply exchanges $S$ and $\Phi(S)$, uniformly scaling the measure of subsets in $S$ or $\Phi(S)$. Notice that since $\Phi$ is an involution, the third condition above implies that for each measurable $E \subseteq \Phi(S)$ we have $\mu(\Phi(E)) = \frac{\mu(S)}{\mu(\Phi(S))}\mu(E)$.

For notational convenience, we let $\rho$ denote the ratio $\frac{\mu(\Phi(S))}{\mu(S)}$. Observe that sets $E \subseteq \Phi(S)$ are scaled by a factor of $\rho^{-1}$ under the application of $\Phi$. The last condition in the definition of a scaling involution can also be described as requiring the Radon-Nikodym derivative $\frac{d\Phi_*\mu}{d\mu}$ equal $\rho$ at each point of $S$ and $\rho^{-1}$ at each point of $\Phi(S)$.

EXAMPLE 1. If $X = [0, 1)$ equipped with the Lebesgue measure $\mu$, then the affine map exchanging $S = [0, 1/3)$ and $\Phi(S) = [1/3, 1)$ is a scaling involution with ratio $\rho = 2$.

By a *reciprocal transformation* we mean the composition $F = \Phi T$ where $T$ is a measure-preserving transformation and $\Phi$ is a scaling involution on a probability space.

REMARK 1. We observe that in defining the reciprocal transformation we made a choice to consider the composition $F = \Phi T$, but could have instead considered the other composition, $G = T\Phi$. We observe, though, that these maps are conjugate to one another with $\Phi$ being the conjugating map:

$$\Phi F = \Phi(\Phi T) = T = (T\Phi)\Phi = G\Phi.$$

Thus dynamical properties of $F$ such conservativity and ergodicity hold if and only if $G$ also has these properties. Throughout this article we will focus on the composition $F = \Phi T$ for concreteness.

EXAMPLE 2. Consider the case when $X = [0, 1)$ is an interval, and $T : X \to X$ is the interval transformation denoted in Figure 4. This transformation is a piecewise translation with four intervals of continuity denoted $A$, $B$, $C$, and $D$. The transformation simply acts as a pair of circle rotations, rotating $A \cup B$ by the length of $B$, and rotating $C \cup D$ by the length of $D$.

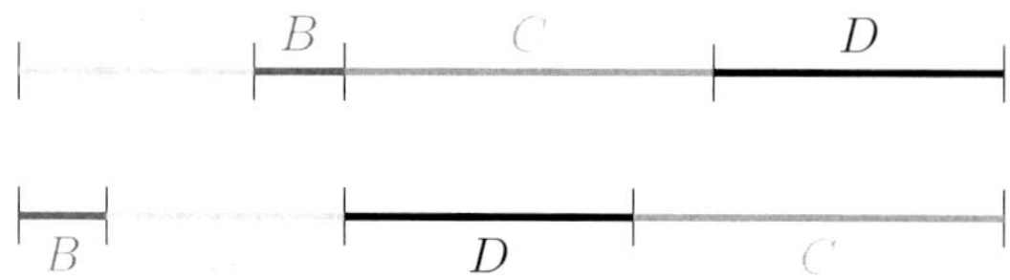


FIGURE 4. An interval exchange transformation $T$ with four intervals permuted by $(A, B)(C, D)$.

Now take $\Phi$ to be the affine piecewise involution which exchanges $S = A \cup B$ and $\Phi(S) = C \cup D$, described in Figure 5.

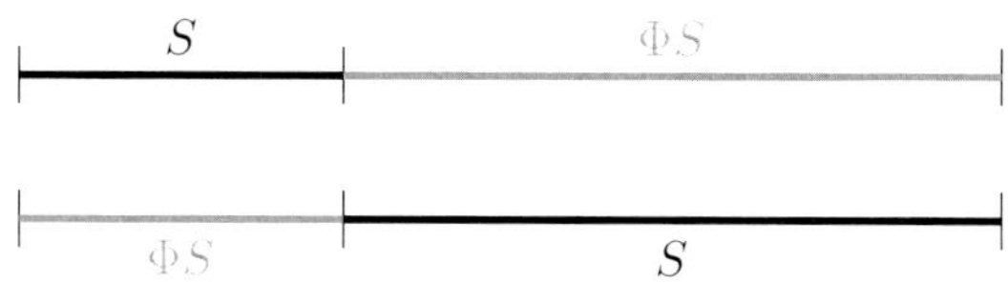


FIGURE 5. A scaling involution $\Phi$ exchanging $S$ and $\Phi(S)$.

The resulting composition $F = \Phi T$ is the affine interval exchange indicated in Figure 6.

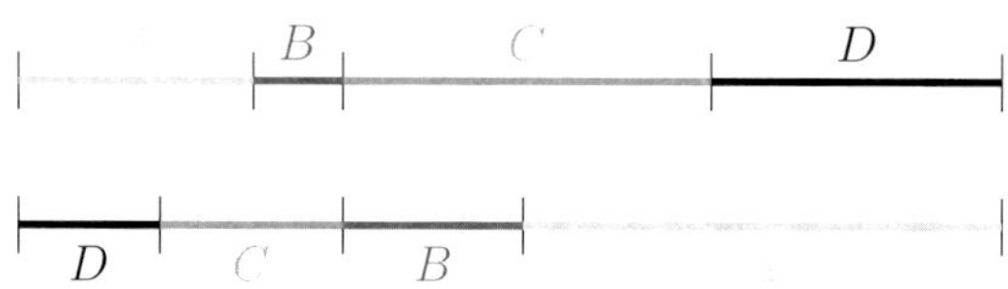


FIGURE 6. An affine interval exchange $F$ obtained by composing an interval exchange $T$ with a scaling involution $\Phi$.

This simple construction gives us a way to build non-singular transformations where we can use information about the dynamics of the measure-preserving $T$ to glean information about the dynamics of $F$. One of our primary goals is to determine what conditions on $\Phi$ and $T$ are sufficient to ensure that the reciprocal transformation $F = \Phi T$ is ergodic.

**3.2. Conservativity and the first-return to $S$.** As we will be interested in the ergodicity of reciprocal transformations and conservativity is prerequisite for ergodicity when $\mu$ is non-atomic, we must first understand when a reciprocal transformation is conservative. To this end it will be helpful to study when the first-return map to $S$ (the smaller of the two sets exchanged by the scaling involution $\Phi$) is defined. Observe that we *are not* currently supposing the reciprocal transformation $F = \Phi T$ is conservative, and so *a priori* the first-return map to an arbitrary subset need not be defined. As we will see, however, the first-return map of a reciprocal transformation to $S$ *is* always defined.

We first observe that when $F$ is applied, any set whose image is in $S$ will have its measure shrink by a factor of $\rho^{-1}$, and any set whose image is in $\Phi(S)$ will have its measure grow by a factor of $\rho$. This is illustrated in Example 3 below.

EXAMPLE 3. Consider the interval exchange $T$ indicated in Figure 7 where each interval has length $1/9$ or $2/9$.

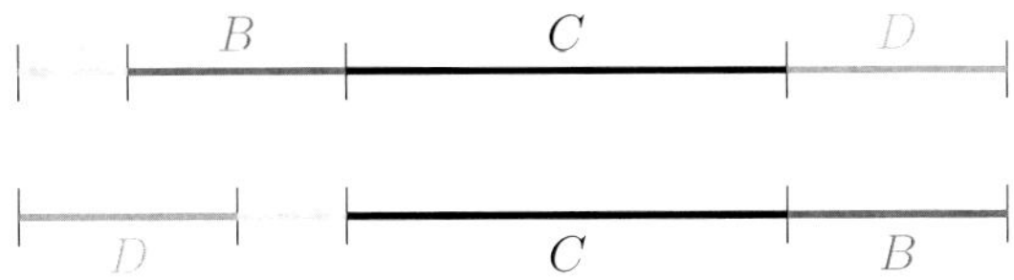


FIGURE 7. An interval exchange transformation.

Composing this map with the $\Phi$ exchanging $[0, 1/3)$ and $[1/3, 1)$ gives us the affine interval exchange indicated $F$ indicated in Figure 8.

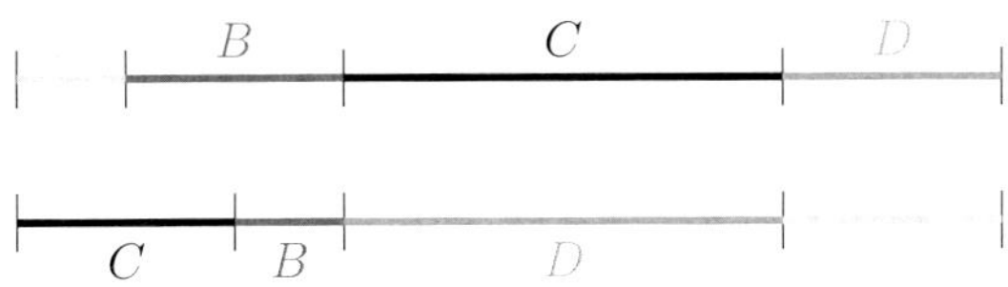


FIGURE 8. A reciprocal transformation.

Observe that the intervals labeled $B$, $C$, and $D$ have image in $S$ and their measures each shrink by a factor of $1/2$. The intervals labeled $A$ and $E$, however, have images in $\Phi(S)$ and their measures grow by a factor of 2.

Now consider the first return map to $S = [0, 1/3)$ which we will denoted $F_S$ and seen in Figure 9.

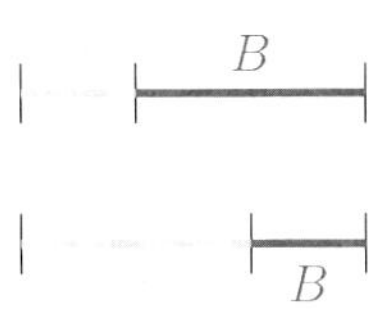


FIGURE 9. The first return map admits a wandering interval.

We observe that the interval $B \setminus F_S(B)$ is wandering under $F_S$, always shrinking and shifting to the right. We also observe that the interval $A$ grows by a factor of 2 under $F_S$, corresponding to the fact $F_S(A) = F^3(A)$; the image of $A$ grows by a factor of 2 under the first two applications of $F$ but then $F^2(A)$ shrinks by a factor of $1/2$ when $F$ is applied again.

Noting that the measure of the image of a set must grow by $\rho$ if that image is not in $S$ is the key observation required for showing almost every point must eventually enter $S$.

LEMMA 1. *Let $F = \Phi T$ be the reciprocal transformation given by composing a measure-preserving bijection $T$ with a scaling involution $\Phi$ of scaling ratio $\rho > 1$. Then for $\mu$-almost every $x \in X$, there exists an $n > 0$ so that $F^n x \in S$.*

PROOF. Let $U$ be the set of points in $X$ which never enter $S$:

$$U = \{x \in X \mid F^n x \notin S \text{ for all } n > 0\}.$$

We claim $U$ must have measure zero. If not, then all iterates of $U$ remain in $\Phi(S)$. However, this means every measurable subset of $U$ has its size increased by a factor of $\rho$ with each application of $F$. Since $\Phi(S)$ has finite measure, though, this is impossible unless $\mu(U) = 0$. □

We emphasize again that throughout the paper all maps are assumed to be measurably bijective unless otherwise explicitly stated. Some results may hold even for transformations which are not bijective, although some proofs given here would need to be modified as we often consider the images instead of preimages in the proofs.

Lemma 1 implies, in particular, that there is a well-defined first-return map from $S$ to itself which we denote $F_S : S \to S$. Notice that dynamical properties of $F$ are reflected in $F_S$ and vice versa when $F_S$ is also a non-singular bijection. Notice, though, that $F_S$ may fail to be surjective in general, and in such situations is not non-singular.

EXAMPLE 4. Consider the reciprocal transformation formed by composing the interval exchange transformation $T$ consisting of four intervals, $A = [0, 1/6)$, $B = [1/6, 1/3)$, $C = [1/3, 1/2)$, and $D = [1/2, 1)$ which are permuted as shown in Figure 10. Composing $T$ with the scaling involution of Example 1, resulting in the transformation shown in Figure 11.

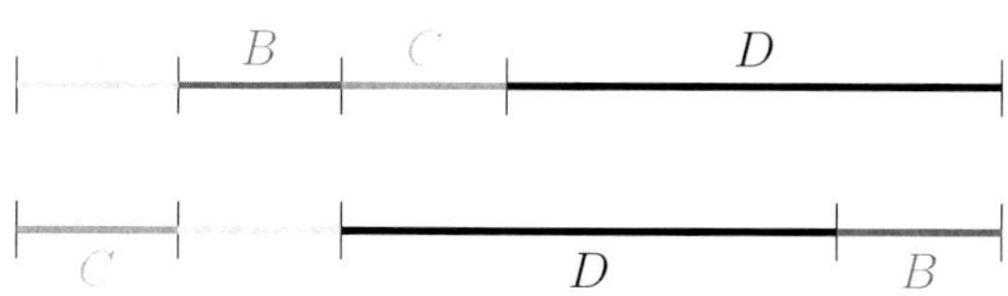


FIGURE 10. An interval exchange $T$ defined on four subintervals of $[0, 1)$.

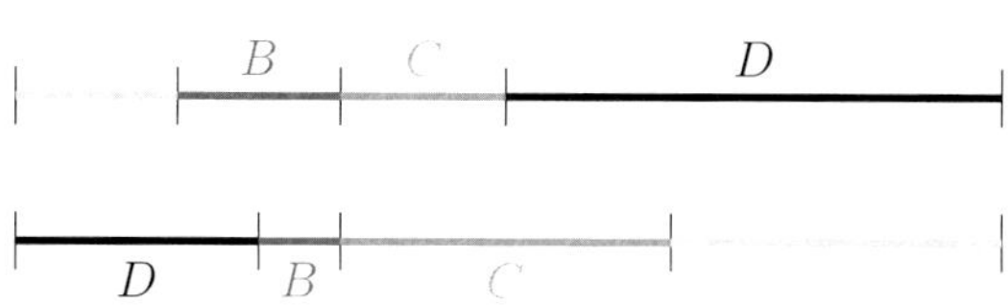


FIGURE 11. The reciprocal transformation $F$ obtained by composing the map $T$ from Figure 10 with the involution $\Phi$ exchanging $[0, 1/3)$ and $[1/3, 1)$.

The first return map to $S = [0, 1/3)$ is then easily seen to be the map in Figure 12. The dashed interval in Figure 12 indicating that $[0, 1/12)$ is not in the image of $F_S$ and so $F_S^{-1}([0, 1/12)) = \emptyset$ and thus $F_S$ in this example is not non-singular.

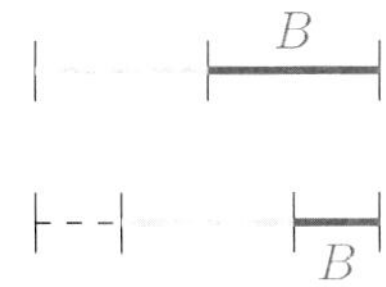


FIGURE 12. The first return map $F_S$ of the map $F$ from Figure 11 to $S = [0, 1/3)$. The dashed segment is not in the image of $F_S$.

LEMMA 2. *If the first-return map $F_S$ is surjective, then it is non-singular.*

PROOF. First note that as $F$ is a bijection, the first-return map $F_S$ is always injective: suppose $s, s' \in S$ with $F_S(s) = F_S(s')$ where $F_S(s) = F^n(s)$ and $F_S(s') = F^{n'}(s')$ where $n, n'$ are the smallest positive integers so that $F^n(s), F^{n'}(s') \in S$. If $n = n'$ then since $F$ is injective, we have $s = s'$. Otherwise suppose without loss of generality that $n > n'$. Applying $F^{-n'}$ to both sides of the equation $F^n(s) = F^{n'}(s')$ yields $F^{n-n'}(s) = s'$, but as $n - n' < n$ this contradicts the definition of $n$.

If $F_S$ is surjective, it is thus bijective. Hence $F$ is non-singular if and only if for each $E \subseteq S$ of positive measure we have that $F(E)$ has positive measure, and this is clear from the decomposition $S = \bigsqcup_{n\in\mathbb{N}} S_n$ with $\frac{dF_{S*}}{d\mu}$ equaling $\rho^{n-2}$ on $S_n$. $\square$

PROPOSITION 3. *The reciprocal transformation $F$ is conservative if and only if the first-return map $F_S$ is conservative.*

PROOF. Suppose $W$ was an $F$-wandering set. By Lemma 1, almost every point of $W$ eventually enters $S$. Let $W_n$ be the subset of $W$ that first enters $S$ after $n$ applications of $F$, for $n \geq 0$. If $W$ has positive measure, then at least one $W_n$ does as well, and $F^n(W_n)$ is an $F_S$-wandering set of positive measure. So, non-conservativity of $F$ implies non-conservativity of $F_S$.

Now suppose $W \subseteq S$ is an $F_S$-wandering set. Then $W$ must also be an $F$-wandering set, for if $F^m(W) \cap F^n(W) \neq \emptyset$, then by assumption $F^m(W) \cap F^n(W) \subseteq \Phi(S)$. There exists some subset $U$ of $F^m(W) \cap F^n(W)$ such that $F^k(U) \subseteq S$ for some $k > 0$ and $F^j(U) \cap S = \emptyset$ for all $0 < j < k$. But then $U' = F^{-(n+k)}(U)$ and $U'' = F^{-(m+k)}(U)$ are subsets of $W \subseteq S$ with the property that $F^{n+k}(U') \cap F^{m+k}(U'')$ is a non-empty subset of $S$, and there exists $n'$ and $m'$ so that $F_S^{n'}(U') \cap F_S^{m'}(U'')$ is non-empty, contradicting that $W$ was wandering. Thus $F_S$-wandering sets are also $F$-wandering sets. Since $F$ and $F_S$ are non-singular, non-conservativity of $F_S$ implies non-conservativity of $F$. $\square$

To study the dynamics of $F_S$, it is helpful to notice that $S$ naturally partitions into a countable collection of subsets. For each $n \geq 1$, we let $S_n$ denote the subset of $S$ which first returns to $S$ after $n$ applications of $F$:

$$S_n := \{x \in S \,|\, F_S(x) = F^n(x)\}.$$

As $F = \Phi T$ where $\Phi$ scales $S$ by $\rho$ and $\Phi(S)$ by $\rho^{-1}$, notice that $\mu(F_S(S_n)) = \rho^{n-2}\mu(S_n)$. That is, $F_S$ shrinks $S_1$ by $\rho^{-1}$, preserves the measure of $S_2$, and enlarges all other $S_n$. As a consequence, if $W$ was an $F_S$-wandering set, it must eventually enter and remain in $S_1$ under iteration of $F_S$. This simple observation establishes the following lemma.

LEMMA 4. *If $\mu(S_1) = 0$, then $F_S$ and $F$ are conservative.*

As we will see, while Lemma 4 easily guarantees that a transformation is conservative, it will be too strong for the applications we have in mind in Section 4; see Example 7.

**3.3. Ergodicity.** We now turn our attention to the ergodicity of reciprocal transformations. First notice that the construction of $F = \Phi T$ as the composition of a scaling involution and measure-preserving transformation places strong restrictions on invariant subsets. As Proposition 5 shows, the proportion of an $F$-invariant set which lives in $S$ (respectively, $\Phi(S)$) must equal the measure of $S$ (resp., $\Phi(S)$).

PROPOSITION 5. *If $E$ is an $F$-invariant set, then*

$$\frac{\mu(E \cap S)}{\mu(E)} = \mu(S) \text{ and } \frac{\mu(E \cap \Phi(S))}{\mu(E)} = \mu(\Phi(S))$$

PROOF. Let $r$ denote the proportion of $E$ which lives in $S$, $r = \frac{\mu(E\cap S)}{\mu(E)}$, and so $1 - r$ is the proportion of $E$ which lives in $\Phi(S)$. Now consider how $\Phi$ distorts the measure of $E$:

$$\begin{aligned}\mu(\Phi(E)) &= \mu(\Phi([E \cap S] \cup [E \cap \Phi(S)])) \\ &= \mu(\Phi(E \cap S)) + \mu(\Phi(E \cap \Phi(S))) \\ &= \rho\mu(E \cap S) + \rho^{-1}\mu(E \cap \Phi(S)) \\ &= \rho r \mu(E) + \rho^{-1}(1 - r)\mu(E).\end{aligned}$$

Notice that as $E$ is $F$-invariant, so $E = F(E) = \Phi T(E)$, and $\Phi$ is an involution we must have $\Phi(E) = T(E)$. As $T$ is measure-preserving, however, this means $\mu(\Phi(E)) = \mu(T(E)) = \mu(E)$. That is, the above equalities become simply

$$\mu(E) = \rho r \mu(E) + \rho^{-1}(1 - r)\mu(E).$$

Dividing by $\mu(E)$ we have $1 = \rho r + \frac{1-r}{\rho}$ or $\rho = \rho^2 r + 1 - r = r(\rho^2 - 1) + 1$. Solving for $r$ yields $r = \frac{\rho-1}{\rho^2-1}$ or simply $r = \frac{1}{\rho+1}$. Keeping in mind $\rho = \frac{\mu(\Phi S)}{\mu(S)}$ this becomes

$$r = \frac{1}{\frac{\mu(\Phi(S))}{\mu(S)} + 1} = \frac{1}{\left(\frac{\mu(\Phi(S))+\mu(S)}{\mu(S)}\right)} = \frac{\mu(S)}{\mu(\Phi(S)) + \mu(S)} = \mu(S).$$

That the proportion of $E$ in $\Phi(S)$ equals $\mu(\Phi(S))$ now easily follows. □

We note that the expression of a reciprocal transformation as a composition $F = \Phi T$ with $\Phi$ a scaling involution and $T$ a probability-preserving transformation allows us the following condition for the ergodicity of $F$.

PROPOSITION 6. *The reciprocal transformation $F = \Phi T$ is ergodic if and only if for every set $E$ which does not contain $S$ (up to a null set), $\Phi(E) \neq T(E)$.*

PROOF. Suppose that $F$ is ergodic and $E$ does not contain all of $S$. Then $E$ can not have full measure and is not an invariant set. That is, $F(E) \neq E$ which is equivalent to $\Phi T(E) \neq E$, and as $\Phi$ is an involution this is equivalent to $T(E) \neq \Phi(E)$.

Suppose now that for every $E$ not containing $S$, $E$ is not $F$-invariant. Thus if a set $U$ were to be $F$-invariant, it must contain $S$. In that case, $U \cap S = S$ and so the conclusion of Proposition 5 becomes

$$\frac{\mu(U \cap S)}{\mu(U)} = \frac{\mu(S)}{\mu(U)} = \mu(S) \implies \mu(U) = 1.$$ □

We can now show that determining whether or not $F$ is ergodic can be rephrased in terms of the ergodicity of $F_S$.

THEOREM 7. *The transformation* $F = \Phi T$ *is ergodic if the first-return map to* $S$, $F_S$, *is ergodic.*

PROOF. Suppose $F_S$ is ergodic and let $E \subseteq X$ be an $F$-invariant set. This would imply that $E \cap S$ is $F_S$-invariant, and so ergodicity of $F_S$ would force $S \subseteq E$. The contrapositive of Proposition 6 then shows $F$ is ergodic. □

EXAMPLE 5. With Theorem 7, we now have a tool that can sometimes be used to easily establish the ergodicity of certain non-singular transformations. In particular, consider affine interval exchanges such as those described in Example 2. The maps described in that example give a 3-parameter family of affine interval exchanges, corresponding to the lengths of the permuted subintervals. It is easy to see that the first-return map, $F_S$, of such a map is itself a circle rotation where points are translated (rotated) by $\mu(B) + \rho^{-1}\mu(D)$, where $\mu$ refers to the Lebesgue measure. Thus this $F_S$, and hence $F$, will be ergodic when $\mu(B) + \rho^{-1}\mu(D)$ is an irrational number. Note however that $F^2$ *is not* ergodic, as each of the rotated circles $A \cup B$ and $C \cup D$ is an invariant subset of $F^2$.

## 4. The discrete Maharam extension

In [**Mah64**], Maharam gave a construction which associates to a non-singular transformation on a probability space a measure-preserving transformation on a space of infinite measure. We now describe a construction which is similar, and can be thought of as a rescaling of the Maharam extension, but restricted to a measure zero subset of the space introduced by the Maharam extension. Note that since we are restricting to a measure zero subset, ergodic theoretic results of Maharam do not instantly carry over to the setting we are about to describe.

Throughout this section we again consider a probability space $(X, \mathcal{B}, \mu)$, a measure-preserving transformation $T$, a scaling involution $\Phi$ with scaling ratio $\rho > 1$, and let $F$ be the reciprocal transformation $F = \Phi T$.

We will place a particular measure on $X \times \mathbb{Z}$ and construct a map $\widetilde{F}$ on $X \times \mathbb{Z}$ which preserves this measure and has $F$ as a factor.

First, let $\pi : X \times \mathbb{Z} \to X$ denote the projection onto the first factor. For any $E \subseteq X \times \mathbb{Z}$ and any $n \in \mathbb{Z}$, let $E_n$ denote the set $E \cap (X \times \{n\})$. We will refer to $E_n$ as the *n-th level of* $E$.

We define a measure $\widetilde{\mu}$ on $X \times \mathbb{Z}$ as follows: for each measurable $E \subseteq X \times \mathbb{Z}$, we declare

$$\widetilde{\mu}(E) = \sum_{n \in \mathbb{Z}} \rho^{-n} \mu(\pi(E_n))$$

where $\rho$ is the scaling ratio of $\Phi$. That is, we imagine $X \times \mathbb{Z}$ as consisting of copies of $X$ scaled by powers of $\rho$ so that "higher" copies of $X$ are smaller than "lower" copies. Note that $\widetilde{\mu}$ is an infinite, but $\sigma$-finite, measure on $X \times \mathbb{Z}$.

Now consider the map $\widetilde{F} : X \times \mathbb{Z} \to X \times \mathbb{Z}$ defined by

$$\widetilde{F}(x, n) = \begin{cases} (F(x), n+1) & \text{if } x \in T^{-1}(S) \\ (F(x), n-1) & \text{if } x \in T^{-1}(\Phi(S)) \end{cases}$$

Notice $F\pi = \pi\widetilde{F}$.

We will refer to $\widetilde{F} : X \times \mathbb{Z} \to X \times \mathbb{Z}$ together with the measure $\widetilde{\mu}$ as the *discrete Maharam extension* of $F : X \to X$ with measure $\mu$. (The $\sigma$-algebra on $X \times \mathbb{Z}$ is simply the product $\sigma$-algebra of the $\sigma$-algebra $\mathcal{B}$ on $X$ with the discrete $\sigma$-algebra, i.e. the power set, on $\mathbb{Z}$.)

EXAMPLE 6. We now describe the discrete Maharam extension of the reciprocal transformation from Example 2. Here $X$ is in the interval $[0, 1)$, and so the levels $X \times \mathbb{Z}$ are scaled versions of the interval; we may think of $X \times \{n\}$ as $[0, \rho^{-n}) = [0, 2^{-n})$. The level $X \times \{n\} = [0, 2^{-n})$ is partitioned into intervals $A_n$, $B_n$, $C_n$, $D_n$, scaled copies of the original $A$, $B$, $C$, $D$, subintervals which $X$ was partitioned into. The transformation $\widetilde{F}$ acts by moving points in $A_n$, $B_n$ into the scaled images of $T(A)$ and $T(B)$ in $X \times \{n+1\}$, and points in $C_n$ and $D_n$ are moved into the scaled images of $T(C)$ and $T(D)$ in $X \times \{n-1\}$. See Figure 13.

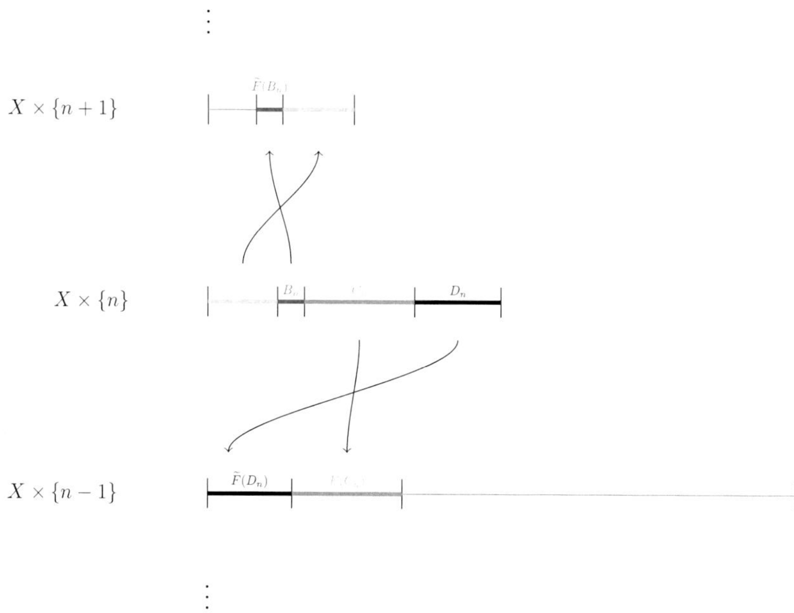


FIGURE 13. The discrete Maharam extension of the reciprocal transformation described in Example 2.

That is, we apply $F$ to each copy of $X$, but then move points up or down a level depending on how $F$ scales at that point (i.e., based on the value of $\frac{dF_*\mu}{d\mu}$). Those points in $T^{-1}(S)$ will first be moved by $T$ into $S$, then scaled by $\rho$ when $\Phi$ is applied. By moving these points up or down one level (to a copy of $X$ whose $\widetilde{\mu}$-measure shrinks by $\rho^{-1}$ or grows by $\rho$), we effectively cancel out the scaling effect of $\Phi$. That is, the measure $\widetilde{\mu}$ we have defined is preserved by $\widetilde{F}$. Before proving this we show that $\widetilde{F}$ is always bijective so that we may consider images of sets instead of their preimages when proving $\widetilde{\mu}$ is preserved.

LEMMA 8. *The discrete Maharam extension $\widetilde{F}$ associated to a reciprocal transformation $F = \Phi T$ is bijective.*

PROOF. First observe that for each integer $m$, the preimage $\widetilde{F}^{-1}(S \times \{m\})$ is $F^{-1}(S) \times \{m+1\}$ and $F^{-1}(S) = (\Phi T)^{-1}(S) = T^{-1}\Phi(S)$. Similarly, $F^{-1}(\Phi(S) \times \{m\}) = F^{-1}(\Phi(S)) \times \{m-1\}$ and $F^{-1}(\Phi(S)) = (\Phi T)^{-1}\Phi(S) = T^{-1}\Phi\Phi(S) = T^{-1}(S)$. As $F$ is a bijection, $\widetilde{F}$ restricted to $T^{-1}\Phi(S) \times \{m+1\}$ is a bijection onto $S \times \{m\}$, and similarly $\widetilde{F}$ restricted to $T^{-1}(S) \times \{m-1\}$ is a bijection onto $\Phi(S) \times \{m\}$. As each level of $X \times \mathbb{Z}$ is decomposted into a disjoint union of $S \times \{m\}$ and $\Phi(S) \times \{m\}$ sets on which $\widetilde{F}$ is bijective, this shows that $\widetilde{F}$ is a bijection. □

PROPOSITION 9. *The map $\widetilde{F}$ preserves the measure $\widetilde{\mu}$.*

PROOF. For any $E \subseteq X \times \mathbb{Z}$ and any $n \in \mathbb{Z}$, we break the $n$-th level $E_n$ up into two pieces: the portion which $\widetilde{F}$ will move up a level (denoted $E_n^+$), and the portion which will move down a level ($E_n^-$). That is,

$$E_n^+ := \left(\pi(E_n) \cap T^{-1}(S)\right) \times \{n\}, \text{ and}$$
$$E_n^- := \left(\pi(E_n) \cap T^{-1}(\Phi(S))\right) \times \{n\}.$$

The $n$-th level of the image $\widetilde{F}(E)$ is $\widetilde{F}(E_{n-1}^+) \sqcup \widetilde{F}(E_{n+1}^-)$, and so has measure

$$\begin{aligned} &\rho^{-n}\left(\mu\left(\pi\left(\widetilde{F}\left(E_{n-1}^+\right)\right)\right) + \mu\left(\pi\left(\widetilde{F}\left(E_{n+1}^-\right)\right)\right)\right) \\ =&\rho^{-n}\left(\mu\left(F\left(\pi\left(E_{n-1}^+\right)\right)\right) + \mu\left(F\left(\pi\left(E_{n+1}^-\right)\right)\right)\right) \end{aligned}$$

However, by definition of the sets $E_{n\pm1}^{\mp}$,

$$\mu\left(\pi\left(F\left(E_{n-1}^+\right)\right)\right) = \rho\mu\left(\pi\left(E_{n-1}^+\right)\right), \text{ and}$$
$$\mu\left(\pi\left(F\left(E_{n+1}^-\right)\right)\right) = \rho^{-1}\mu\left(\pi\left(E_{n+1}^-\right)\right).$$

Thus the $n$-th level of $\widetilde{F}(E)$ has measure

$$\rho^{-(n-1)}\mu\left(\pi\left(E_{n-1}^+\right)\right) + \rho^{-(n+1)}\mu\left(\pi\left(E_{n+1}^-\right)\right).$$

That is, the terms of

$$\widetilde{\mu}\left(\widetilde{F}(E)\right) = \sum_{n\in\mathbb{Z}} \rho^{-n}\mu\left(\pi\left(\widetilde{F}(E_n)\right)\right)$$

are the same as the terms of

$$\widetilde{\mu}(E) = \sum_{n\in\mathbb{Z}} \rho^{-n}\mu\left(\pi\left(E_n\right)\right),$$

just slightly reordered. (Note that the terms of all our series are non-negative, and so these series either diverge to infinity or are absolutely convergent. In either case, reordering the terms of the series does not change the series' value.) □

A natural question to consider is whether ergodicity of a reciprocal transformation $F$ is sufficient to ensure the ergodicity of its discrete Maharam extension $\widetilde{F}$. Using Lemma 10 and Proposition 11 below, Example 7 gives a counterexample to the reasonable-seeming conjecture that ergodicity of $F$ implies ergodicity of $\widetilde{F}$.

LEMMA 10. *Suppose $F = \Phi T$ is a reciprocal transformation where some iterate $F^n$ is measure-preserving. Then the associated discrete Maharam extension can not be ergodic.*

Proof. As $F^n$ is measure-preserving and each iterate of $F$ modifies the measure of a subset by a factor of $\rho$ or $\rho^{-1}$, for each $0 \leq m \leq n$ the Radon-Nikodym derivative $\frac{dF_*^m \mu}{d\mu}$ is bounded below by $\rho^{-n}$ and above by $\rho^n$. Thus any point of $X \times \{0\}$ can not visit any level $X \times \{k\}$ for $|k| > n$, as it must return to $X \times \{0\}$ at the $n$-th iterate. □

Proposition 11. *If $F = \Phi T$ is a reciprocal transformation such that $\mu(S_1)$ (in the notation of Lemma 4) has measure zero, then the discrete Maharam extension $\widetilde{F}$ is not ergodic.*

Proof. Recall $\mu(F_S(E)) = \rho^{n-2}\mu(E)$ for each $E \subseteq S_n$. If $\mu(S_1) = 0$, then subsets of $S$ can never decrease in measure, as entering into $S_1$ is the only way that the image of a set under $F_S$ can decrease its measure. By Lemma 4, $F_S$ is conservative, and so if some $S_n$ for $n > 2$ had positive measure, almost every point in $S_n$ must return to $S_n$ infinitely-often. Thus if some $S_n$, $n > 2$, had positive measure, any subset of $S_n$ of positive measure must have its measure increase by a factor of $\rho^{n-2}$ infinitely often. However, $S$ has finite measure, and so if $\mu(S_1) = 0$ then $S = S_2$ up to a null set. This implies $F^2$ is in fact a measure-preserving transformation, and by Lemma 10, the associated discrete Maharam extension can not be ergodic. □

Example 7. Consider our running example of a reciprocal transformation from Example 2 and Example 5, and its associated discrete Maharam extension in Example 6. As previously noted, when $\mu(B) + \rho^{-1}\mu(D)$ is irrational, the reciprocal transformation $F$ is ergodic. Notice, in the notation of Lemma 4, that $S_1 = \emptyset$ and $S_2 = S$. Hence the associated discrete Maharam extension from Example 6 is not ergodic.

We thus see that ergodicity of $\widetilde{F}$ requires a more restrictive condition than just ergodicity of $F$. In order to relate the ergodicity of $\widetilde{F}$ to that of the reciprocal transformation $F = \Phi T$, we must first describe the "Krieger type" and ratio set of non-singular transformations.

The study of Krieger types, the closely related notion of the ratio set, and non-singular ergodic theory in general has its origin in a simple question about whether a given measurable transformation necessarily has any invariant measures. The work of Murray and von Neumann had established a connection between ergodic theory and operator algebras [**MvN43**], which was utilized by Dye in his study of orbit equivalence [**Dye59**]. Shortly afterwards, Ornstein provided the first example of a measurable transformation not preserving any equivalent measure [**Orn60**]. This initiated the study of the various types of non-singular transformations that could exist, and would lead to Krieger's classification of non-singular transformations motivated by the existing classification of von Neumann algebras [**Kri69**].

Associated to each non-singular transformation $F$ is a collection of non-negative real numbers called the transformation's *ratio set*, denoted $r(F)$. The number $\lambda$ is an element of $r(F)$ if for every set $E$ of positive measure and every $\varepsilon > 0$ there exists a subset $E' \subseteq E$ of positive measure and an integer $n$ so that for every $x \in E'$, $F^n(x) \in E'$ and $\left|\frac{dF_*^n \mu}{d\mu}(x) - \lambda\right| < \varepsilon$. That is, the ratio set essentially measures all of the proportions by which all sets get scaled. Computing the ratio set of a transformation is an extremely difficult problem in general, though can sometimes be accomplished in certain special settings; e.g., see [**HO81**], [**CHP87**] [**Dan04**], [**DHR11**], and [**Fur14**].

Non-singular transformations are characterized into various *Krieger types* based on their ratio set. We say an ergodic transformation $F$ has *(Krieger) type $II$* if its ratio set is $r(F) = \{1\}$, and *(Krieger) type $III_\lambda$* if its ratio set is $r(F) = \{0\} \cup \{\lambda^n \mid n \in \mathbb{Z}\}$ for a real number $\lambda > 0$. It was shown in [**Kri70**] that type $II$ transformations always admit an equivalent invariant measure, and correspondingly there are two subtypes: type $II_1$ transformations admit an invariant probability measure, whereas type $II_\infty$ transformations admit an infinite invariant measure. Type $III$ transformations do not admit any invariant measures, and similarly have a breakdown into subtypes.

The subtypes of type $III$ transformations are described as being type $III_\lambda$ where $\lambda \in [0,1]$. Of these subtypes there are three particular cases corresponding to $r(F) = \{0,1\}$ (type $III_0$), $r(F) = [0,\infty)$ (type $III_1$), and $r(F) = \{\lambda^n \mid n \in \mathbb{Z}\}$ with $\lambda \in (0,1)$. Theorem 12 below shows that $\widetilde{F}$ is ergodic (with respect to $\widetilde{\mu}$) precisely when $F$ is ergodic (with respect to $\mu$) is type $III_\lambda$ where $\lambda = \frac{1}{\rho}$.

THEOREM 12. *Let $T : X \to X$ be a measure-preserving transformation of a probability space $(X, \mathscr{B}, \mu)$, $\Phi : X \to X$ a scaling involution with scaling ratio $\rho > 1$, $F = \Phi T$ their composition, and $\widetilde{F} : X \times \mathbb{Z} \to X \times \mathbb{Z}$ the corresponding discrete Maharam extension with measure $\widetilde{\mu}$. The map $\widetilde{F}$ is $\widetilde{\mu}$-ergodic if and only if $F$ is ergodic and type $III_{1/\rho}$.*

PROOF. Suppose that $F$ is ergodic and type $III_{1/\rho}$. To establish that $\widetilde{F}$ is ergodic, we will show that for every subset of $X \times \mathbb{Z}$ of positive measure, $\widetilde{\mu}$-almost every point $(x_0, n_0) \in X \times \mathbb{Z}$ eventually enters the set.

Let $E \subseteq X \times \mathbb{Z}$ be any set of positive $\widetilde{\mu}$ measure. As $E = \bigsqcup_{k\in\mathbb{Z}} E_k$ with $E_k = E \cap (X \times \{k\})$, there must exist some $k$ with $\widetilde{\mu}(E_k) > 0$. Let $U \subseteq X \times \mathbb{Z}$ consist of those points $(x,n)$ so that $\widetilde{F}^N(x,n) \notin E_k$ for all $N$. Observe that if $U$ has positive $\widetilde{\mu}$-measure, then $\pi(U)$ must have positive $\mu$-measure. By the assumption that $F$ is ergodic, for $\mu$-almost every point $x \in \pi(U)$ there would exist some integer $m$ so that $F^m(x) \in \pi(E_k)$.

For each integer $j \geq 0$, let $V_j \subseteq \pi(U)$ consist of those points $x \in \pi(U)$ so that $F^j(x) \in \pi(E_k)$ but $F^\ell(x) \notin \pi(E_k)$ for each $\ell < j$. We may further decompose each $V_j$ into subsets $V_j^q$ where $\frac{dF_*^j\mu}{d\mu}$ equals $\rho^q$ for $\mu$-a.e. $x \in V_j^q$. Observe if $x \in V_j^q$, then $\widetilde{F}^j(x, k-q) \in E_k$. As $\pi(U) = \bigcup_{j,q} V_j^q$ is a countable union, at least one $V_j^q$ has positive measure. This would mean $U \cap \pi^{-1}(V_j^q)$ has positive measure, and so a positive measure subset of $U$ enters $E_k$, contradicting the definition of $U$. Thus $U$ must have had zero measure, and so almost every point of $X \times \mathbb{Z}$ eventually enters every set of positive measure, and $\widetilde{F}$ is ergodic.

Suppose now that $\widetilde{F}$ is ergodic. We will first show that $F$ must be ergodic as well by showing an invariant subset $E \subseteq X$ has full measure or zero measure. Notice that $\widetilde{E} = \pi^{-1}(E)$ is $\widetilde{F}$-invariant. Hence $\widetilde{E}$ or its complement has zero measure. Thus either $E$ or its complement has zero measure as well, since $\pi$ is non-singular.

Now we must establish that $F$ is type $III_{1/\rho}$. Let $E$ be any subset of $X$ of positive measure, and consider $E \times \{0\}$. By ergodicity of $\widetilde{F}$, for each integer $q$ and each $x \in E \times \{0\}$ there exists an $m$ so that $\widetilde{F}^m(x,0) \in E \times \{q\}$. As $F$ is semiconjugate to $\widetilde{F}$, this means $F^m(x) \in E$ and $\frac{dF_*\mu}{d\mu}(x) = \rho^q$. Thus all powers of $\rho$ are in the ratio set of $F$, and $F$ is type $III_{1/\rho}$. $\square$

## 5. Directions for future work

As previously noted, determining the ratio set of a given transformation seems to be a very difficult problem in general. In order for Theorem 12 to be useful, we would like to have some easy-to-check condition that will guarantee a reciprocal transformation has Krieger type $III_{1/\rho}$. One possible avenue which the author is currently pursuing has to do with the case that the reciprocal transformation $F$ is "affinely self-similar" to its first-return map $F_S$. To be precise, suppose that $X$ is an interval, $T$ is an interval exchange, and $\Phi$ is a piecewise affine involution of $X$. The reciprocal transformation $F = \Phi T$ is then an affine interval exchange, and it is not difficult to see that the first-return map $F_S$ is also an affine interval exchange transformation. In the special case that $F_S : S \to S$ is simply a rescaled copy of $F : X \to X$ (this is what we mean by "affinely self-similar"), it may be possible to show that $F$ has Krieger type $III_{1/\rho}$. In particular, the self-similarity allows us to obtain scaled copies of the smaller set $S$ into which almost every point of $X$ must enter. It may be that this existence of smaller and smaller subsets, each scaled by a factor of $\rho^{-1}$, will imply that all sets have subsets that must eventually scale by $\rho^{-n}$ as the points visit these scaled copies of $S$.

At the moment the author does not know of an example of an interval exchange $T$ and scaling involution $\Phi$ whose associated reciprocal transformation $F = \Phi T$ would exhibit this self-similarity. It is possible that in fact no such transformation exists, and so simply establishing the existence or non-existence of these affinely self-similar transformations seems interesting. To this end, it would be helpful to have a better understanding of the moduli space of reciprocal transformations, at least in the case when $X$ is an interval and $T$ is an interval exchange with some fixed number of intervals of continuity. Once this moduli space is understood, a transformation on that space similar to Rauzy-Veech induction for traditional interval exchanges might be helpful to search for self-similar reciprocal transformations. To date no such procedure exists for general affine interval exchanges, but perhaps in this restricted case such a procedure could be defined.

## References


[Aar97] Jon Aaronson, *An introduction to infinite ergodic theory*, Mathematical Surveys and Monographs, vol. 50, American Mathematical Society, Providence, RI, 1997, DOI 10.1090/surv/050. MR1450400

[CHP87] J. R. Choksi, J. M. Hawkins, and V. S. Prasad, *Abelian cocycles for nonsingular ergodic transformations and the genericity of type $III_1$ transformations*, Monatsh. Math. **103** (1987), no. 3, 187–205, DOI 10.1007/BF01364339. MR894170

[Dan04] Alexandre I. Danilenko, *Infinite rank one actions and nonsingular Chacon transformations*, Illinois J. Math. **48** (2004), no. 3, 769–786. MR2114251

[DHR11] A. H. Dooley, J. Hawkins, and D. Ralston, *Families of type $III_0$ ergodic transformations in distinct orbit equivalent classes*, Monatsh. Math. **164** (2011), no. 4, 369–381, DOI 10.1007/s00605-010-0258-0. MR2861592

[DS23] Alexandre I. Danilenko and Cesar E. Silva, *Ergodic theory: nonsingular transformations*, Ergodic theory, Encycl. Complex. Syst. Sci., Springer, New York, 2023, pp. 233–292, DOI 10.1007/978-1-0716-2388-6_183. MR4647080

[Dye59] H. A. Dye, *On groups of measure preserving transformations. I*, Amer. J. Math. **81** (1959), 119–159, DOI 10.2307/2372852. MR131516

[Fur14] Joanna Furno, *Orbit equivalence of p-adic transformations and their iterates*, Monatsh. Math. **175** (2014), no. 2, 249–276, DOI 10.1007/s00605-014-0645-z. MR3260869

[HO81] Toshihiro Hamachi and Motosige Osikawa, *Ergodic groups of automorphisms and Krieger's theorems*, Seminar on Mathematical Sciences, vol. 3, Keio University, Department of Mathematics, Yokohama, 1981. MR617740

[Kri69] Wolfgang Krieger, *On non-singular transformations of a measure space. I, II*, Z. Wahrscheinlichkeitstheorie und Verw. Gebiete **11** (1969), 83–97; ibid. 11 (1969), 98–119, DOI 10.1007/BF00531811. MR240279

[Kri70] Wolfgang Krieger, *On the Araki-Woods asymptotic ratio set and non-singular transformations of a measure space*, Contributions to Ergodic Theory and Probability (Proc. Conf., Ohio State Univ., Columbus, Ohio, 1970), Lecture Notes in Math., Vol. 160, Springer, Berlin-New York, 1970, pp. 158–177. MR414823

[Mah64] D. Maharam, *Incompressible transformations*, Fund. Math. **56** (1964), 35–50, DOI 10.4064/fm-56-1-35-50. MR169988

[MvN43] F. J. Murray and J. von Neumann, *On rings of operators. IV*, Ann. of Math. (2) **44** (1943), 716–808, DOI 10.2307/1969107. MR9096

[Orn60] Donald S. Ornstein, *On invariant measures*, Bull. Amer. Math. Soc. **66** (1960), 297–300, DOI 10.1090/S0002-9904-1960-10478-8. MR146350

[Pet89] Karl Petersen, *Ergodic theory*, Cambridge Studies in Advanced Mathematics, vol. 2, Cambridge University Press, Cambridge, 1989. Corrected reprint of the 1983 original. MR1073173

[Via06] Marcelo Viana, *Ergodic theory of interval exchange maps*, Rev. Mat. Complut. **19** (2006), no. 1, 7–100, DOI 10.5209/rev_REMA.2006.v19.n1.16621. MR2219821

[Vor96] Ya. B. Vorobets, *Plane structures and billiards in rational polygons: the Veech alternative* (Russian), Uspekhi Mat. Nauk **51** (1996), no. 5(311), 3–42, DOI 10.1070/RM1996v051n05ABEH002993; English transl., Russian Math. Surveys **51** (1996), no. 5, 779–817. MR1436653

[Wal82] Peter Walters, *An introduction to ergodic theory*, Graduate Texts in Mathematics, vol. 79, Springer-Verlag, New York-Berlin, 1982. MR648108

[Wri15] Alex Wright, *Translation surfaces and their orbit closures: an introduction for a broad audience*, EMS Surv. Math. Sci. **2** (2015), no. 1, 63–108, DOI 10.4171/EMSS/9. MR3354955

[Zor06] Anton Zorich, *Flat surfaces*, Frontiers in number theory, physics, and geometry. I, Springer, Berlin, 2006, pp. 437–583. MR2261104

Department of Mathematics and Computer Science, Western Carolina University, Cullowhee, North Carolina 28723

*Email address*: ccjohnson@gmail.com

Contemporary Mathematics
Volume **839**, 2026
https://doi.org/10.1090/conm/839/16729

# Concatenations of periodic words in subshifts defined by linear orders and poles of the Artin-Mazur zeta function

Chenxi Wu

ABSTRACT. For certain pairs of unimodal maps on the interval with periodic critical orbits, it is known that one can combine them to create another interval map with periodic critical points whose topological entropy is close to one while the poles of Artin-Mazur $\zeta$ function outside the unit circles are close to the other. We provided a formulation and proof of this result in a more general, symbolic dynamical setting, which allow us to generalize this fact to certain families of maps on finite trees.

## 1. Introduction

The Milnor-Thurston kneading theory [**7**] allows one to establish connections between interval maps and symbolic dynamics, and prior works by Tiozzo [**11, 12**], Calegari-Koch-Walker [**3**], Lindsey-Tiozzo-Wu [**5**], Bray-Davis-Lindsey-Wu [**2, 6**] and others applied this idea to the study of entropy of certain families of post-critically finite interval maps, as well as core entropies of some superattracting polynomials. In particular, in [**2, 5**], a key result on the dynamics of these families of maps is the following "persistence theorem":

THEOREM 1.1 ([**2, 5**]). *Let $\mathcal{F}$ be a family of maps, which can be:*

- *Unimodal maps on intervals with periodic critical orbits.*
- *Maps on Hubbard trees of a superattracting parameter on a principal vein on the Mandelbrot set, with periodic critical orbits.*

*Every map $f \in F$ admits a finite Markov decomposition. Let $M_f$ be its incidence matrix, $E_f$ the set of eigenvalues of $M_f$, and $h_f$, which is the* log *of the leading eigenvalue of $M_f$, the topological entropy.*

*If $f_1, f_2, f_3 \in \mathcal{F}$, $h_{f_1} < h_{f_2} < h_{f_3}$, then for any $\epsilon > 0$, any $z$ which is an eigenvalue of $M_{f_1}$ such that $|z| < 1$, there is some $f' \in \mathcal{F}$, such that $h_{f_2} < h_{f'} < h_{f_3}$ and $M_{f'}$ has an eigenvalue $z'$ which is $\epsilon$-close to $z$.*

As a consequence, the "master teapots" (see [**10**], and Figure 1) corresponding to these families of maps, defined as

$$T_{\mathcal{F}} = \overline{\{(z, \lambda) : z \in E_f, \lambda = e^{h_f}, f \in \mathcal{F}\}}$$

2020 *Mathematics Subject Classification.* Primary 37B10, Secondary 37B40, 37E05, 37E25, 37F10.

FIGURE 1. The intersection of Thurston's master teapot with the unit cylinder, plotting the half where the imaginary part is non negative.

when intersecting with the unit cyliner $\{(z, y) : |z| < 1\}$, consists of vertical line segments that end at the same height. This also provides a connection between horizontal slices of these "teapots" with the Mandelbrot set (in the sense of Solomyak [**9**]) of parametrized families of graph directed systems, c.f. [**6**].

The rough idea behind these kind of "persistence" results is the following: one can use Milnor-Thurston kneading theory to relate maps in $\mathcal{F}$ and the forward itinerary of the critical value of these maps, and such itineraries satisfies the condition of *admissibility* which means that when deleting any prefix, it becomes no larger than the original itinerary under some linear order. Now for two such maps and corresponding periodic itineraries, such that one has larger entropy, one can find another map in $\mathcal{F}$ with periodic itinerary such that the prefix of the period is close to the itinerary of the map with higher entropy, and the suffix is close to the itinerary of the map with lower entropy. As entropies are controlled by the prefix of the period, while eigenvalues of $M_f$ which are inside the unit circle are controlled by the suffix, one gets the "persistence" result.

In this paper, we will examine under exactly what conditions on the linear order we can carry out these kind of "concatenation" argument as in [**2**, **5**]. From this we will provide a symbolic dynamics formulation of the concept of "persistence", which can be generalized to other settings, which will be stated as Theorem 6.1. In Section 2 we will set the notations, in Section 3-5 we will develop the technical tools of tuning and dominance, in Section 6 we will state and prove the main theorem, and discuss its consequence on Artin-Mazur $\zeta$ functions. Finally, in Section 7 we will provide applications that recovers prior works as special cases as well as applied to families of tree maps.

In an upcoming paper with Kathryn Lindsey and Ethan Farber, we will use a slight variant of the technique in this paper to show "persistence" result for a family of real polynomial maps with 2 critical points.

## 2. Notations

Let $\Gamma = (V, E)$ be a finite directed graph with no parallel edges, that is fully connected and aperiodic. The vertex set is $V = \{0, 1, 2, \ldots, N\}$. Let $O : V \to \{-1, 1\}$ be a function. Let $\Sigma = \{a \in V^{\mathbb{N}} : \text{for all } n \in \mathbb{N}, (a_n, a_{n+1}) \in E\}$ be the set of infinite paths on $\Gamma$, then we have a shift map $\sigma : \Sigma \to \Sigma$ defined as $(\sigma(a))_n = a_{n+1}$. Here we always assume $0 \in \mathbb{N}$.

DEFINITION 2.1. We define a linear order in $\Sigma$, called the *twisted lexicographic order*, as follows: we say $a < b$ if and only if there is some $n \in \mathbb{N}$, such that $a_k = b_k$ for $k < n$, and $\prod_{k=0}^{n-1} O(a_k)(a_n - b_n) < 0$. We call an element in $\Sigma$ *admissible*, if and only if $a_0 = N$ and $\sigma^n(a) \leq a$ for all $n \in \mathbb{N}$.

REMARK 2.2. Let $a \in \Sigma$ be an admissible element, $\Sigma_a = \{b \in \Sigma : b \leq a\}$, then $\sigma(\Sigma_a) \subseteq \Sigma_a$.

Let $w$ be a finite word with alphabet $V$, i. e. $w \in V^n$ for some $n$. Denote the length of $w$ as $|w|$, and let $\mathbf{sgn}(w) = \prod_k O(w_k)$ be the *sign* of $w$. We define a partial order $\leq$ on the set of finite words as follows: if $u, v$ are two finite words, we say $u < v$ iff there is some $0 \leq j < \min(|u|, |v|)$, such that $u_k = v_k$ for all $k < j$ and $\prod_{k<j} O(u_k)(u_j - v_j) < 0$. The length $k$ prefix and suffix of a word $w$ is denoted by $\mathbf{Pre}_k(w)$ and $\mathbf{Suff}_k(w)$ respectively.

Some of the following notations follow that of Tiozzo [**11**].

DEFINITION 2.3. Let $w$ be a finite word starting with $N$. We call $w$ *periodic* if $w^\infty \in \Sigma$, *irreducible* if $w$ can not be written as a concatenation of more than one copies of another word, *extremal* if $w$ is periodic and $w^\infty \in \Sigma$ is admissible, *admissible* if $w$ is both extremal and has positive **sgn**, *dominant* if $w$ is periodic, and for any $2 \leq k \leq |w|$, $\mathbf{Suff}_k(wN) < \mathbf{Pre}_k(w)$. Here $wN$ is the concatenation of $w$ with the word $N$.

REMARK 2.4. A finite word $w$ satisfies $w^\infty \in \Sigma$ iff it satisfies the following two conditions:

(1) For any $0 \leq i \leq |w| - 2$, $(w_i, w_{i+1}) \in E$.
(2) $(w_{|w|-1}, w_0) \in E$.

REMARK 2.5. A finite word $w$ starting with $N$ is extremal iff $w^\infty \in \Sigma$, and for any $k < |w|$, $\mathbf{Suff}_{|w|-k}(w)\mathbf{Pre}_k(w) \leq w$. Also, if $w$ is dominant then it is both extremal and irreducible.

Let $a, b \in \Sigma$, $a < b$, the **open interval** $(a, b)$ is the elements in $\Sigma$ that lies between $a$ and $b$, and the closed interval is the open interval union with $\{a, b\}$, in other words, $[a, b] = \{c \in \Sigma : a \leq c \leq b\}$. Let $I_k$ be the set of elements in $\Sigma$ that starts with $k$.

REMARK 2.6. If $w$ is extremal, then $\Sigma_{w^\infty}$ admits a finite Markov decomposition as follows: Let $a_1 > a_2 > \cdots > a_m$ be distinct elements in the set $\{\sigma^k(w^\infty) : k \in \mathbb{N}\}$. Let $a_{m+1}$ be the smallest element in $\Sigma$. Now it is easy to see that $\mathcal{P} = \{[a_{i+1}, a_i] \cap I_j : 1 \leq j \leq N, 1 \leq i \leq m+1, [a_{i+1}, a_i] \cap I_j \neq \emptyset\}$ form a Markov decomposition of $\Sigma_{w^\infty}$. Hence the Artin-Mazur $\zeta$-function of $\sigma|_{\Sigma_{w^\infty}}$ is rational with integer coefficients, and the topological entropy is the logrithm of an algebraic integer. Furthermore, because if $w_1^\infty < w_2^\infty$ then $\Sigma_{w_1^\infty} \subseteq \Sigma_{w_2^\infty}$, the topological entropy of $\Sigma_{w_2^\infty}$ is no less than the topological entropy of $\Sigma_{w_1^\infty}$. In other words, the topological entropy of $\Sigma_{w^\infty}$ is monotone with respect to $w^\infty$ under the linear order $<$.

## 3. Concatenation of dominant and admissible words

The purpose of this section is to prove a generalization of Proposition 5.1 of [**2**], which says that one can create admissible words by concatenating dominant and admissible words. The proof follows closely to that of [**2**] as well.

Firstly some elementary lemmas, which are generalizations of arguments in Carminati-Tiozzo [**4**]. See also [**2**, **11**].

LEMMA 3.1. *Let $a, b$ be two periodic words, in other words, $a^\infty, b^\infty \in \Sigma$, and both start at $N$. Then*

(1) *$(ab)^\infty$ and $(ba)^\infty$ are both in $\Sigma$.*
(2) *$a^\infty > b^\infty$ iff $ab > ba$*

PROOF. Part (1) follows from the observation in Remark 2.4.

Suppose $ab = ba$, then $a$ and $b$ must be common multiples of the same word, which implies $a^\infty = b^\infty$. Similarly $a^\infty = b^\infty$ also implies $ab = ba$. Now to show (2) one only need to show that it is not possible to have both $a^\infty > b^\infty$ and $ab < ba$. Suppose $a^\infty > b^\infty$ and $ab < ba$. We shall now look at where $ab$ and $ba$ first differ. Suppose they differ for the first time in the $k$-th letter, i.e. $\mathbf{Pre}_{k-1}(ab) = \mathbf{Pre}_{k-1}(ba)$, and $\mathbf{Pre}_k(ab) \neq \mathbf{Pre}_k(ba)$.

- Case 1: If $k \leq \min(|a|, |b|)$, this implies $a < b$, hence $a^\infty < b^\infty$, a contradiction.
- Case 2: If $|a| = |b|$, then it is easy to see that $k \leq |a| = |b|$, which reduces to case 1.
- Case 3: If $|a| < |b|$, and $k > |a|$. This means that $b = a^j c$ for some word $c$ which does not have $a$ as a prefix, and $j > 0$ a positive integer. Hence $a^{j+1}c < a^j ca$. Now if $k \leq (j+1)|a|$, we can similarly get $a^\infty < b^\infty$. If $k > (j+1)|a|$, then $c$ must be a prefix of $a$, hence $a^{j+1}c$ is a prefix of $a^\infty$ while $a^j ca$ is a prefix of $b^\infty$, this also implies that $a^\infty < b^\infty$.
- Case 4: If $|b| < |a|$, and $k > |b|$. This can be shown to be impossible via the same argument as in Case 3.

□

An immediate consequence of Lemma 3.1 is the following:

LEMMA 3.2. *If $a^\infty$ and $b^\infty$ are both in $\Sigma$ and both starts with $N$, then $a^\infty > b^\infty$ iff $ab^\infty > b^\infty$ iff $ba^\infty < a^\infty$.*

PROOF. Apply Lemma 3.1 to $a$ and $b^n$ or $a^n$ and $b$, and set $n \to \infty$. □

LEMMA 3.3. *Let $w$ be an extremal and irreducible word, $0 < k < |w|$. If $\boldsymbol{Pre}_k(w) = \boldsymbol{Suff}_k(w)$, then $\boldsymbol{sgn}(\boldsymbol{Pre}_k(w)) = -1$.*

PROOF. Let $v = \mathbf{Pre}_k(w)$, then $w$ can be written as $w = vx = yv$, and extremality and irreducibility of $w$ implies that $vx > xv$, $yv > vy$.

Now assume that $\mathbf{sgn}(v) = 1$, then $vx = yv > vy$ hence $x > y$, and $yv = vx > xv$ hence $y \geq x$, a contradiction. □

Now we state and prove the main theorem for this Section:

THEOREM 3.4. *Let $w$ be a dominant word, $v$ an admissible and irreducible word, $v^\infty < w^\infty$, and $n$ a positive integer such that $n|w| < 2|v|$, then $w^n v^n$ is also admissible.*

PROOF. By Remark 2.4, $w^n v^n$ is periodic, and it is evident that it has positive $\mathbf{sgn}$. Now we only need to show that $\sigma^k((w^n v^n)^\infty) \leq (w^n v^n)^\infty$, for $1 \leq k \leq n|w| + n|v| - 1$.

(1) If $k = t|w|$ for $1 \leq t \leq n$, remove the common prefix $w^{n-t}$, then compare the $n|w| + n|v|$ prefixes by applying the second part of Lemma 3.1.
(2) If $k = n|w| + t|v|$ for $0 \leq t \leq n-1$, compare the $(n-t)|v| + n|w|$ prefixes by applying the second part of Lemma 3.1.
(3) If $k = t|w| + r$, where $0 \leq t \leq n-1$, $0 < r < |w|$, compare the $|w| + 1 - r$ prefixes, by making use of the assumption that $w$ is dominant.
(4) If $k = n|w| + t|v| + r$ where $0 \leq t < n-1$, $0 < r < |v|$.
    (a) If the $|v| - r$ prefixes are different,

$$\mathbf{Pre}_{|v|-r}((w^n v^n)^\infty) = \mathbf{Pre}_{|v|-r} w^\infty$$

$$\mathbf{Pre}_{|v|-r}(\sigma^k (w^n v^n)^\infty) = \mathbf{Suff}_{|v|-r} v \leq \mathbf{Pre}_{|v|-r} v \leq \mathbf{Pre}_{|v|-r} w^\infty$$

    The first $\leq$ due to $v$ being admissible, and the second $\leq$ due to $v^\infty < w^\infty$. Hence $\sigma^k(w^n v^n)^\infty < (w^n v^n)^\infty$.
    (b) If the $|v| - r$ prefixes are the same, by Lemma 3.3, this common $|v| - r$ prefix has negative cumulative sign. Deleting this common prefix, we reduce it to case (1) or case (3).

□

## 4. Tuning and renormalizability

DEFINITION 4.1. Let $W_n$ be the set of periodic words of length $n$. Then the $\leq$ defined in Section 2 gives a linear order on the finite set $W_n$. For any $w \in W_n$, let $\mathbf{Next}(w)$ be the element in $W_n$ which is right above $w$ under $\leq$, and $\mathbf{Prev}(w)$ the element that is right below $w$.

DEFINITION 4.2. We say a pair of elements $a, a'$ in $W_n$ form a *tuning pair*, if:

- $a$ is admissible, $a'$ is extremal and $\mathbf{sgn}(a') = -1$.
- $a' = \mathbf{Next}(a)$.

DEFINITION 4.3. We call the tuple $(\Gamma = (V, E), O)$ to be *tunable* if

(1) There is a periodic word $w_{\min}$, with only a single $N$, and $w_{\min} N$ is the minimal element in $W_{|w_{\min}|+1}$.
(2) If $a \in W_n$ is extremal and irreducible, and $\mathbf{sgn}(a) = -1$, then either $a = w_{\min}^k$ for some odd number $k$, or $a$ lies in a tuning pair.

EXAMPLE 4.4. If

$$\Gamma_2 = (\{0, 1\}, \{(0, 0), (0, 1), (1, 0), (1, 1)\})$$

$O_2$ sends $0$ to $1$ and $1$ to $-1$. This is the symbolic dynamics corresponding to unimodal interval maps [**2, 6, 12**].

Now we show that $(\Gamma_2, O_2)$ is tunable. Firstly it is easy to see that $w_{\min} = 1$ satisfies the first condition. To check the second condition, suppose $a \neq 1^n$ and $a$ is extremal, irreducible and $\mathbf{sgn}(a) = -1$. Let $b$ be $a$ with the last letter removed, then $\mathbf{sgn}(b) = -\mathbf{sgn}(a_{n-1})$, hence

$$a' = \begin{cases} b0 & a_{n-1} = 1 \\ b1 & a_{n-1} = 1 \end{cases}$$

equals $\mathbf{Prev}(a)$ and $\mathbf{sgn}(a') = 1$. Now we only need to show that $a'$ is also extremal. Suppose otherwise, there are $u$, $v$ such that $a' = uv$, and $vu > uv$, hence $vu \geq a$,

$$v \geq \mathbf{Pre}_{|v|}(a') = \mathbf{Pre}_{|v|}(a) \geq \mathbf{Suff}_{|v|}(a)$$

and $v \neq \mathbf{Suff}_{|v|}(a)$ as they differ at the last digit, hence $v > \mathbf{Suff}_{|v|}(a)$. this, together with the fact that $uv = a' < a = u\mathbf{Suff}_{|v|}(a)$, imply that $\mathbf{sgn}(u) = -1$, hence $\mathbf{sgn}(v) = -1$, and $\mathbf{sgn}(\mathbf{Suff}_{|v|}(a)) = -1$. Because $v$ and $\mathbf{Suff}_{|v|}(a)$ differ only at the last letter, there can be no word in between, hence $\mathbf{Pre}_{|v|}(a') = \mathbf{Pre}_{|v|}(a)$ must equals either $v$ or $\mathbf{Suff}_{|v|}(a)$. The latter case contradicts with Lemma 3.3. In the former case, deleting the common $|v|$-prefix of $vu$ and $a$ one gets $u \leq \mathbf{Suff}_{|u|}(a)$, which, due to the extremality of $a$, can only mean that $u = \mathbf{Suff}_{|u|}(a)$, i.e. $vu = a$. Yet $\mathbf{sgn}(a) = -1$, $\mathbf{sgn}(vu) = \mathbf{sgn}(uv) = \mathbf{sgn}(a) = 1$, a contradiction.

REMARK 4.5. The argument above means that to check a pair is tunable, one can first check part 1, and next show that for any $w \in W_n$, $\mathbf{Prev}(w)$ is $w$ with one letter changed, where the letter is either the last one or is followed by a suffix that is shared by all periodic words.

From now on we always assume that our $(\Gamma, O)$ is tunable.

DEFINITION 4.6. Let $(w, w')$ be a tuning pair, $v$ an admissible word in the setting of $(\Gamma_2, O_2)$ as in Example 4.4. Then by the *tuning* of $w$ by $v$ we mean the word

$$w * v = w_0 w_1 \dots w_{|v|-1}$$

such that

$$w_i = \begin{cases} w & v_i = 0 \\ w' & v_i = 1 \end{cases}$$

If an admissible word can not be written as a tuning we call it *nonrenormalizable*. Similarly one can define tuning of admissible infinite sequences, where $v$ is an admissible infinite sequence in the setting of $(\Gamma_2, O_2)$.

LEMMA 4.7. *If $w$ is extremal but not admissible (i.e. $w^\infty$ admissible and **sgn**$(w) = -1$), and $w \neq w_{min}^n$, then $w$ can be written as a tuning, and there are $w'$, $v$ such that $w = w' * v$, and $w'$ is nonrenormalizable.*

PROOF. $w$ can always be written as $w = w_1^k$, where $k$ is an odd integer, $w_1$ is irreducible, extremal and $\mathbf{sgn}(w_1) = -1$. Hence tunability of $(\Gamma, O)$ implies that $(\mathbf{Prev}(w_1), w_1)$ is a tuning pair, and $w = \mathbf{Prev}(w_1) * 1^k$.

Consider the non-empty set $\{(w', v) : w = w' * v\}$, pick $(w'_0, v_0)$ such that $w'_0$ is the shortest. Suppose $w'_0 = w'' * v''$, where $|v''| > 1$, then by the definition of the order and the fact that $\mathbf{Next}(v_0)$ is always $v_0$ with the last digit flipped (see Example 4.4), $\mathbf{Next}(w'_0) = w'' * \mathbf{Next}(v'')$, and $w = w'' * (v'' * v_0)$, a contradiction. ∎

REMARK 4.8. It is easy to see that $v < v'$ iff $w * v < w * v'$. Furthermore, if $w$ is admissible, $a$ is nonrenormalizable, and $w = a * v$, then the intervals in Example 2.6 that are of the form $[\sigma^i(w^\infty), \sigma^{i+k|a|}(w^\infty)]$ where $k \in \mathbb{Z}$ must be sent to one another, in other words the Markov decomposition incidence matrix can not be irreducible.

DEFINITION 4.9. We say that the tuple $(\Gamma = (V, E), O)$ has *unique maximal tuning*, if the $w'$ in Lemma 4.7 is always *minimal*, in the sense that if there is some $w''$, $v''$ such that $w = w'' * v''$, then the length of $w''$ is a multiple of the length of $w'$.

REMARK 4.10. One can show that tuples that satisfies Remark 4.5 also has unique maximal tuning.

From now on we always assume that our $(\Gamma, O)$ has unique maximal tuning.

## 5. Construction of dominant words

The main goal of this Section is to prove the following, which is a generalization of Proposition 10.5 of [**11**]:

THEOREM 5.1. *Suppose $w$ is an admissible, irreducible and non-renormalizable word, then either $w$ is dominant, or for any $n$, there is some word $b$ such that $w^n b$ is dominant.*

Following the notation in [**11**], for any admissible word $w$, let $PS(w)$ be the set of common prefixes and suffixes of $w$, i.e.

$$PS(w) = \{v : 0 < |v| < |w|, w = yv = vz, z_0 = N\}$$

And

$$RS(w) = \{z : w = vz, v \in PS(w)\}$$

It is evident that elements of $PS(w)$ and $RS(w)$ are all periodic, and an admissible word $w$ is dominant iff $PS(w) = \emptyset$.

Firstly we need the following four lemmas:

LEMMA 5.2. *Let $w$ be an admissible and nonrenormalizable word which is not dominant. Let $z_0$ be the element of $RS(w)$ that maximizes $z_0^\infty$. Let $v$ be the word such that $w = vz_0$, then $v$ is extremal.*

PROOF. Consider any decomposition of $v$ into $v = ab$, we only need to show $ba \leq ab$. By definition of $RS(w)$ and $PS(w)$, $v$ is a suffix of $w$ as well, hence $\sigma^{|w|-|b|}(w^\infty) = bw^\infty < w^\infty$. Comparing the $|v|$-prefix of both sides we see that $ba \leq ab$. □

LEMMA 5.3. *Let $a$ be a admissible word such that $(a, a')$ form a tuning pair, $\beta$ an admissible infinite sequence such that $a * 0^\infty \leq \beta \leq a * 10^\infty$, then there is an admissible infinite sequence $s$ in the setting of $(\Gamma_2, O_2)$ such that $\beta = a * s$.*

PROOF. We only need to show that each section of $\beta$ from the $k|a|+1$-th place to the $(k+1)|a|$ place, where $k$ is arbitrary natural number, is either $a$ or $a'$.

By admissibility of $\beta$, this section is no more than the $|a|$-prefix of $\beta$, which is no more than $a'$. Now the prior sections, by inductive hypothesis, are all either $a$ or $a'$. If before this section there is some $a'a^j$ ($j$ may be 0), do $\sigma^{(k-1-j)|a|}$ and compare the $(j+2)|a|$ prefix with $\beta$ (which is no more than $a * 10$ by assumption), we see that this section can not be less than $a$. If the previous sections are all $a$, by $\beta \geq a^\infty$ we see that this section is no less than $a$ as well. □

LEMMA 5.4. *Let $w$ be an admissible and nonrenormalizable word which is not dominant. Let $z_0$ and $v$ be defined as in Lemma 5.2, and $v = \alpha * t$ for some nonrenormalizable word $\alpha$. Then $\alpha^\infty > z_0^\infty$.*

PROOF. Let $\alpha' = \mathbf{Next}(\alpha)$. Because $10^\infty$ is the largest $(\Gamma_2, O_2)$-sequence, while any $(\Gamma_2, O_2)$-sequence is larger than $0^\infty$,

$$\alpha'\alpha^\infty = \alpha * 10^\infty > \alpha * t0^\infty = v\alpha^\infty$$

$$v^\infty = \alpha * t^\infty > \alpha * 0^\infty = \alpha^\infty$$

Now let $y$ be a word such that $w = yv$, then $y^\infty > v^\infty = (v^n)^\infty$ for any positive integer $n$, hence $yv^n > v^n y$ for any $n$. Let $n$ goes to infinity we have $yv^\infty = wv^\infty > v^\infty$ which implies that $w^\infty > v^\infty > \alpha^\infty$.

Now we proceed by looking at various cases:

- Case 1: $(z_0 v)^\infty \geq \alpha^\infty$. We have

$$w^\infty = v(z_0 v)^\infty \leq v\alpha^\infty < \alpha'\alpha^\infty$$

  which implies that $w = \alpha * t'$ for some $(\Gamma_2, O_2)$-word $t'$ by Lemma 5.3. Apply Definition 4.9 we see that this contradicts with the assumption that $w$ is non renormalizable.
- Case 2: $(z_0 v)^\infty < \alpha^\infty$. There are two subcases:
  - Case 2(a): The $|z_0|$-prefix of $\alpha^\infty$ is strictly larger than $z_0$: by looking at the $|z_0|$-prefix we have $\alpha^\infty > z_0^\infty$.
  - Case 2(b): The $|z_0|$-prefix of $\alpha^\infty$ equals $z_0$. This means that $z_0$ is of the form $z_0 = \alpha^k p$ where $p$ is a proper prefix of $\alpha$. By Lemma 3.3, $\mathbf{sgn}(z_0) = \mathbf{sgn}(v) = -1$, so $\mathbf{sgn}(p) = -1$. Now delete from $\alpha^\infty$ and $z_0^\infty$ the common prefix $z_0$, by admissibility and irreducibility of $\alpha$ we see that $\alpha^\infty > z_0^\infty$.

□

LEMMA 5.5. *Let $w$ be an admissible and nonrenormalizable word which is not dominant. Let $u$ be a dominant word, such that*

- *$w > u$.*
- *For any $z \in RS(w)$, $u^\infty > z^\infty$.*

*Then for sufficiently large $n$, $w^n u^n$ is dominant.*

PROOF. We need to show that for any $k \geq 1$, the $k+1$-prefix of $w^n u^n$ is strictly larger than the $k$ suffix of $w^n u^n$ concatenated with $N$.

- Case 1: $k \leq n|u|$: the $k$ suffix concatenated with $N$ either start with a proper suffix of $u$ followed by $N$, which due to the dominance of $u$, is strictly smaller than the corresponding prefix of $u$, which in turn is no larger than the corresponding prefix of $w$; or it may start with a $u$, which is smaller than $w$ by assumption.
- Case 2: $k \geq n|u| + |w|$: the $k$-suffix either starts with a shift of $w$, or is a (less than $n$) multiple of $w$ followed by $u^n$. In the former case, admissibility and irreducibility of $w$ implies that this suffix is strictly smaller than the corresponding prefix. In the latter case, the suffix is strictly smaller than the corresponding prefix because $u < w$.

- Case 3: $n|u| < k < n|u| + |w|$. In this case the $k$-suffix is of the form $du^n$ where $d$ is a proper suffix of $w$.
  - Case 3a: If $dN$ is not also a prefix of $w$, $dN$ must be strictly smaller than the corresponding prefix of $w$ due to admissibility of $w$, hence this $k$-suffix must be strictly smaller than the corresponding prefix.
  - Case 3b: $dN$ is also a prefix of $w$. Let $w = de$, then $e \in RS(w)$ by definition. Because $w^k = (de)^k$ is admissible, we have $(w^{k-1}d)^\infty > e^\infty$. Let $k$ be sufficiently large we have $w^\infty \geq e^\infty$. By Lemma 3.3, $sgn(e) = sgn(d) = -1$, hence $ew^\infty \leq e^\infty < u^\infty$, $dew^\infty = w^\infty > du^\infty$, so when $n$ is large enough, $du^nN$ is strictly smaller than the corresponding prefix in $w^\infty$.

□

Now we are ready to prove the main theorem:

PROOF OF THEOREM 5.1. Suppose the theorem is not true, let $w$ be the shortest word which is admissible, nonrenormalizable, non-dominant, and there exists $n \in \mathbb{N}$ such that $w^n r$ is not dominant for any $r$.

Let $z_0 = \arg\max_{z \in RS(w)} z^\infty$, and let $v$ be a word such that $w = vz_0$. By Lemma 3.3, $\mathbf{sgn}(v) = -1$, and by Lemma 5.2, $v$ is extremal. Hence by Lemma 4.7, either of the two cases below are true:

i) $v = \alpha * t$ for some nonrenormalizable word $\alpha$.
ii) $v = w_{\min}^k$, $k$ is odd, $\mathbf{sgn}(w_{\min}) = -1$.

In case i), by Lemma 5.4, $\alpha^\infty > z_0^\infty$. Because $|\alpha| < |w|$ and $w$ is the shortest word where the claim does hold hold, either $\alpha$ itself is dominant, or for any integer $n$ there is some word $r'$ such that $\alpha^n r'$ is dominant. We now have a dominant word $u$, which equals $\alpha$ if $\alpha$ is dominant itself, and $\alpha^n r'$ for $n$ sufficiently large if otherwise, such that $w > u$, $u^\infty > z^\infty$ for any $z \in RS(w)$. Here $u^\infty > z^\infty$ is because $u^\infty$ shares a long common prefix with $a^\infty$, which is greater than $z_0^\infty$. Hence by Lemma 5.5, when $M \gg 0$, $w^M u^M$ is dominant. So let $r = w^{M-n}u^M$, we get a contradiction.

In case ii), because $w^\infty \geq \sigma^{|w_{\min}|}(w^\infty)$, $w$ must equals $w_{\min}^k$, which contradicts with the assumption of $w$ being irreducible. □

## 6. Topological entropy and poles of the Artin-Mazur zeta function

Combining the results from the previous 3 sections, we get the following:

THEOREM 6.1. *Suppose $(\Gamma, O)$ is tunable and has unique maximal tuning. For any nonrenormalizable word $a$, any admissible word $b$, any positive integer $n$, there is a word $c$ such that $a^n cb^n$ is admissible.*

PROOF. This follows immediately from Theorem 3.4 and Theorem 5.1. □

Let $\Gamma$ be a fully connected and aperiodic directed graph, with vertices labeled $0, \ldots, N$, and let $w$ be an admissible word. Then by Remark 2.6, the subshift $(\Sigma_{w^\infty}, \sigma)$ admits a Markov decomposition, and the poles of the Artin-Mazur $\zeta$ function are reciprocals of the eigenvalues of the Markov decomposition incidence matrix. These eigenvalues can also be written as roots of a polynomial as below:

By Gaussian elimination, there are polynomials $p_0(\lambda), \ldots, p_N(\lambda)$, such that for each $0 \leq i \leq N-1$,

$$\lambda p_i(\lambda) = \sum_{(i,j)\in E(\Gamma)} p_j(\lambda)$$

and $\gcd(p_0, \ldots, p_N) = 1$. Now for any edge $(i,j)$ in $\Gamma$, define

$$F_{i,j}(x) = \begin{cases} \lambda x - \sum_{(i,k)\in E(\Gamma), k<j} p_k(\lambda) & O(i) = 1 \\ p_j(\lambda) - \lambda(p_i(\lambda) - x) + \sum_{(i,k)\in E(\Gamma), k>j} p_k(\lambda) & O(i) = -1 \end{cases}$$

And let

$$F_w(\lambda) = F_{w_{|w|-1},N} \circ \cdots \circ F_{w_0,w_1}(p_N(\lambda)) - p_N(\lambda)$$

With exactly the same argument as in Theorem 1.2 of [**5**], one can show that the roots of $F_w$ that are off unit circle are exactly the eigenvalues of the Markov decomposition matrix $M$ which are off the unit circle, and furthermore by construction the coefficients of $F_w$ are uniformly bounded with the bound only dependent on $(\Gamma, O)$, while the lowest non-zero degree is bounded as well. Hence, by Rouche's theorem, the eigenvalues outside of the unit circle are controlled by prefix of $w$ while the eigenvalues inside the unit circle are controlled by suffix of $w$. This observation, together with Theorem 6.1 and Remark 4.8, implies that:

THEOREM 6.2. *Let $(\Gamma, O)$ be tunable and has unique maximal tuning. Let $w$ be an admissible word such that the Markov decomposition matrix of $\Sigma_{w^\infty}$ defined in Remark* 2.6 *is irreducible, and $v$ an admissible word such that $v^\infty < w^\infty$, then, for any $\epsilon > 0$, any pole $z$ of the Artin-Mazur $\zeta$ function of $\Sigma_{v^\infty}$ outside the unit circle, there is some admissible word $w'$ such that the entropy of $\Sigma_{w'^\infty}$ is within $\epsilon$ to the entropy of $\Sigma_{w^\infty}$, and a pole of the Artin-Mazur $\zeta$ function of $\Sigma_{w'^\infty}$ is within $\epsilon$ distance to $z$.*

□

REMARK 6.3. Similar to [**6**], Theorem 6.2 also implies that when $w$ is a non-renormalizable admissible word, the suffix of all possible admissible words $w'$ such that $w'^\infty \leq w^\infty$, when reversed order and taken closure, is a subshift of finite type. Furthermore, the closure of the set of all Markov matrix eigenvalues for such $\Sigma_{w'^\infty}$ which are inside the unit circle, can be written as the limit set of graph directed system, hence can be studied via the technique of Solomyak [**9**].

## 7. Applications to dynamics and further questions

EXAMPLE 7.1. By Milnor-Thurston kneading theory [**7**], and Remark 4.4, Theorem 6.2 implies Theorem 1 in [**2**] for $\lambda > \sqrt{2}$. By Branner-Hubbard surgery [**1,5,8**] one can obtain the "persistence" result in [**5**] as well.

EXAMPLE 7.2. Let $\Gamma$ be a graph with vertices $\{0, 1, \ldots, n\}$, $O(k) = (-1)^k$, and there is a directed edge from any vertex to any vertex. This $(\Gamma, O)$ is tunable with $w_{\min} = n$, and $\mathbf{Prev}(w)$ is $w$ with the last letter $i$ replaced by either $i+1$ or $i-1$. This corresponds to continuous interval maps that have $n$ critical points and 2 critical values, where the smaller critical value is also a fixed point.

EXAMPLE 7.3. Let $\Gamma$ be a graph with vertices $\{0, 1, 2, 3\}$, and there are edges from $i$ to $j$ except for $i = 1$ or $2$ and $j = 3$, let $O(k) = (-1)^k$. Then we can check that $(\Gamma, O)$ is tunable:

(1) $w_{\min} = 3$.
(2) If $w$ is an extremal word that is not admissible, $\mathbf{Prev}(w)$ is $w$ with the last 0 replaced with 3 or vice versa. Now apply Remark 4.5.

EXAMPLE 7.4. This example is inspired by an Madison Experimental Math project I supervised, with Robert Argus, Beining Mu, Anvit Thekkatte, and Zihan Zhao. Consider the following family of continuous maps on a tree: Let $T$ be a metric graph which is also a tree with 3 branches, two of which denoted as $I_0$, $I_1$, the third divided into two segments $I_2$ and $I_3$ such that $I_2$ has the center of the tree as an end point. Consider the family of continuous maps that fixes the center, send $I_0$ bijectively to $I_2 \cup I_3$, $I_2$ to $I_1$, $I_1$ to $I_0$, and $I_3$ to a sub-segment of $I_0 \cup I_1$.

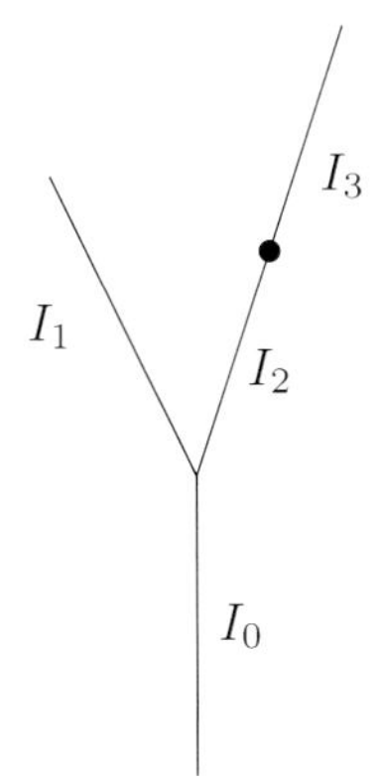


FIGURE 2. The tree in Example 7.4

Such map corresponds to the directed graph and sign function as follows:

$$\Gamma = \{\{0, 1, 2, 3\}, \{(0, 2), (0, 3), (1, 0), (2, 1), (3, 0), (3, 1)\}$$

$$O(0) = O(1) = O(3) = -1, O(2) = 1$$

We can also show that this $(\Gamma, O)$ is tunable: $w_{\min} = 3210$, and if $w$ is extremal but not admissible. Because $w^\infty \in \Sigma$, its suffix can only be 210 or 310 or 30. If an extremal word ends with 30 it has to be $(30)^n$ hence admissible. If it ends with 210 or 310, $\mathbf{Prev}(w)$ is $w$ with the other 3-suffix. Now apply Remark 4.5. Figure 3 is a plot of the "teapot" for this example, one can see that the "persistence" behavior is also illustrated in the plot.

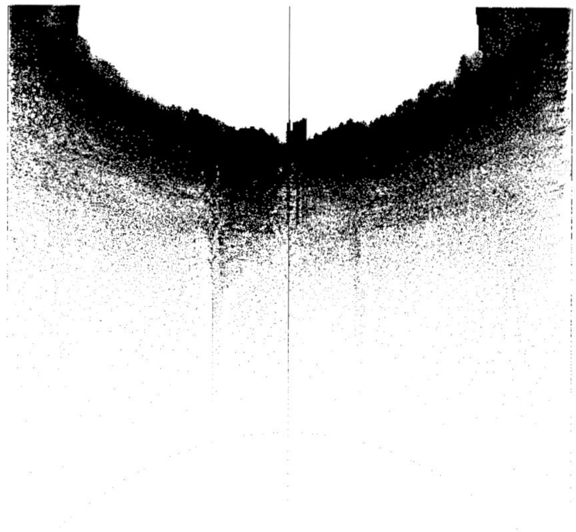

FIGURE 3. The teapot for Example 7.4, with the unit cylinder, plotting the half where the imaginary part is non negative.

A question that remains is the following:

QUESTION 7.5. Can the assumptions for tunable and unique maximal tuning be further relaxed? In particular, can Theorem 5.1 be strengthened to made sufficient to cover persistent results for all veins of the Mandelbrot set, thereby strengthening the main result in [**5**].

## References


[1] Bodil Branner and Adrien Douady, *Surgery on complex polynomials*, Holomorphic dynamics (Mexico, 1986), Lecture Notes in Math., vol. 1345, Springer, Berlin, 1988, pp. 11–72, DOI 10.1007/BFb0081395. MR980952

[2] Harrison Bray, Diana Davis, Kathryn Lindsey, and Chenxi Wu, *The shape of Thurston's master teapot*, Adv. Math. **377** (2021), Paper No. 107481, 32, DOI 10.1016/j.aim.2020.107481. MR4182912

[3] Danny Calegari, Sarah Koch, and Alden Walker, *Roots, Schottky semigroups, and a proof of Bandt's conjecture*, Ergodic Theory Dynam. Systems **37** (2017), no. 8, 2487–2555, DOI 10.1017/etds.2016.17. MR3719268

[4] Carlo Carminati and Giulio Tiozzo, *A canonical thickening of $\mathbb{Q}$ and the entropy of $\alpha$-continued fraction transformations*, Ergodic Theory Dynam. Systems **32** (2012), no. 4, 1249–1269, DOI 10.1017/S0143385711000447. MR2955313

[5] Kathryn Lindsey, Giulio Tiozzo, and Chenxi Wu, *Master teapots and entropy algorithms for the Mandelbrot set*, Trans. Amer. Math. Soc. **378** (2025), no. 5, 3297–3348, DOI 10.1090/tran/9346. MR4891377

[6] Kathryn Lindsey and Chenxi Wu, *A characterization of Thurston's master teapot*, Ergodic Theory Dynam. Systems **43** (2023), no. 10, 3354–3382, DOI 10.1017/etds.2022.73. MR4637157

[7] John Milnor and William Thurston, *On iterated maps of the interval*, Dynamical systems (College Park, MD, 1986), Lecture Notes in Math., vol. 1342, Springer, Berlin, 1988, pp. 465–563, DOI 10.1007/BFb0082847. MR970571

[8] Johannes Riedl, *Arcs in multibrot sets, locally connected Julia sets and their construction by quasiconformal surgery*, Ph.D. thesis, Techniche Universität München, Zentrum Mathematik, 2001.

[9] Boris Solomyak, *On the 'Mandelbrot set' for pairs of linear maps: asymptotic self-similarity*, Nonlinearity **18** (2005), no. 5, 1927–1943, DOI 10.1088/0951-7715/18/5/003. MR2164725

[10] William P. Thurston, *Entropy in dimension one*, Frontiers in complex dynamics, Princeton Math. Ser., vol. 51, Princeton Univ. Press, Princeton, NJ, 2014, pp. 339–384. MR3289916

[11] Giulio Tiozzo, *Topological entropy of quadratic polynomials and dimension of sections of the Mandelbrot set*, Adv. Math. **273** (2015), 651–715, DOI 10.1016/j.aim.2014.12.033. MR3311773

[12] Guilio Tiozzo, *Galois conjugates of entropies of real unimodal maps*, International Mathematics Research Notices **2020** (2020), no. 2, 607–640.

Department of Mathematics, University of Wisconsin–Madison, Van Vleck Hall, Madison, Wiconsin 53703

*Email address*: cwu367@wisc.edu

Contemporary Mathematics
Volume **839**, 2026
https://doi.org/10.1090/conm/839/16730

# Dimensions and metric dyadic cubes

Efstathios-K. Chrontsios-Garitsis

ABSTRACT. In this note, we provide equivalent definitions for fractal geometric dimensions through dyadic cube constructions. Given a metric space $X$ with finite Assouad dimension, i.e., satisfying the doubling property, we show that the construction of systems of dyadic cubes by Hytönen-Kairema is compatible with many dimensions. In particular, the Hausdorff, Minkowski, and Assouad dimensions can be equivalently expressed solely using dyadic cubes in the aforementioned system. The same is true for the Assouad spectrum, a collection of dimensions introduced by Fraser-Yu.

## 1. Introduction

For the past three decades, fractals have gained significant attention within both pure and applied mathematics. The study of dimension notions is a fundamental part of fractal geometry, which facilitates the understanding of fractal objects. Two notions that have been popular since the early 20th century are the Hausdorff and the Minkowski dimension, which have found applications within many fields of research. For instance, these dimensions often show up in the context of partial differential equations (PDEs) [**12**], number theory [**29**], signal processing [**28**] and mathematical physics [**4**]. We refer to the book of Falconer [**14**] for a thorough exposition on these dimensions. Another notion of dimension that has been enjoying a lot of interest for the past decade is the Assouad dimension. Initially introduced under a different name by Assouad in [**2**], in order to investigate embeddability properties of abstract metric spaces, the Assouad dimension has been recently tied to a lot of areas through its fractal-geometric properties. Moreover, this notion has even motivated the introduction of dimension spectra, such as the Assouad spectrum introduced by Fraser-Yu [**19**], which interpolate between known dimensions (see [**16**] for a modern exposition and applications).

All three notions of dimension have proved to be extremely useful, and are often studied simultaneously in situations where they differ. However, their original definitions can be quite complicated for the sake of calculations. The main idea behind the Hausdorff dimension $\dim_H E$ of a set $E \subset \mathbb{R}^n$ is that the dimension of the set is essentially the critical exponent $s \geq 0$ for which the $s$-dimensional Hausdorff measure of $E$, denoted by $H^s(E)$, changes from being infinite to being 0. This is motivated by the property of the Lebesgue measure that, for instance, assigns infinite length and zero volume to a 2-dimensional object. On the other

2020 *Mathematics Subject Classification.* Primary 28A80; Secondary 30L99, 31E05.

hand, the Minkowski dimension $\dim_B E$ of a bounded set $E \subset \mathbb{R}^n$ is intuitively the exponent $s > 0$ for which $N(E, r)$, i.e. the minimum number of sets of diameter at most $r$ needed to cover $E$, is approximately $r^{-s}$, for all small $r > 0$. The Assouad dimension $\dim_A E$ of $E$ is defined by utilizing the idea of the Minkowski dimension after "zooming in" (or "zooming out") at all points of $E$. For $\theta \in (0,1)$, the $\theta$-Assouad spectrum $\dim_A^\theta E$ of $E$ is defined similarly to the Assouad dimension, with a restriction on the allowed zoom scales that depends on $\theta$. We refer to Section 2 for rigorous definitions.

In the Euclidean setting, one feature of $\mathbb{R}^n$ that has facilitated the study of dimensions, especially within applied areas, has been the simplification of the aforementioned definitions using dyadic cubes. In particular, the Hausdorff dimension of $E$ is equal to the critical exponent $s \geq 0$ for which the $s$-dimensional cubic measure $M^s(E)$ changes from being infinite to being equal to 0. The cubic measure $M^s(E)$ is defined using dyadic cubes, unlike the Hausdorff measure which uses arbitrary sets. Similarly, for the definition of the Minkowski and Assouad dimension, if $E$ is contained in a ball $B(x, R)$ and its circumscribed cube $Q(x, R)$, one can replace the number $N(E, r)$ by the smallest number of dyadic sub-cubes of $Q(x, R)$ of level $m$ needed to cover $E$, denoted by $D(E, m)$, and ask that $D(E, m) \simeq 2^{ms}$, for all large $m \in \mathbb{N}$. Such simplifications have been an invaluable tool in many applications of fractal geometry. In particular, in the area of dimension distortion under mapping classes, the simplified definitions have been used extensively (for instance [**20**, **22**, **26**]), including in previous work of the author in [**7**, **8**], and jointly by Tyson and the author in [**10**, **11**].

Given the purely metric nature of the Hausdorff dimension, Minkowski dimension, Assouad dimension, and Assouad spectrum, it would be helpful to have similar reductions of the original definitions in the setting of more abstract spaces. In fact, the origin and popularity of the Assouad dimension within the analysis on metric spaces community [**2**, **21**], and the recent embeddability theorem of Troscheit and the author [**9**] emphasize this need for simplified definitions in higher generality. In arbitrary metric spaces, however, there are various generalizations of dyadic cube constructions. One of the first manuscripts addressing this idea was by David [**13**], while one of the first explicit constructions of a system of dyadic cubes is due to Christ [**5**]. See also [**1**], [**25**], [**30**], which is not an exhaustive list.

The main result of this note is to express the aforementioned dimensions equivalently in the metric spaces setting using the dyadic cube systems constructed by Hytönen-Kairema in [**24**]. Specifically, we show that the Hausdorff measure $H^s(E)$ and minimal covering number $N(E, r)$ can be replaced by the cubic measure $M^s(E)$ and dyadic cube covering number $D(E, m)$, respectively, as defined through the Hytönen-Kairema dyadic cube systems (see Section 2 for definitions).

THEOREM 1.1. Let $(X, d)$ be a doubling metric space. There is a finite collection of $\delta$-dyadic cube systems $\mathcal{S} = \{\mathcal{D}_1, \ldots, \mathcal{D}_K\}$ of $X$, such that for every non-empty $E \subset X$, the following hold:

(i) For any $t \in \{1, \ldots, K\}$, we have

$$\dim_H E = \inf\{s \geq 0 : M^s(E) = 0\} = \sup\{s \geq 0 : M^s(E) = \infty\},$$

where $M^s(E)$ is the cubic measure of $E$ with respect to $\mathcal{D}_t$.

(ii) If there are $x \in X$, $R > 0$ with $E \subset B(x,R)$, and $m_E \in \mathbb{N}$ with $\delta^{m_E} \le |E|$, then

$$\dim_B E = \inf\{\alpha \ge 0 : \exists C > 0 \text{ s.t. } D(E,m) \le C\delta^{-m\alpha} \text{ for all } m \ge m_E\},$$

where $D(E,m)$ is with respect to the circumscribed cube $Q(x,R)$ of the ball $B(x,R)$, which lies in $\mathcal{D}_t$ for some $t \in \{1,\dots,K\}$.

(iii) There is a constant $\widetilde{C} > 0$, independent of $E$, such that for any $\theta \in (0,1)$, the Assouad spectrum $\dim_A^\theta E$ is equal to the infimum of all $\alpha > 0$ for which there is $C > 0$ with

$$D(E \cap B(x,R), m) \le C\delta^{-m\alpha},$$

for all $x \in E$, $R > 0, m \in \mathbb{N}$ with $0 < \widetilde{C}\delta^m R \le R^{1/\theta} < R < 1$, and $D(E \cap B(x,R), m)$ with respect to the circumscribed cube $Q(x,R)$.

(iv) There is a constant $\widetilde{C} > 0$, independent of $E$, such that the Assouad dimension $\dim_A E$ is equal to the infimum of all $\alpha > 0$ for which there is $C > 0$ with

$$D(E \cap B(x,R), m) \le C\delta^{-m\alpha},$$

for all $x \in E$, $R > 0$, $m \in \mathbb{N}$ with $\widetilde{C}\delta^m R \le R$, and $D(E \cap B(x,R), m)$ with respect to the circumscribed cube $Q(x,R)$.

This paper is organized as follows. Section 2 reviews the background on the relevant dimensions and introduces the quantities $M^s(E)$, $D(E,m)$ for $E \subset X$, using the dyadic cube systems of Hytönen-Kairema for the metric space $X$. In Section 3 we prove our main result, Theorem 1.1, by employing properties of the dyadic cubes of Hytönen-Kairema that resemble those of the Eucledian systems. Section 4 contains remarks and future directions motivated by this work.

## 2. Background

**2.1. Dimension notions.** Let $(X,d)$ be a metric space. We use the Polish notation $|x-y| := d(x,y)$ for all $x,y \in X$ and denote the open ball centered at $x$ of radius $r > 0$ by $B(x,r) := \{z \in X : \ |x-z| < r\}$. Given a ball $B = B(x,r) \subset X$, we denote by $\lambda B$ the ball $B(x,\lambda r)$, for $\lambda > 0$. Given a subset $U$ of $X$, we denote by $|U|$ the diameter of $U$. While for every $x \in X$, $r > 0$, we have $|B(x,r)| \le 2r$, the radius of a ball in a metric space is not uniquely determined. However, for the rest of this manuscript we make the convention that for every ball $B$ that contains at least two distinct points, all radii of $B$ that we consider do not exceed $2|B|$. Thus, the notation $B(x,r)$ always implies that

$$2^{-1}|B(x,r)| \le r \le 2|B(x,r)|. \tag{2.1}$$

We recall the dimension notions that we are focusing on in this note (see [**14**,**16**] for more details). For $s > 0$, $r > 0$, and a subset $E$ of $X$, the *s-dimensional r-approximate Hausdorff measure* of $E$ is defined as

$$H^s_r(E) = \inf\left\{\sum_i |U_i|^s : \{U_i\} \text{ is a countable cover of } E \text{ with } |U_i| \le r\right\}.$$

The *s-dimensional Hausdorff (outer) measure* of $E$ is the limit

$$H^s(E) = \lim_{r\to 0} H_r^s(E),$$

and the Hausdorff dimension of $E$ is defined as

$$\dim_H E = \inf\{s \geq 0 : H^s(E) = 0\} = \sup\{s \geq 0 : H^s(E) = \infty\}.$$

While the Hausdorff dimension stems from the idea of a dimensional threshold between null and infinite measure for the given set, other dimension notions rely on the behavior of the cardinality of coverings by small sets. Let $E$ be a bounded subset of $X$. For $r > 0$, denote by $N(E, r)$ the smallest number of sets of diameter at most $r$ needed to cover $E$. The *(upper) Minkowski dimension* of $E$ is defined as

$$\overline{\dim_B} E = \limsup_{r\to 0} \frac{\log N(E,r)}{\log(1/r)}.$$

This notion is also known as *upper box-counting dimension*, which justifies the notation with the subscript 'B' typically used in the literature (see [**14**], [**16**]). We drop the adjective 'upper' and the bar notation throughout this paper as we will make no reference to the lower Minkowski dimension. For any fixed $r_0 \leq |E|$, an equivalent formulation is

$$\dim_B E = \inf\{\alpha > 0 \,:\, \exists C > 0 \text{ s.t. } N(E,r) \leq Cr^{-\alpha} \text{ for all } 0 < r \leq r_0\}.$$

This formulation reveals the relation of the Minkowski dimension to other notions.

For an arbitrary (not necessarily bounded) set $E \subset X$, the *Assouad dimension* of $E$ is

$$\dim_A E = \inf\left\{\alpha > 0 : \begin{array}{c} \exists C > 0 \text{ s.t. } N(E \cap B(x,R), r) \leq C(R/r)^{\alpha} \\ \text{for all } 0 < r \leq R \text{ and all } x \in E \end{array}\right\}.$$

The idea behind the Assouad dimension is that we are allowed to zoom-in and zoom-out at all points of the set, trying to find the largest dimension possible. This notion first appeared (under a different name) in a 1983 paper of Assouad on metric embedding problems [**2**]. However, it has recently gained a lot of popularity in the fractal geometry and dynamics community, due to its various applications (see [**16**] for a modern exposition).

It is easy to see that for a fixed bounded set $E \subset X$, the relation between the aforementioned dimensions is

$$\dim_H E \leq \dim_B E \leq \dim_A E,$$

with the inequalities being strict in various cases (see for instance [**6**]). The potential "gap" between the Minkowski and the Assouad dimension received increased attention by Fraser and Yu [**19**], who defined a collection of dimension notions that reside in that gap. For $0 < \theta < 1$ and a set $E \subset X$, define

$$\dim_A^\theta E = \inf\left\{\alpha > 0 : \begin{array}{c} \exists C > 0 \text{ s.t. } N(B(x,R) \cap E, r) \leq C(R/r)^{\alpha} \\ \text{for all } 0 < r \leq R^{1/\theta} < R < 1 \text{ and all } x \in E \end{array}\right\}.$$

Thus, $\dim_A^\theta E$ is defined by the same process as $\dim_A E$, but with the restriction that the two scales $r$ and $R$ involved in the definition of the latter are related by the inequality $r^\theta \le R$. The function $\theta \mapsto \dim_A^\theta E$ is called the *(regularized)*[1] *Assouad spectrum* of $E$. We often abuse the terminology and refer to a value $\dim_A^\theta E$ for some $\theta \in (0,1)$ as the Assouad spectrum of $E$.

We emphasize that the Assouad spectrum naturally (and continuously [**16**]) interpolates between the Minkowski and Assouad dimensions. For a fixed $E$, the limit $\lim_{\theta\to 0^+} \dim_A^\theta E$ exists and equals $\dim_B E$. Moreover, $\lim_{\theta\to 1^-} \dim_A^\theta E$ coincides with the so-called *quasi-Assouad dimension* of $E$, denoted by $\dim_{qA} E$. While there are instances where the quasi-Assouad dimension differs from the Assouad dimension, in many natural situations they coincide (see [**17, 18, 31**]). Furthermore, if $E$ is bounded, for all $\theta \in (0,1)$ we have

$$\dim_H E \le \dim_B E \le \dim_A^\theta E \le \dim_{qA} E \le \dim_A E,$$

with the first two inequalities being strict in various cases (see for instance [**16**, Theorem 3.4.7] and [**6**]).

**2.2. Metric dyadic cubes.** In the context of the typical Euclidean metric spaces $\mathbb{R}^n$, it is quite elementary to equivalently express the dimensions $\dim_H E$ and $\dim_B E$ using dyadic cubes instead of arbitrary sets, by modifying the quantities $H_r^s(E)$ and $N(E,r)$. This idea was also extended for the Assouad dimension and the Assouad spectrum (see [**11**, Proposition 2.5] for a proof). In this subsection we present and define the appropriate dyadic cube notion in the context of metric spaces, in order to prove similar expressions for the aforementioned dimensions.

We say that $(X,d)$ is a *doubling metric space* if there is a *doubling constant* $C_d \ge 1$ such that for all $x \in X$ and $r > 0$, the smallest number of balls of radius $r$ needed to cover $B(x,2r)$ is at most $C_d$. The property of a metric space $X$ being doubling is in fact equivalent to $\dim_A X < \infty$ (see for instance [**21**]). Note that the doubling property implies that $X$ is separable.

As noted in the Introduction, the dyadic cube systems that we employ are those constructed by Hytönen and Kairema.

THEOREM A (Hytönen, Kairema [**24**]). *Suppose $(X,d)$ is a doubling metric space. Let $0 < c_0 \le C_0 < \infty$ and $\delta \in (0,1)$ with $12C_0\delta \le c_0$. For any $k \in \mathbb{N}$ and collection of points $\{z_i^k\}_{i\in I_k}$ with*

$$|z_i^k - z_j^k| \ge c_0\delta^k, \quad \text{for } i \ne j, \tag{2.2}$$

*and*

$$\min_i |z_i^k - x| < C_0\delta^k, \quad \text{for all } x \in X, \tag{2.3}$$

*where $I_k$ is an index set, we can construct a collection of sets $\{Q_i^k\}_{i\in I_k}$ such that*

(i) *if $l \ge k$ then for any $i \in I_k$, $j \in I_l$ either $Q_j^l \subset Q_i^k$ or $Q_j^l \cap Q_i^k = \emptyset$,*
(ii) *$X$ is equal to the disjoint union $\bigcup_{i\in I_k} Q_i^k$, for every $k \in \mathbb{N}$,*
(iii) *$B(z_i^k, c_0\delta^k/3) \subset Q_i^k \subset B(z_i^k, 2C_0\delta^k) =: B(Q_i^k)$ for every $k \in \mathbb{N}$,*
(iv) *if $l \ge k$ and $Q_j^l \subset Q_i^k$, then $B(Q_j^l) \subset B(Q_i^k)$.*

---

[1] This definition can also be found as *upper* Assouad spectrum in the literature, while the original spectrum in [**19**] was defined with $r = R^{1/\theta}$. Due to the relation $r \le R^{1/\theta}$ ensuring that the spectrum is monotone in $\theta$ for a fixed $E$, thus resulting in a more regular function of $\theta$, the term "regularized" was suggested by Tyson and the author. See also the discussion in [**11**].

For $k \in \mathbb{N}$, we call the sets $Q_i^k$ from the construction of Theorem A ($\delta$-)*dyadic cubes* of level $k$ of $X$, and the collection $\{Q_i^k : k \in \mathbb{N}, i \in I_k\}$ a *dyadic cube system*.

Fix a doubling metric space $(X, d)$ with doubling constant $C_d$, and constants $\delta < 1/100$, $c_0$ and $C_0$, as in Theorem A, for the rest of the paper. Moreover, for every $k \in \mathbb{N}$ we fix a collection of points $\{z_i^k\}_{i \in I_k}$ satisfying (2.2), (2.3). To see why such a collection of points exists, consider the covering $\{B(z, c_0\delta^k) : z \in X\}$ of $X$ and apply the $5B$-covering lemma (see [**21**]). By separability of $X$ and by choosing $c_0$ and $C_0$ so that $5c_0\delta^k < C_0\delta^k$, the existence of centers $\{z_i^k\}_{i \in I_k}$ is ensured.

By [**24**, Theorem 4.1], we can fix finitely many dyadic cube systems of $X$ satisfying Theorem A, say $\mathcal{S} = \{\mathcal{D}_1, \dots, \mathcal{D}_K\}$ with $K = K(C_d, \delta) \in \mathbb{N}$, so that for every ball $B(x, R)$, $x \in X$, $R > 0$, which contains at least two points, there are a constant $C_\delta > 0$ and a $\delta$-dyadic cube $Q(x, R) \in \mathcal{D}_t$, for some $t = t_{x,R} \in \{1, \dots, K\}$, so that $B(x, R) \subset Q(x, R)$ and

$$C_\delta^{-1} R \leq |Q(x, R)| \leq C_\delta R. \tag{2.4}$$

We call $Q(x, R)$ the *circumscribed cube* of $B(x, R)$. This is the particular property of the dyadic systems from [**24**] that ensures their compatibility with Assouad-like dimensions.

Henceforth, if two quantities $A, B > 0$ are related by $A \leq CB$, for some uniform constant $C > 0$ that depends only on intrinsic constants of $X$ and the systems in $\mathcal{S}$, such as $C_d, \delta, c_0, C_0$, we write $A \lesssim B$ and call $C$ the *comparability constant* of the relation. Similarly, if there is a uniform constant $C > 0$ such that $A \geq CB$, we write $A \gtrsim B$. Lastly, if there are uniform constants $C_1, C_2 > 0$ such that $C_1 B \leq A \leq C_2 B$, we write $A \simeq B$. Note that the relation $A \simeq B$ is equivalent to $C^{-1}B \leq A \leq CB$, for some uniform $C > 0$, which we also call the comparability constant of the relation.

We have introduced all the necessary notions to rigorously define the cubic measure and dyadic cube covering number in the setting of metric spaces, which are stated in Theorem 1.1. We further assume without loss of generality that all balls considered in what follows contain at least two points, even if not explicitly stated, since all covering number conditions in the definitions of the Assouad dimension and spectrum are trivial otherwise. Given $x \in X$, $R > 0$, $m \in \mathbb{N}$, denote by $W_m(Q(x, R)) \subset \mathcal{D}_t$ the dyadic cubes of level $L_R + m$ that are contained in $Q(x, R) \in \mathcal{D}_t$, where $L_R$ is the level of the circumscribed cube $Q(x, R)$ of the ball $B(x, R)$. Note that by the convention $|B(x, R)| \geq R/2$, Theorem A (iii) and (2.4), the level $L_R$ of $Q(x, R)$ does not depend on the point $x$. Given a set $E \subset X$, we denote by $D(E \cap B(x, R), m)$ the smallest number of cubes in $W_m(Q(x, R))$ needed to cover $E \cap B(x, R)$. If $E \subset B(x, R)$, we simplify the notation and write $D(E, m)$.

Given $\mathcal{D} \in \mathcal{S}$, $s > 0$, $r > 0$ and $E \subset X$, the *$s$-dimensional $r$-approximate cubic measure* of $E$ with respect to $\mathcal{D}$ is defined as

$$M_r^s(E) = \inf \left\{ \sum_i |Q_i|^s : \begin{array}{l} \{Q_i\} \subset \mathcal{D} \text{ is a cover of } E \text{ by dyadic cubes} \\ \text{of level } m \text{ with } 4C_0\delta^m \leq r \end{array} \right\}. \tag{2.5}$$

The *$s$-dimensional cubic measure* of $E$ with respect to $\mathcal{D}$ is then defined as

$$M^s(E) = \lim_{r \to 0} M_r^s(E).$$

The fact that $M_r^s(E)$ is decreasing in $r$ ensures that the above limit exists.

## 3. Proof of equivalent expressions

**3.1. Hausdorff dimension and cubic measure.** In this subsection we show that the threshold dimension of the cubic measure $M^s(E)$ is indeed equal to the Hausdorff dimension $\dim_H E$.

PROOF OF THEOREM 1.1 (I). Let $s > 0$, $r \in (0,1)$, and $E \subset X$. Fix a dyadic system $\mathcal{D} \in \mathcal{S}$ of $X$, and consider the cubic measure $M_r^s(E)$ with respect to $\mathcal{D}$. We first show that $M_r^s(E) \simeq H_r^s(E)$, with comparability constant independent of $r$.

One direction is trivial, namely, $H_r^s(E) \leq M_r^s(E)$, due to the cubes used in the definition of $M_r^s(E)$ having diameter at most $r$, by Theorem A (iii). For the other direction, let $\{U_i\}$ be a cover of $E$ with $|U_i| \leq r$. For every $i$, fix some $x_i \in U_i$ and set $m_i$ to be the unique integer such that

$$4C_0\delta^{m_i+1} < |U_i| \leq 4C_0\delta^{m_i}. \tag{3.1}$$

Note that we assume that $|U_i| > 0$, otherwise it would not contribute to the sum. Due to (2.2) and the one-to-one correspondence of cubes of level $m_i$ and centers $z_\ell^{m_i}$, along with $\dim_A X < \infty$ imply that there can be at most $C' > 0$ cubes of level $m_i$ intersecting the ball $B(x_i, 4C_0\delta^{m_i})$, for some constant $C'$ that depends only on $\delta$ and the doubling constant $C_d$ of $X$. By (2.3), these cubes in fact cover the ball $B(x_i, 4C_0\delta^{m_i})$. An additional application of the doubling condition of $X$ yields that every dyadic cube of level $m_i$ intersecting $B(x_i, 4C_0\delta^{m_i})$ can be decomposed into at most $C'' > 0$ cubes of level $m_i + 1$, with $C''$ depending only on $\delta$ and $C_d$. Denote this collection of cubes of level $m_i + 1$ by $\{Q_j^{m_i+1}\}_j$. Note that by the right-hand side of (3.1) we have $U_i \subset B(x_i, 4C_0\delta^{m_i})$, for every $i$. Thus, since $\{U_i\}_i$ is a cover of $E$, we have

$$E \subset \bigcup_{i,j} Q_j^{m_i+1},$$

with $|Q_j^{m_i+1}| \leq 4C_0\delta^{m_i+1} \leq r$ by Theorem A (iii), (3.1), and choice of $\{U_i\}$. In addition, by (3.1) we have

$$\sum_{i,j} |Q_j^{m_i+1}|^s \leq \sum_{i,j} (4C_0\delta^{m_i+1})^s \leq C'' \sum_i (4C_0\delta^{m_i+1})^s \lesssim \sum_i |U_i|^s,$$

where the comparability constant does not depend on $r$. Since the collection of cubes $\{Q_j^{m_i+1}\}_{i,j}$ satisfies the conditions for a cube cover of $E$ in (2.5), the above inequality implies

$$M_r^s(E) \lesssim \sum_i |U_i|^s.$$

But the cover $\{U_i\}$ is arbitrary, which implies by the above that

$$M_r^s(E) \lesssim H_r^s(E)$$

as needed. As a result, it is shown that

$$M_r^s(E) \simeq H_r^s(E),$$

with the comparability constant independent of $r$, which allows for $r \to 0$, proving that $M^s(E) \simeq H^s(E)$. By the definition of $\dim_H E$, this is enough to complete the proof. □

**3.2. Dyadic covering numbers and dimensions.** Recall that for a fixed, but arbitrary dyadic cube system of $X$, the author proved in [**8**] that using dyadic cubes instead of arbitrary sets indeed yields the Minkowski dimension.

PROOF OF THEOREM 1.1 (II). Suppose $E \subset X$ is bounded, with $E \subset B(x, R)$, for some $x \in X, R > 0$. Let $\mathcal{D}_t \in \mathcal{S}$ be the dyadic cube system in which the circumscribed cube $Q(x, R)$ lies, and fix $m_E \in \mathbb{N}$ so that $\delta^{m_E} \leq |E|$. Thus, for any integer $m \geq m_E$, the covering number $D(E, m)$ is simply the minimum number of cubes of level $L_R + m$ in $\mathcal{D}_t$ needed to cover $E$. The desired relation for $\dim_B E$ follows by [**8**, Proposition 2.1] (where $D(E, m)$ essentially corresponds to the covering number $N_{L_R+m}(E)$ in the notation of Proposition 2.1, for the fixed given dyadic cube system $\mathcal{D}_t$). □

REMARKS 3.1.

(i) After a closer analysis of the above proof, it is evident that the statement of Theorem 1.1 (ii) can be slightly strengthened by including potentially more levels of cubes in the expression of $\dim_B E$. In particular, the number $m_E$ can be replaced by the minimal integer so that $\delta^{L_R+m_E} \leq |E|$, and the statement still holds.

(ii) It should be noted that in the proofs of Theorem 1.1 (i) and (ii), the fact that the dyadic system $\mathcal{D}$ is one of those lying in $\mathcal{S}$ is not used. Hence, the Hausdorff and Minkowski dimensions can in fact be expressed using *any* $\delta$-dyadic system with the properties listed in Theorem A.

We proceed with the proofs regarding the Assouad spectrum and the Assouad dimension.

PROOF OF THEOREM 1.1 (III). We first pick the desired uniform constant $\widetilde{C} > 0$ to be

$$\widetilde{C} = \frac{12 C_0 C_\delta}{c_0}, \tag{3.2}$$

where $c_0, C_0$ are the constants as in Theorem A and $C_\delta$ is the uniform constant in (2.4). Let $\theta \in (0, 1)$, and set

$$A_\theta = \left\{ \alpha > 0 : \begin{array}{c} \exists C > 0 \text{ s.t. } N(B(x,R) \cap E, r) \leq C(R/r)^\alpha \\ \text{for all } 0 < r \leq R^{1/\theta} < R < 1 \text{ and all } x \in E \end{array} \right\},$$

so that $\dim_A^\theta E = \inf A_\theta$ by definition, and

$$B_\theta = \left\{ \alpha > 0 : \begin{array}{c} \exists C > 0 \text{ s.t. } D(E \cap B(x,R), m) \leq C\delta^{-m\alpha} \\ \text{for all } x \in E, m \in \mathbb{N}, R > 0 \text{ with } 0 < \widetilde{C}\delta^m R \leq R^{1/\theta} < R < 1 \end{array} \right\},$$

for which we need to show that $\dim_A^\theta E = \inf B_\theta$. We do so by proving that in fact $A_\theta = B_\theta$. We also emphasize that even if we temporarily consider balls $B = B(x, R)$ in the definitions of the sets $A_\theta, B_\theta$ against the convention (2.1), i.e. with $R > 2|B(x, R)|$, then due to $N(B(x, R) \cap E, r) = N(B(x, 2|B|) \cap E, r)$ the proof still reduces to considering only balls according to the convention.

Let $x \in E$, $R > 0$, $m \in \mathbb{N}$ and a cube $Q_i \in W_m(Q(x,R)) \in \mathcal{D}$, for some $\mathcal{D} \in \mathcal{S}$ where the circumscribed cube $Q(x,R)$ is contained. By definition of $W_m(Q(x,R))$, the cube $Q_i$ is of level $L_R + m$, where $L_R$ is the level of $Q(x,R)$. Since $|Q(x,R)| \geq 3^{-1}c_0\delta^{L_R}$ by Theorem A (iii) and (2.1), we have by (2.4) and another application of Theorem A (iii) on $Q_i$ that

$$|Q_i| \leq 4C_0\delta^{L_R+m} = 4C_0\delta^m\delta^{L_R} \leq 4C_0\delta^m 3c_0^{-1}|Q(x,R)| \leq \widetilde{C}\delta^m R. \tag{3.3}$$

Hence, the cubes used in the definition of $B_\theta$, i.e., cubes in $W_m(Q(x,R))$ for $m \in \mathbb{N}$ such that $\widetilde{C}\delta^m R \leq R^{1/\theta}$, are the appropriate cubes to use in coverings of $E$ for the Assouad spectrum $\dim_A^\theta E$, since the diameter of these cubes are at most $R^{1/\theta}$ by (3.3). This trivially leads to the relation

$$N(B(x,R) \cap E, \widetilde{C}\delta^m R) \leq D(E \cap B(x,R), m), \tag{3.4}$$

for all $m \in \mathbb{N}$ such that $\widetilde{C}\delta^m R \leq R^{1/\theta}$.

Let $\alpha \in B_\theta$, and $0 < r \leq R^{1/\theta}$. Set $m_r \in \mathbb{N}$ to be the unique integer so that

$$\widetilde{C}\delta^{m_r} R \leq r < \widetilde{C}\delta^{m_r - 1} R.$$

Since the cubes in $W_{m_r}(Q(x,R))$ are of diameter at most $\widetilde{C}\delta^{m_r} R \leq r \leq R^{1/\theta}$, we have by (3.4) that

$$N(B(x,R) \cap E, r) \leq N(B(x,R) \cap E, \widetilde{C}\delta^{m_r} R) \leq D(E \cap B(x,R), m_r).$$

In addition, by $\alpha \in B_\theta$ and by choice of $m_r$, the above implies

$$N(B(x,R) \cap E, r) \leq C\delta^{-m_r\alpha} \leq (\widetilde{C}\delta^{-1})^\alpha \left(\frac{R}{r}\right)^\alpha.$$

Since $x \in E$ and the scales $r, R$ are arbitrary, we have $\alpha \in A_\theta$, which proves the inclusion $B_\theta \subset A_\theta$.

For the other inclusion we need to show a comparability relation similar to (3.4) with opposite direction. Let $\{U_i\}_i$ be a cover of $E$ with $|U_i| \leq \widetilde{C}\delta^m R$. Recall by (2.4) that $R \simeq \delta^{L_R}$, which implies $|U_i| \lesssim \delta^{L_R+m}$. By a similar argument to that in the proof of Theorem 1.1 (i), it can be shown that there are at most $M_0$ cubes of level $L_R + m$ needed to cover $U_i$, for some uniform constant $M_0$ that does not depend on $i$, $m$, $R$. This establishes the relation

$$D(E \cap B(x,R), m) \leq M_0 N(B(x,R) \cap E, \widetilde{C}\delta^m R), \tag{3.5}$$

for all $m \in \mathbb{N}$ such that $\widetilde{C}\delta^m R \leq R^{1/\theta}$.

Let $\alpha \in A_\theta$, and $m \in \mathbb{N}$ with $\widetilde{C}\delta^m R \leq R^{1/\theta}$. For $r = \widetilde{C}\delta^m R \leq R^{1/\theta}$, by (3.5) and definition of $A_\theta$ we have

$$D(E \cap B(x,R), m) \leq M_0 N(B(x,R) \cap E, r) \leq M_0 C\left(\frac{R}{\widetilde{C}\delta^m R}\right)^\alpha = M_0 C\widetilde{C}^{-\alpha}\delta^{-m\alpha},$$

which implies that $\alpha \in B_\theta$. Therefore, $A_\theta = B_\theta$, and the proof is complete.

(iv) While the statement does not follow by (iii), due to the limit of $\dim_A^\theta E$ as $\theta \to 1^-$ being equal to the quasi-Assouad dimension $\dim_{qA} E \leq \dim_A E$, the proof is nonetheless identical to that of (iii), by replacing $\theta$ with the number 1 in all arguments. □

## 4. Final remarks

The construction of dyadic cube systems of Hytönen-Kairema in [**24**] is actually given for quasimetric spaces. As a result, Theorem 1.1 is also true if $X$ is a quasimetric doubling space. The proofs are almost identical, with the only difference being the dependence of a few of the uniform constants on the quasimetric constant of the space. Another generalization of Theorem 1.1 would be to state it for all $\delta \in (0,1)$ small enough that satisfy the relation $12C_0\delta \leq c_0$ in Theorem A, which follows with identical arguments to those in Section 3.

While the cube systems defined in Theorem A have been used extensively in various areas [**3**, **23**, **27**], there are many different constructions of dyadic cubes in the metric setting. The main advantage of the Hytönen-Kairema systems is the existence of a circumscribed cube for every ball in $X$. This is a fundamental property that is necessary for the arguments in the proof of Theorem 1.1 (iii), (iv). This is evident already by the representation of the Assouad dimension and spectrum using dyadic cubes in $\mathbb{R}^n$, as shown in [**11**, Proposition 2.5]. It would be interesting to establish similar expressions for the dimensions in Theorem 1.1 using a different construction of dyadic cube systems on metric spaces, for instance the notion defined in [**25**]. We expect the corresponding proofs for the Hausdorff and Minkowski dimension to be almost identical to those in Section 3. However, the Assouad dimension and Assouad spectrum need to be treated differently, due to the lack of circumscribed cubes for arbitrary balls.

Moreover, Falconer, Fraser and Kempton introduced in [**15**] the *intermediate dimensions*, a collection of dimensions that interpolate between the Hausdorff and Minkowski dimension, similarly to how the Assouad spectrum interpolates between the Minkowski and Assouad dimension. It would be an interesting result to represent those dimensions using dyadic cube systems in the metric setting as well. Since a circumscribed cube is only necessary in the arguments in Section 3 due to the local "zooming in" nature of the Assouad dimension and spectrum, we expect to have similar representations to those in Theorem 1.1 for the intermediate dimensions, using any of the systems from [**24**, **25**]

## Acknowledgments

I wish to thank Alex Rutar for the interesting conversations regarding this note. I am also grateful to the anonymous referee, whose comments improved the exposition of the note.

## References

[1] H. Aimar, A. Bernardis, and B. Iaffei, *Comparison of Hardy-Littlewood and dyadic maximal functions on spaces of homogeneous type*, J. Math. Anal. Appl. **312** (2005), no. 1, 105–120, DOI 10.1016/j.jmaa.2005.03.034. MR2175208

[2] P. Assouad, *Plongements lipschitziens dans* $\mathbf{R}^n$ (French, with English summary), Bull. Soc. Math. France **111** (1983), no. 4, 429–448. MR763553

[3] M. Cao, J. J. Marín, and J. M. Martell, *Extrapolation on function and modular spaces, and applications*, Adv. Math. **406** (2022), Paper No. 108520, 87, DOI 10.1016/j.aim.2022.108520. MR4440073

[4] V. V. Chepyzhov and M. I. Vishik, *Attractors for equations of mathematical physics*, American Mathematical Society Colloquium Publications, vol. 49, American Mathematical Society, Providence, RI, 2002, DOI 10.1051/cocv:2002056. MR1868930

[5] M. Christ, *A $T(b)$ theorem with remarks on analytic capacity and the Cauchy integral*, Colloq. Math. **60/61** (1990), no. 2, 601–628, DOI 10.4064/cm-60-61-2-601-628. MR1096400

[6] E.-K. Chrontsios-Garitsis, *On concentric fractal spheres and spiral shells*, Nonlinearity **38** (2025), no. 3, Paper No. 035019, 21, DOI 10.1088/1361-6544/adb530. MR4871234

[7] E.-K. Chrontsios-Garitsis, *Quasiregular distortion of dimensions*, Conform. Geom. Dyn. **28** (2024), 165–175, DOI 10.1090/ecgd/398. MR4840245

[8] E.-K. Chrontsios-Garitsis, *Sobolev mappings on metric spaces and Minkowski dimension*, Proc. Amer. Math. Soc. **153** (2025), no. 1, 223–237, DOI 10.1090/proc/17016. MR4840271

[9] E.-K. Chrontsios-Garitsis and S. Troscheit, *Minkowski weak embedding theorem*, Houston J. Math. **50** (2024), no. 2, 259–273. MR4877071

[10] Chrontsios Garitsis, E. K., and Tyson, J. T. *On the Assouad spectrum of Hölder and Sobolev graphs*, arXiv:2309.07783, 2023.

[11] E. K. Chrontsios Garitsis and J. T. Tyson, *Quasiconformal distortion of the Assouad spectrum and classification of polynomial spirals*, Bull. Lond. Math. Soc. **55** (2023), no. 1, 282–307, DOI 10.1112/blms.12727. MR4568342

[12] P. Constantin and C. Foias, *Navier-Stokes equations*, Chicago Lectures in Mathematics, University of Chicago Press, Chicago, IL, 1988. MR972259

[13] G. David, *Morceaux de graphes lipschitziens et intégrales singulières sur une surface* (French), Rev. Mat. Iberoamericana **4** (1988), no. 1, 73–114, DOI 10.4171/RMI/64. MR1009120

[14] K. Falconer, *Fractal geometry: Mathematical foundations and applications*, 3rd ed., John Wiley & Sons, Ltd., Chichester, 2014. MR3236784

[15] K. J. Falconer, J. M. Fraser, and T. Kempton, *Intermediate dimensions*, Math. Z. **296** (2020), no. 1-2, 813–830, DOI 10.1007/s00209-019-02452-0. MR4140764

[16] J. M. Fraser, *Assouad dimension and fractal geometry*, Cambridge Tracts in Mathematics, vol. 222, Cambridge University Press, Cambridge, 2021, DOI 10.1017/9781108778459. MR4411274

[17] J. M. Fraser, K. E. Hare, K. G. Hare, S. Troscheit, and H. Yu, *The Assouad spectrum and the quasi-Assouad dimension: a tale of two spectra*, Ann. Acad. Sci. Fenn. Math. **44** (2019), no. 1, 379–387, DOI 10.5186/aasfm.2019.4419. MR3919144

[18] J. M. Fraser and S. Troscheit, *The Assouad spectrum of random self-affine carpets*, Ergodic Theory Dynam. Systems **41** (2021), no. 10, 2927–2945, DOI 10.1017/etds.2020.93. MR4308161

[19] J. M. Fraser and H. Yu, *New dimension spectra: finer information on scaling and homogeneity*, Adv. Math. **329** (2018), 273–328, DOI 10.1016/j.aim.2017.12.019. MR3783415

[20] F. W. Gehring and J. Väisälä, *Hausdorff dimension and quasiconformal mappings*, J. London Math. Soc. (2) **6** (1973), 504–512, DOI 10.1112/jlms/s2-6.3.504. MR324028

[21] J. Heinonen, *Lectures on analysis on metric spaces*, Universitext, Springer-Verlag, New York, 2001, DOI 10.1007/978-1-4613-0131-8. MR1800917

[22] S. Hencl and P. Honzík, *Dimension of images of subspaces under Sobolev mappings*, Ann. Inst. H. Poincaré C Anal. Non Linéaire **29** (2012), no. 3, 401–411, DOI 10.1016/j.anihpc.2012.01.002. MR2926241

[23] G. Hong, B. Liao, and S. Wang, *Noncommutative maximal ergodic inequalities associated with doubling conditions*, Duke Math. J. **170** (2021), no. 2, 205–246, DOI 10.1215/00127094-2020-0034. MR4202493

[24] T. Hytönen and A. Kairema, *Systems of dyadic cubes in a doubling metric space*, Colloq. Math. **126** (2012), no. 1, 1–33, DOI 10.4064/cm126-1-1. MR2901199

[25] A. Käenmäki, T. Rajala, and V. Suomala, *Existence of doubling measures via generalised nested cubes*, Proc. Amer. Math. Soc. **140** (2012), no. 9, 3275–3281, DOI 10.1090/S0002-9939-2012-11161-X. MR2917099

[26] R. Kaufman, *Sobolev spaces, dimension, and random series*, Proc. Amer. Math. Soc. **128** (2000), no. 2, 427–431, DOI 10.1090/S0002-9939-99-05383-6. MR1670383

[27] J. Li, C.-W. Liang, C.-Y. Shen, and B. D. Wick, *Muckenhoupt-type weights and quantitative weighted estimates in the Bessel setting*, Math. Z. **309** (2025), no. 1, Paper No. 13, 29, DOI 10.1007/s00209-024-03641-2. MR4832561

[28] S. Mallat, *A wavelet tour of signal processing*, Academic Press, Inc., San Diego, CA, 1998. MR1614527

[29] M. Pollicott and K. Simon, *The Hausdorff dimension of $\lambda$-expansions with deleted digits*, Trans. Amer. Math. Soc. **347** (1995), no. 3, 967–983, DOI 10.2307/2154881. MR1290729

[30] E. Sawyer and R. L. Wheeden, *Weighted inequalities for fractional integrals on Euclidean and homogeneous spaces*, Amer. J. Math. **114** (1992), no. 4, 813–874, DOI 10.2307/2374799. MR1175693

[31] S. Troscheit, *The quasi-Assouad dimension of stochastically self-similar sets*, Proc. Roy. Soc. Edinburgh Sect. A **150** (2020), no. 1, 261–275, DOI 10.1017/prm.2018.112. MR4065082

Department of Mathematics, University of Tennessee, Knoxville, 1403 Circle Dr, Knoxville, Tennessee 37966

*Email address*: echronts@utk.edu, echronts@gmail.com

Contemporary Mathematics
Volume **839**, 2026
https://doi.org/10.1090/conm/839/16731

# Asymptotic homotopical complexity of an infinite sequence of dispersing $2D$ billiards

Nandor Simanyi

ABSTRACT. We investigate the large scale chaotic, topological structure of the trajectories of an infinite sequence of dispersing, hence ergodic, $2D$ billiards with the configuration space $Q_n = \mathbb{T}^2 \setminus \bigcup_{i=0}^{n-1} D_i$, where the scatterers $D_i$ ($i = 0, 1, \dots, n-1$) are disks of radius $r << 1$ centered at the points $(i/n, 0)$ mod $\mathbb{Z}^2$. We get effective lower and upper radial bounds for the rotation set $R$. Furthermore, we also prove the compactness of the admissible rotation set $AR$ and the fact that the rotation vectors $v$ corresponding to admissible periodic orbits form a dense subset of $AR$. We also obtain asymptotic lower and upper estimates for the sequence $h_{top}(n)$ of topological entropies and precise asymptotic formulas for the metric entropies $h_\mu(n, r)$.

## 1. Introduction

The concept of rotation number finds its origin in the study of the average rotation around the circle $S^1$ per iteration, as classically defined by H. Poincaré in the 1880's [**Poi96**], when one iterates an orientation-preserving circle homeomorphism $f : S^1 \to S^1$. This is equivalent to studying the average displacement $(1/n)(F^n(x) - x)$ ($x \in \mathbb{R}$) for the iterates $F^n$ of a lifting $F : \mathbb{R} \to \mathbb{R}$ of $f$ on the universal covering space $\mathbb{R}$ of $S^1$. The study of fine homotopical properties of geodesic lines on negatively curved, closed surfaces goes back at least to Morse [**Mor24**]. As far as we know, the first appearance of the concept of homological rotation vectors (associated with flows on manifolds) was the paper of Schwartzman [**Sch57**], see also Boyland [**Boy00**] for further references and a good survey of homotopical invariants associated with geodesic flows. Following an analogous pattern, in [**BMS06**] we defined the (still commutative) rotation numbers of a $2D$ billiard flow on the billiard table $\mathbb{T}^2 = \mathbb{R}^2/\mathbb{Z}^2$ with one convex obstacle (scatterer) $\mathbf{O}$ removed. Thus, the billiard table (configuration space) of the model in [**BMS06**] was $\mathbf{Q} = \mathbb{T}^2 \setminus \mathbf{O}$. Technically speaking, we considered trajectory segments $\{x(t) | 0 \le t \le T\} \subset \mathbf{Q}$ of the billiard flow, lifted them to the universal covering space $\mathbb{R}^2$ of $\mathbb{T}^2$ (not of the configuration space $\mathbf{Q}$), and then systematically studied the rotation vectors as limiting vectors of the average displacement $(1/T)(\tilde{x}(T) - \tilde{x}(0)) \in \mathbb{R}^2$ of the lifted orbit segments $\{\tilde{x}(t) | 0 \le t \le T\}$ as $T \to \infty$.

2020 *Mathematics Subject Classification.* Primary 37D05.

These rotation vectors are still "commutative", for they belong to the vector space $\mathbb{R}^2$.

Despite all the advantages of the homological (or "commutative") rotation vectors (i. e. that they belong to a real vector space, and this provides us with useful tools to construct actual trajectories with prescribed rotational behaviour), in our current view the "right" lifting of the trajectory segments $\{x(t)|0 \le t \le T\} \subset \mathbf{Q_n}$ is to lift these segments to the universal covering space of $\mathbf{Q_n}$, not of $\mathbb{T}^2$. This, in turn, causes a profound difference in the nature of the arising rotation "numbers", primarily because the fundamental group $\pi_1(\mathbf{Q_n})$ of the configuration space $\mathbf{Q_n}$ is a highly complex hyperbolic group.

After a bounded modification, trajectory segments $\{x(t)|0 \le t \le T\} \subset \mathbf{Q_n}$ give rise to closed loops $\gamma_T$ in $\mathbf{Q_n}$, thus defining an element $g_T = [\gamma_T]$ in the fundamental group $\pi_1(\mathbf{Q_n})$. The limiting behavior of $g_T$ as $T \to \infty$ will be investigated, quite naturally, from two viewpoints:

(1) The direction "$e$" is to be determined, in which the element $g_T$ escapes to infinity in the hyperbolic group $\pi_1(\mathbf{Q_n})$ or, equivalently, in its Cayley graph $\mathbf{G}$. All possible directions $e$ form the horizon or the so called ideal boundary $\mathrm{Ends}(\pi_1(\mathbf{Q_n}))$ of the group $\pi_1(\mathbf{Q_n})$, see [**CP93**].
(2) The average speed $s = \lim_{T\to\infty}(1/T)\,\mathrm{dist}(g_T, 1)$ is to be determined, at which the element $g_T$ escapes to infinity, as $T \to \infty$. These limits (or limits $\lim_{T_n\to\infty}(1/T_n)\,\mathrm{dist}(g_{T_n}, 1)$ for sequences of positive reals $T_n \nearrow \infty$) are nonnegative real numbers.

The natural habitat for the two limiting data $(s, e)$ is the infinite cone

$$C = ([0, \infty) \times \mathrm{Ends}(\pi_1(\mathbf{Q_n}))/(\{0\} \times \mathrm{Ends}(\pi_1(\mathbf{Q_n}))$$

erected upon the set $\mathrm{Ends}(\pi_1(\mathbf{Q_n}))$, the latter supplied with the usual Cantor space topology. Since the homotopical "rotation vectors" $(s, e) \in C$ (and the corresponding homotopical rotation sets) are defined in terms of the non-commutative fundamental group $\pi_1(\mathbf{Q_n})$, these notions will be justifiably called homotopical or noncommutative rotation numbers and sets.

The rotation set arising from trajectories obtained by the arc-length minimizing variational method will be the so called admissible homotopical rotation set $AR \subset C$. The homotopical rotation set $R$ defined without the restriction of admissibility will be denoted by $R$. Plainly, $AR \subset R$ and these sets are closed subsets of the cone $C$.

In this paper we study the large scale chaotic topological (actually, homotopic) structure of the trajectories of an infinite sequence of dispersing, hence ergodic, $2D$ billiards with a configuration space $Q_n = \mathbb{R}^2/\mathbb{Z}^2 \setminus \cup_{i=0}^{n-1} D_i = \mathbb{T}^2 \setminus \cup_{i=0}^{n-1} D_i$, where the scatterers $D_i$ $(i = 0, 1, \dots, n-1)$ are disks of radius $r << 1$ centered at $(i/n, 0)$ modulo $\mathbb{Z}^2$. In Theorems 2.1 and 2.2 we provide lower and upper radial estimates for the homotopical rotation set of the system. Namely, in Theorem 2.1 we prove

$$R \subset B(0, 2\sqrt{2}),$$

whereas Theorem 2.2 states that

$$B\left(0, \frac{1}{\sqrt{5}} - \mathcal{O}(\frac{1}{n})\right) \subset AR.$$

Furthermore, in Theorems 3.1 and 3.2 we also prove the compactness and the convexity of the admissible rotation set $AR$ and the property that the set of rotation vectors $v$ corresponding to periodic admissible orbits form a dense subset of $AR$. For all of the above notions, please see the first three pages (up to the formulation of Theorem 2.2) of [**BMS06**]. In particular, admissibility of a trajectory segment $S^{[0,T]}(x_0) = \{S^t(x_0) = (x(t), x'(t)) \,|\, 0 \le t \le T\}$, lifted to the universal covering space $\tilde{\mathbf{Q}}_\mathbf{n}$ of the configuration space $\mathbf{Q}_\mathbf{n}$, means two things:

(i) any two, consecutively visited obstacles (in the universal covering space of the configuration space, as always) are always different and the convex hull of them does not intersect any other obstacle in $\tilde{\mathbf{Q}}_\mathbf{n}$;

(ii) the obstacle visited at the $k$-th collision $k = 2, 3, 4, \ldots$ does not intersect the convex hull of the obstacles visited at the $k-1$-st and $k+1$-st collisions.

For a fixed admissible sequence of obstacles $(\sigma_0, \sigma_1, \ldots, \sigma_n)$ from $\tilde{\mathbf{Q}}_\mathbf{n}$, consider all piecewise smooth curves

$$\{x(t) \,|\, 0 \le t \le A\}$$

in $\tilde{\mathbf{Q}}_\mathbf{n}$ with the arc-length parametrization and with the following property:

(P) There are numbers $0 = t_0 < t_1 < t_2 < \cdots < t_n = A$ such that $x(t_i) \in \partial\sigma_i$ ($i = 0, 1, 2, \ldots, n$), and for all other values of $t$ the point $x(t)$ is in the interior of $\tilde{\mathbf{Q}}_\mathbf{n}$.

Elementary inspection shows that amongst the above curves there is a unique shortest one, and the shortest curve is actually a billiard trajectory. Furthermore, for the shortest curve the collisions at times $t_0 = 0$ and $t_n = A$ are perpendicular to the boundaries of the involved scatterers $\sigma_0$ and $\sigma_n$, therefore the minimal curve $\{x(t) \,|\, 0 \le t \le A\}$ is actually half of the periodic orbit $\{x(t) \,|\, 0 \le t \le 2A\}$, where $x(t) = x(2A - t)$ for $A \le t \le 2A$.

The above observation is often called in the literature, in particular in [**BMS06**], the variational principle, or variational method for the construction of admissible orbits.

For the topological entropy $h_{top}(n)$ of the $n$-th system, in Theorem 4.1 we obtain

$$\frac{1}{\sqrt{5}} \le \liminf_{n\to\infty} \frac{h_{top}(n)}{\log n} \le \limsup_{n\to\infty} \frac{h_{top}(n)}{\log n} \le 2\sqrt{2}.$$

Finally, for the metric entropy $h_\mu(n, r)$ of the flow, in Section 5 we prove that for any fixed $n$,

$$h_\mu(n, r) = \mathcal{O}(-r \log r)$$

as $r \to 0$, and for $r = \mathcal{O}(1/n)$

$$h_\mu(n, T) = \mathcal{O}(\log(n)),$$

as $n \to \infty$ for the Poincare section map $T$.

## 2. Radial size estimates for the admissible rotation set

As described in Section 2 of [**GS11**] (see page 4 in that paper, in particular Figures 1 and 2), the free generators of the fundamental group $\pi_1(Q_n) \cong F_{n+1}$ are $a$, $b_1$, $\ldots$, $b_n$, so that each of these generators correspond to a specific wall crossing of a trajectory: The generator $a$ corresponds to a trajectory crossing the wall

$$\{0\} \times [r, 1 - r]$$

(modulo $\mathbb{Z}^2$) in the positive direction, whereas the generator $b_i$ means that the trajectory crosses the wall

$$\left[\frac{i-1}{n}+r,\ \frac{i}{n}-r\right]\times\{0\}$$

(taken again modulo $\mathbb{Z}^2$) in the positive direction. Needless to say, crossing these walls in the opposite, negative directions means the inverses of the above mentioned elements of $\pi_1(Q_n)$.

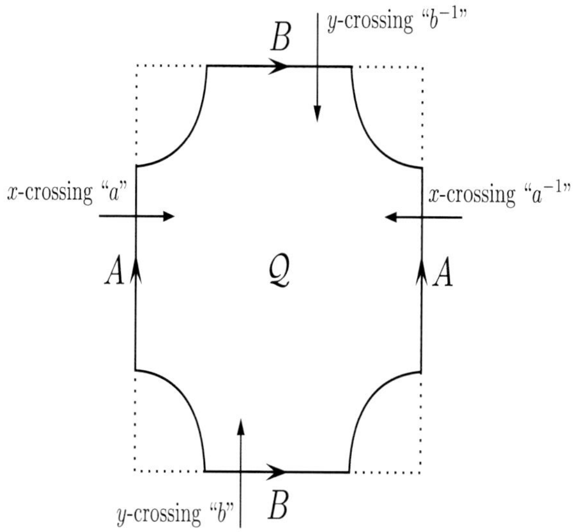


FIGURE 1

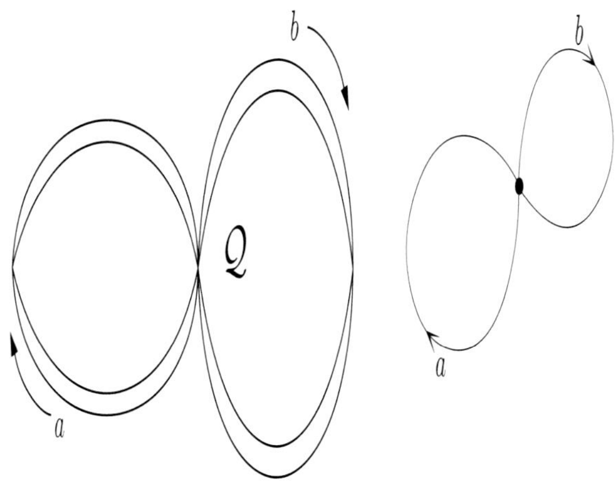


FIGURE 2

First we prove the upper radial size estimate:

THEOREM 2.1. $R \subset B(0, 2\sqrt{2})$

PROOF. Let us investigate the symbolic wall-crossing sequence, or associated word, $W$ of a trajectory segment $S^{[0,T]}x_0$ of the billiard flow. We write $W$ as the shortest product of generators $a, b_1, b_2, \dots, b_n$ and their inverses, i.e. only the exponents 1 and $-1$ are permitted. In such a writing, a maximal subsequence of consecutive symbols $a$ or $a^{-1}$ of $W$ will be called an $a$-block $B^a$, whereas a maximal subsequence of consecutive symbols $b_j$ or $b_j^{-1}$ $(j = 1, 2, \dots, n)$ of $W$ will be called a $b$-block $B^b$. To be clear, in a $b$-block, distinct symbols $b_i$ and $b_j$ $(i \neq j)$ and their inverses are allowed to be elements of the same $b$-block. For just simplifying the notations, we assume that $W$ begins with an $a$-block $B_1^a$, and it ends with a $b$-block $B_s^b$, so that $W$ is of the form

$$W = B_1^a B_1^b B_2^a B_2^b \dots B_s^a B_s^b.$$

At the end of proving the upper radial size estimate $2\sqrt{2}$ for such words, we will be clearly pointing out the minor differences in the argument to cover the other cases for the word $W$.

Let us denote the total number of the symbols $a$ and $a^{-1}$ in $W$ by $k$, whereas the total number of the symbols $b_j^{\pm 1}$ $(j = 1, 2, \dots, n)$ is denoted by $m$. Clearly, $s \leq \min\{k, m\}$.

Observe that for our trajectory segment

$$S^{[0,T]}x_0 = \{(x(t), x'(t)) \mid 0 \leq t \leq T\}$$

the estimate

$$\int_0^T |x_1'(t)|\, dt \geq k - s \tag{2.1}$$

holds true. We note that here, and everywhere else in this paper, $(x(t), x'(t))$ denotes the phase point at time $t$, where $x(t)$ is the position of the moving particle, already lifted to the universal covering space $\tilde{Q}_n$ of the configuration space $Q_n = \mathbb{R}^2/\mathbb{Z}^2 \setminus \cup_{i=0}^{n-1} D_i = \mathbb{T}^2 \setminus \cup_{i=0}^{n-1} D_i$, and $x'(t) = \dfrac{d}{dt}x(t)$ is its time derivative. Furthermore, $x_1(t)$ and $x_2(t)$ denote the first and second coordinates of $x(t)$, and similar notational convention holds true for the time derivative $x'(t)$.

Indeed, for each $a$-block $B_j^a$ the horizontal position changes at least by the amount $|B_j^a| - 1$, where $|B_j^a|$ is the number of symbols in $B_j^a$. Taking the sum of all these lower estimates, we obtain (2.1).

The situation with $b$-blocks $B_j^b$ and the corresponding vertical move

$$\int_0^T |x_2'(t)|\, dt$$

is somewhat different. Namely, we consider not only the block $B_j^b$, but also the subsequent two blocks $B_{j+1}^a$ and $B_{j+1}^b$. Elementary inspection shows that, after having at least the amount of $|B_j^b| - 1$ of vertical motion with $B_j^b$, in order to switch from $B_j^b$ to $B_{j+1}^b$ through the block $B_{j+1}^a$, the orbit has to make at least a unit of vertical motion betwen $B_j^b$ and $B_{j+1}^b$. (More precisely, this motion is at least $2 - \mathcal{O}(1/n)$, if one of $B_j^b$ and $B_{j+1}^b$ corresponds to an upward motion and the other one to a downward motion.)

Thus, by taking the sum of all these lower estimates, one obtains the lower bound

$$\int_0^T |x_2'(t)|\,dt \geq m-1. \tag{2.2}$$

Taking the sum of (2.1) and (2.2), one ends up with the estimates

$$m+k-s-1 \leq \int_0^T \|x'(t)\|_1\,dt \leq \sqrt{2}T, \tag{2.3}$$

which, in turn, implies that

$$\max\{m,k\}-1 \leq \sqrt{2}T,$$

thus

$$|W| = m+k \leq 2\sqrt{2}T+2. \tag{2.4}$$

Dividing (2.4) by $T$ and passing to the limit $T\to\infty$ yields

$$R \subset B(0, 2\sqrt{2}).$$

As said earlier, here we review the minor changes in the proof that are necessary to make, if the first block of $W$ is not an $a$-block, or the last block of it is not a $b$-block. Then, by simply truncating the unwanted block(s) at the two ends of $W$ (thus also truncating the trajectory segment under investigation), we can reduce the study to the above case. Note that, in this way, we make a bounded error in the right-hand-side of (2.4), but that error vanishes after division by $T$ and passing to the limit $T\to\infty$. This completes the proof of the theorem. □

THEOREM 2.2. *The lower radial estimate*

$$B\left(0, \frac{1}{\sqrt{5}} - \mathcal{O}(\frac{1}{n})\right) \subset AR$$

*holds for the admissible rotation set AR.*

PROOF. For any given infinite word $W_\infty = w_0w_1w_2\dots$ (where each $w_j$ either a generator or the inverse of a generator, and $W_\infty$ is in the shortest form) we are going to construct an infinite sequence

$$\left(S^{[0,T_m]}x_m\right)_{m=1}^{\infty}$$

of admissible orbit segements with $\lim_{m\to\infty} T_m = \infty$ and such that the symbolic wall crossing sequence of $S^{[0,T_m]}x_m$ is $w_0w_1\dots w_m$. In this process of constructing admissible orbit segments we use the arc-length minimizing variational method, described earlier. We will try to economize time and make $T_m$ as small as possible, thus making the escape speed

$$\liminf_{m\to\infty} \frac{m}{T_m}$$

as large as possible. We want to prove that

$$\liminf_{m\to\infty} \frac{m}{T_m} \geq \frac{1}{\sqrt{5}} - \mathcal{O}(\frac{1}{n}) \tag{2.5}$$

can be achieved. Then, by "slowing down" the speed of escape to infinity by inserting "idle runs", just as in the proofs of Theorems 3.1 and 3.3 in [**MS17**], one gets that any speed $s$, $0 \le s \le \frac{1}{\sqrt{5}} - \mathcal{O}(\frac{1}{n})$, is achievable while escaping to infinity in the given (but arbitrary!) direction of $W_\infty$, thus proving the desired set inequality

$$B\left(0, \frac{1}{\sqrt{5}} - \mathcal{O}(\frac{1}{n})\right) \subset AR.$$

In order to prove (2.5), it is enough to show that, when constructing longer and longer orbit segments with the symbolic sequence $w_0 w_1 \dots w_m$, in order to construct the passage $w_j \to w_{j+1}$, it is sufficient to spend at most the time $\sqrt{5} + \mathcal{O}(1/n)$. We consider all possible, mutually non-isomorphic passages $w_j \to w_{j+1}$ by excluding geometrically symmetric cases and cases equivalent after time reversal. After this reduction, here is the list of the remaining four cases for the passages:

**2.1. Case 1. Passage $ab_i$.** Direct inspection shows that here the worst case scenario is when $i = n$ and the admissible orbit (under construction) enters the unit fundamental cell at the upper end of the left vertical $a$-wall. Then, the most efficient admissible construction bounces back at the disk with index $\left\lfloor \frac{n}{2} \right\rfloor$ at the lower edge of the unit cell, before leaving it upwards through the wall $b_n$. The time spent to do this is at most $\sqrt{5} + \mathcal{O}(1/n^2)$.

REMARK 2.3. The possible error $\mathcal{O}(1/n^2)$ may be caused by the occurence of an odd value of $n$, $n = 2q + 1$, and having to select $D_q$ or $D_{q+1}$ at the bottom side of the unit cell.

**2.2. Case 2. Passage $b_i b_j^{-1}$ $(i \neq j)$.** Here again, direct inspection shows that the least favorable (i.e. the most time consuming) situation is when $|i - j|$ is maximal, i.e. $i = 1$ and $j = n$. (Or the other way around.) Then the most time efficient construction of the admissible trajectory in the unit fundamental cell $[0, 1] \times [0, 1]$ is to have it bounce back at the disk with index $\lfloor n/2 \rfloor$ at the ceiling of the unit cell. This takes time $\sqrt{5} + \mathcal{O}(1/n)$.

REMARK 2.4. As opposed to the previous case, the reason why the error is of order $\mathcal{O}(1/n)$ (and not $\mathcal{O}(1/n^2)$) is that, in the recursive construction of the passages of the admissible trajectory, it can happen that the incoming wall crossing $b_1$ takes place while bouncing back at the scatterer $D_1$. In order to facilitate the admissible passage to $D_{\lfloor n/2 \rfloor}$ at the ceiling of the unit fundamental cell, the orbit needs to visit the scatterer $D_0$ first, before shooting towards $D_{\lfloor n/2 \rfloor}$ at the ceiling. This takes an extra time of $\mathcal{O}(1/n)$.

**2.3. Case 3. Passage $aa$.** By excluding symmetric cases, we may assume that the orbit enters the unit fundamental cell by bouncing back at $D_0$ at the top of the left wall "$a$". Then we can have it exit the fundamental unit cell through the right wall "$a$" by bouncing back from $D_0 = D_n$ at the bottom of this wall. This takes time $\sqrt{2}$.

**2.4. Case 4. Passage $b_i b_j$.** Here again the least favorable case is when $|i - j|$ is maximal, i.e. $i = 1$ and $j = n$ (or, vice versa). Then an admissible trajectory construction in the unit fundamental cell can diagonally traverse the cell, similarly to Case 3, thus spending time not more than $\sqrt{2} + \mathcal{O}(1/n)$ in the cell. The $\mathcal{O}(1/n)$ error term should be included here for the same reason as in Case 2, but this does not matter, since $\sqrt{2} + \mathcal{O}(1/n) < \sqrt{5}$.

Summarizing all the above, when constructing the admissible orbit segments $S^{[0,T_m]}x_m$ with the symbolic sequence $W_m = w_0w_1 \dots w_m$ (being the $m$-th truncation of the given, infinite admissible word $W_\infty = w_0w_1 \dots$) we can achieve that

$$\liminf_{m\to\infty} \frac{m}{T_m} \geq \frac{1}{\sqrt{5}+\mathcal{O}(1/n)} = \frac{1}{\sqrt{5}} - \mathcal{O}(1/n).$$

This completes the proof of the theorem. □

## 3. Geometric properties of the admissible rotation set $AR$

THEOREM 3.1. *The set $AR$ is a convex, compact subset of the infinite cone $C$ erected upon the topological Cantor set* $\mathrm{Ends}(\pi_1(Q_n))$.

*(Here the set* $\mathrm{Ends}(\pi_1(Q_n))$ *denotes the set of all ends, i.e. the horizon of, the hyperbolic group $\pi_1(Q_n)$, see (1) on page 2 of* [**MS17**].*)*

PROOF. First we note that the convexity of a compact subset $K$ of the cone $C$ is equivalent to the star shaped property of $K$. This is a consequence of the fact that the basis $\mathrm{Ends}(\pi_1(Q_n))$ of the cone is totally disconnected. Indeed, the shortest path (geodesic line) connecting two points $(s_1, e_1)$ and $(s_2, e_2)$ of $C$ ($s_1 > 0$, $s_2 > 0$, $e_1 \neq e_2$) is the curve $\gamma$ that first connects $(s_1, e_1)$ with the vertex $(0, e_1)$ by a linear change in the radial speed coordinate $s$, then it connects the vertex with $(s_2, e_2)$ by also a linear change in the $s$ coordinate.

We follow the ideas of the proof of Theorem 3.3 of [**MS17**]. Indeed, the cone $C$ is a totally disconnected, Cantor set-type family of infinite rays that are glued together at their common endpoint, the vertex of the cone. Therefore, the convexity of $AR$ means that for any $(s, e) \in AR$ and for any $t$ with $0 \leq t \leq s$ we have $(t, e) \in AR$. However, this immediately follows from our construction, since we can always insert a suitable amount of idle runs into an admissible orbit segment to be constructed, hence slowing it down to the asymptotic speed $t$, as required. □

THEOREM 3.2. *The rotation vectors $(s, e) \in AR$ that correspond to periodic admissible trajectories form a dense subset of $AR$.*

PROOF. We adopt the main ideas of the proof of Theorem 3.4 of [**MS17**]. Consider an arbitrary trajectory segment

$$S^{[0,T]}x_0 = \left\{ S^t x_0 = (x_t, x'_t) \middle|\, 0 \leq t \leq T \right\}$$

with $x(0), x(T) \in \partial Q_n$ that bounces back from the scatterers with centers at $C_0, \dots, C_n$ (so that $\|C_i - x(t_i)\| = r$ for $0 = t_0 < t_1 < t_2 < \dots < t_n = T$) with an admissible symbolic sequence $(C_0, C_1, \dots, C_n)$. We want to find a periodic, admissible trajectory near $S^{[0,T]}x_0$ such that its rotation vector is close to the rotation vector of $S^{[0,T]}x_0$.

A direct inspection shows that the symbolic sequence $(C_0, C_1, \dots, C_n)$ has a bounded extension to a longer admissible sequence $(C_0, C_1, \dots, C_m)$, i.e. $m > n$ and $m - n$ has an upper bound $K$, independent of $S^{[0,T]}x_0$ and $n$, such that

(1) $C_m = C_0 + (a, b), \quad a, b \in \mathbb{Z}$,
(2) $(C_{m-1}, C_m, C_1 + (a, b))$ is admissible.

Then, by anchoring the trajectory segment at $C_0$ and $C_m$ with the configuration points $x(0)$ (with $\|x(0) - C_0\| = r$) and $x(T) = x(0) + (a, b)$, and performing the usual length minimization, we obtain a periodic, admissible orbit $S^{[0,T']}x'_0$ whose rotation vector is close to the rotation vector of $S^{[0,T]}x_0$. □

## 4. Topological entropy

Here we prove

THEOREM 4.1.
$$\frac{1}{\sqrt{5}} \le \liminf_{n\to\infty} \frac{h_{top}(n)}{\log(n)} \le \limsup_{n\to\infty} \frac{h_{top}(n)}{\log(n)} \le 2\sqrt{2}.$$

PROOF. It turns out that the estimates of this theorem are corollaries of Theorems 2.1 and 2.2. Indeed, following the ideas of Section 5 of [**MS17**], we construct a generating partition $\mathcal{P}$ for the topological entropy of the billiard flow, as follows:

We define the following subsets of the full phase space $M = M_n$ of the studied system with $n$ scatterers:

$$D_k^+ = \left\{(q, v) \in M \middle| \ \frac{k}{n} < q_1 < \frac{k+1}{n},\ \{q_2\} < \epsilon_0\right\}, \tag{4.1}$$

$$D_k^- = \left\{(q, v) \in M \middle| \ \frac{k}{n} < q_1 < \frac{k+1}{n},\ 1 - \epsilon_0 < \{q_2\}\right\} \tag{4.2}$$

for some $0 < \epsilon_0 << r$ (i.e. $\epsilon_0 = o(r)$ is small ordo of $r$), $k = 0, 1, \dots, n-1$,

$$S^+ = \left\{(q, v) \in M \middle| \ \{q_1\} < \epsilon_0\right\}, \tag{4.3}$$

$$S^- = \left\{(q, v) \in M \middle| \ \{q_1\} > 1 - \epsilon_0\right\}, \tag{4.4}$$

where $\{x\} = x - \lfloor x \rfloor$ denotes the fractional part of $x$. By definition, the partition $\mathcal{P}$ consists of all sets in (4.1), (4.2), (4.3), (4.4) and the rest of the phase space

$$B = M \setminus \left( \bigcup_{k=0}^{n-1} D_k^+ \cup \bigcup_{k=0}^{n-1} D_k^- \cup S^+ \cup S^- \right). \tag{4.5}$$

Just like in Section 5 of [**MS17**], an easy geometric inspection shows that the $\mathcal{P}$-itinerary $(P_1, P_2, \dots, P_k)$ of a trajectory segment $S^{[a,b]}x_0$ uniquely determines the homotopy type of $S^{[a,b]}x_0$.

Similarly, if one considers the discrete-time dynamics, i.e. the collision map, then the full $\mathcal{P}$-itinerary of a phase point uniquely determines the phase point, that is, the partition $\mathcal{P}$ is a generating partition for the topological entropy.

The lower radial estimate of Theorem 2.2 allows us to construct admissible orbit segments $S^{[0,T]}x$, with a given time span $T$, with at least as many as

$$(2n+1)^{(5^{-1/2} - \mathcal{O}(1/n))T} \tag{4.6}$$

different $\mathcal{P}$-itineraries. By taking the natural logarithm of (4.6), dividing by $T$, and passing to the limit as $T \to \infty$, one concludes that

$$h_{top}(n) \ge (5^{-1/2} - \mathcal{O}(1/n)) \log(2n+1).$$

Finally, dividing by $\log(n)$, then passing to the limit inferior as $n \to \infty$, we ontain the lower bound of Theorem 4.1.

On the other hand, Theorem 2.1 shows that the orbit segments $S^{[0,T]}x$, with a given time span $T$, cannot form more than

$$(2n+1)^{2\sqrt{2}T}$$

different $\mathcal{P}$-itineraries. Again, by taking natural logarithm, dividing by $T$, passing first to the limit $T \to \infty$, then to the limit superior as $n \to \infty$, we obtain the upper bound of Theorem 4.1. $\square$

## 5. Concluding remarks

In this brief section we make a quick comparison between the topological entropy $h_{top}(n)$ of the flow and its metric entropy $h_\mu(n) = h_\mu(n, r)$, where $\mu = \mu_n$ is the unique, absolutely continuous invariant measure of the billiard flow. As it turns out, if the radius $r = r(n)$ of the disks is a sufficiently fast decreasing function of $n$, than $h_\mu(n)$ is dramatically smaller than $h_{top}(n)$. Indeed, first fix the value of $n$, and study the behaviour of the metreic entropy $h_\mu(n, r)$ of the flow as a function of $r$, when $r \to 0$. As it follows from the Marklof-Strömbergsson theory of the distribution of the free path length in a dilute Lorentz gas [**MS10**], the expected value of the free path length is $\mathcal{O}(1/r)$. After a typical collision with a scatterer of radius $r$, an expanding (convex) local orthogonal manifold collects a curvature of the order of $\mathcal{O}(1/r)$, and undergoes an expansion of the order $\mathcal{O}(1/r^2)$ until the next collision. Thus the positive Lyapunov exponent will be of the order $\mathcal{O}(-r\log(r))$, therefore $h_\mu(n, r) = \mathcal{O}(-r\log(r))$ for any fixed $n$, while $r \to 0$. This immediately implies that, if $r = r(n)$ is a sufficiently fast decreasing function of $n$, then the ratio $\dfrac{h_\mu(n)}{h_{top}(n)}$ can be made to converge to zero as fast as we wish, as $n \to \infty$.

On the other hand, if $r = \mathcal{O}(1/n)$ and we consider the billiard map $T$ instead of the flow $\{S^t\}$, then the metric entropy $h_\mu(n, T)$ will be of the order $\mathcal{O}(\log(n))$, i.e. of the order of the topological entropy of the map. Indeed, the iterate $T^k$ produces asymptotically the same amount of dilation of the local unstable manifolds (as $k \to \infty$) as the iterate $S^{k/r} \approx S^{kn}$ of the flow so, after rescaling, we gain an extra multiplying factor $n$ in the metric entropy, i.e. $h_\mu(n, T) = \mathcal{O}(-nr\log(r)) = \mathcal{O}(\log(n))$.

## References


[Boy00] Philip Boyland, *New dynamical invariants on hyperbolic manifolds*, Israel J. Math. **119** (2000), 253–289, DOI 10.1007/BF02810671. MR1802657

[BMS06] Alexander Blokh, Michał Misiurewicz, and Nándor Simányi, *Rotation sets of billiards with one obstacle*, Comm. Math. Phys. **266** (2006), no. 1, 239–265, DOI 10.1007/s00220-006-0014-4. MR2231972

[CP93] Michel Coornaert and Athanase Papadopoulos, *Symbolic dynamics and hyperbolic groups*, Lecture Notes in Mathematics, vol. 1539, Springer-Verlag, Berlin, 1993. MR1222644

[GS11] Lee M. Goswick and Nándor Simányi, *Homotopical complexity of 2D billiard orbits*, Studia Sci. Math. Hungar. **48** (2011), no. 4, 540–562, DOI 10.1556/SScMath.48.2011.4.1191. MR3058358

[MS10] Jens Marklof and Andreas Strömbergsson, *The distribution of free path lengths in the periodic Lorentz gas and related lattice point problems*, Ann. of Math. (2) **172** (2010), no. 3, 1949–2033, DOI 10.4007/annals.2010.172.1949. MR2726104

[Mor24] Harold Marston Morse, *A fundamental class of geodesics on any closed surface of genus greater than one*, Trans. Amer. Math. Soc. **26** (1924), no. 1, 25–60, DOI 10.2307/1989225. MR1501263

[MS17] Caleb C. Moxley and Nandor J. Simanyi, *Homotopical complexity of a 3D billiard flow*, Dynamical systems, ergodic theory, and probability: in memory of Kolya Chernov, Contemp. Math., vol. 698, Amer. Math. Soc., Providence, RI, 2017, pp. 169–180, DOI 10.1090/conm/698/13981. MR3716092

[MS19] Caleb C. Moxley and Nandor J. Simanyi, *Homotopical complexity of a billiard flow on the 3D flat torus with two cylindrical obstacles*, Ergodic Theory Dynam. Systems **39** (2019), no. 4, 1071–1081, DOI 10.1017/etds.2017.62. MR3916847

[Poi96] Henri Poincaré, *Œuvres. Tome VII: Masses fluides en rotation. Principes de mécanique analytique. Problème des trois corps. [Rotating fluid masses. Principles of analytic mechanics. Three-body problem]* (French), Les Grands Classiques Gauthier-Villars. [Gauthier-Villars Great Classics], Éditions Jacques Gabay, Sceaux, 1996. With a preface by Jacques Lévy; Reprint of the 1952 edition. MR1401348

[Sch57] Sol Schwartzman, *Asymptotic cycles*, Ann. of Math. (2) **66** (1957), 270–284, DOI 10.2307/1969999. MR88720

Department of Mathematics, University of Alabama at Birmingham, 1402 10th Avenue South, Birmingham, Alabama 35294-1241

*Email address*: simanyi@uab.edu

Contemporary Mathematics
Volume **839**, 2026
https://doi.org/10.1090/conm/839/16732

# Operator theory of diffusion equations governed by self-similar measures

Palle E. T. Jorgensen and James Tian

*Dedicated to the memory of Professor Ole Barndorff-Nielsen.*

ABSTRACT. For classes of self-affine fractal measures, we offer a new analysis of the associated diffusions, Markov processes, semigroups, and Dirichlet forms. With view to an harmonic analysis of fractal measures, we introduce particular dual-pair systems of operators. The starting point is a fixed measure $\mu$. We then introduce associated transforms which allow a comparison of $\mu$ to Lebesgue measure $\lambda$. Our dual pairs are carried by the two Hilbert spaces, $L^2(\mu)$ and $L^2(\lambda)$. A dual-pair analysis consist of a choice of a pair of Hilbert spaces, and two densely defined (unbounded) operators, each one contained in the adjoint of the other. Applications include the Krein-Feller operator, and the associated $\mu$ diffusion. Our approach goes beyond earlier related results in the general theory, in several ways: For example, our new duality framework is versatile, and it accommodates the setting of general measure spaces, thus going beyond that of classical domains, or IFSs, in ambient $\mathbb{R}^k$. Key tools for our analysis are associated reproducing kernel Hilbert spaces (RKHSs), time-change, and Gaussian fields. We further present an application to an inverse problem from finance, specifically, determining probability distributions of underlying stochastic assets arising in option-price formulas.

## CONTENTS



---

2020 *Mathematics Subject Classification.* Primary 60G20, 60H05, 60H30, 60J35, 60J60, 60J65; Secondary 47B32, 47B25, 47E05, 46E22, 46N20.

*Key words and phrases.* Hilbert space, reproducing kernel Hilbert space, harmonic analysis, Gaussian free fields, transforms, covariance, Krein-Feller operators, unbounded operators, dual pairs, semigroup, diffusion, Ito lemma.

## 1. Introduction

The purpose of the present paper is to prove new results for certain stochastic processes which arise from self-similar fractals. We accomplish this by making a number of new links between following two settings: (i) the harmonic analysis of the $\nabla_\mu$ operators for general positive non-atomic Borel measures, and (ii) more standard harmonic analysis questions for the case of Lebesgue measure $\lambda$ on an interval. One benefit of this is that we will then be able to "transport" facts from (the easier case of) (ii) to (i). The measures $\mu$ in (i) will typically be fractal type, either of a general kind, or IFS-measures, or, more particularly, Cantor measures. Our task will be carried out below with the use of specific intertwining operators; (when $\mu$ is fixed) from $L^2(\mu)$ onto $L^2(\lambda)$. Background references for this include [**Jor04**, **JT18**, **JT20**, **CP68**].

Our paper is interdisciplinary, and a second purpose for us is to illustrate an intriguing interplay between recent and new ideas from separate areas. The list includes the following areas: operator theory (commutative and non-commutative; bounded and unbounded), a harmonic analysis of iterated function system (IFS) measures and the role they play in a study of diffusion processes, systems of transition probability, a certain class of stochastic processes and their dynamics. This interplay of areas serves in turn to combine and add to our understanding and insight into an important families of diffusion processes. The latter are stochastic processes that are often said to be generated by classes of Krein-Feller operators. Our presentation is therefore aimed at diverse audiences. And to facilitate outreach, we have included below some background material that may be familiar to some but perhaps not all potential readers.

**Organization.** Starting with a positive non-atomic Borel measure $\mu$, our first results deal with the operator theory, and the spectral resolution, of an associated $\mu$-derivative, called $\nabla_\mu$, as well as a second order operator, called the Krein-Feller operator $K_F$, formally $K_F := \nabla_\mu d/dx$. Our focus is the case of singular measures $\mu$, especially measures arising as iteration-limits of iterated function systems (IFS), i.e., measures having a system of scales reflecting affine self-similarity. (Cantor-measures are special cases of IFS-measures.) We further study the diffusion process corresponding to $K_F$. To this end, we introduce a Gaussian process $W^{(\mu)}$ as a generalized Brownian motion. The quadratic variation of $W^{(\mu)}$ yields back the measure $\mu$. It will be convenient for us to present this Gaussian process $W^{(\mu)}$ in a more general framework. This is detailed in section 3. Section 4 deals with a duality relation for the operator systems introduced in section 3. An application of the results in the paper is given in the final section: It entails an inverse problem, and deals with the price formulas for derivative securities in financial mathematics; more specifically, determining the probability distribution of underlying random variables or stochastic processes.

This paper is a continuation of an earlier one [**JT22a**] by the co-authors. Our present focus is the presentation of new results, and a new duality framework. We further demonstrate its versatility with new applications to problems from analysis of fractals, and to the associated diffusion processes. For the purpose of readability, we have chosen to recall below some facts from [**JT22a**] as they will be needed. See also the papers [**Min20**, **AN06**, **Fre11**, **KLLW21**, **GL20**, **GLQ19**, **KLW17**, **DL17**].

## 2. Analysis of Cantor measures

The section below is subdivided as follows: It first presents some of the key geometric features of iterated function system (IFS)-measures $\mu$, stressing those Hilbert space properties that will be needed; thus, extending results from classical analysis to the context of fractals (and singular measures). We then turn to transformation of measure results needed in our results regarding the $\mu$ derivative operator, denoted $\nabla_\mu$. This, in turn, will serve as key tools in our spectral analysis of $\nabla_\mu$, realized as a skew symmetric operator with dense domain in $L^2(\mu)$.

Fix a non-atomic measure $\mu$ on $[0,1]$. Let

$$\mathscr{D}_1 := \left\{ f : f(x) = f(0) + \int_0^x f^{(\mu)} d\mu,\ f^{(\mu)} \in L^2(\mu), \text{ for all } x \right\}. \tag{2.1}$$

Then $\mathscr{D}_1 \subset L^2(\mu) \cap C([0,1])$.

Define

$$\nabla_\mu f = f^{(\mu)}, \quad \forall f \in \mathscr{D}_1. \tag{2.2}$$

In the lemma below we express eq (2.2) for the operator $\nabla_\mu$ (acting on functions $f$) in terms of associated Stieltjes measures. This point is summarized best in eq (2.5) in Lemma 2.1 below, where $df$ then denotes the *Stieltjes measure* corresponding to some function $f$. In the sequel we shall reserve the notation $df$ for Stieltjes measure, (not to be confused with notions of differential.) Recall that for the Stieltjes measure $df$ to make sense, the function $f$ must be assumed to be locally of bounded variation.

In our discussions and applications below, we make use of an important extension of the more familiar Radon-Nikodym derivative. This extension is based on an $L^2(\mu)$ analysis, and is therefore different from more traditional Radon-Nikodym considerations. For our present purpose, we make use of the main result in the paper [**AJ22**] where this extension is proved, and is referred to as a Krein-Feller derivative. Readers are directed to [**AJ22**] for the justification of why the premise is satisfied in our present considerations, and hence why the use of the Krein-Feller derivative is justified.

LEMMA 2.1. *Let $f$ be a function on $\mathbb{R}$, assumed to be locally of bounded variation, so that the Stieltjes measure $df$ is well defined. Let $\mu$ be a positive measure defined on the Borel $\sigma$-algebra $\mathscr{B}$, and assume that*

$$df \ll \mu, \tag{2.3}$$

*i.e., that we have the implication* (2.4) *below holds:*

$$\mu(B) = 0 \Longrightarrow df(B) = 0. \tag{2.4}$$

*Let $f^{(\mu)}$ be the corresponding Krein-Feller derivative (also denoted $f^{(\mu)} = \nabla_\mu f$), then*

$$df = f^{(\mu)} d\mu. \tag{2.5}$$

PROOF. The assertion in (2.5) amounts to the identity

$$df(B) = \int_B f^{(\mu)} d\mu, \tag{2.6}$$

for all $B \in \mathscr{B}$. But since $f$ is locally of bounded variation, (2.6) follows from the corresponding assumption for intervals, i.e., $B = [x, y]$ for all $x < y$; so

$$f(y) - f(x) = \int_x^y f^{(\mu)} d\mu. \tag{2.7}$$

But condition (2.7) in turn is equivalent to the definition of $f^{(\mu)} = \nabla_\mu f$ given in (2.1) above. □

If $F$ is absolute continuous then $F'$ exists a.e. and

$$F(y) - F(x) = \int_x^y F'(s)\, ds. \tag{2.8}$$

This is false if $F$ is not absolutely continuous.

Definition 2.2. We say a function $F$ is absolutely continuous if, $\forall \epsilon > 0\ \exists \delta > 0$ s.t.

$$\sum_i |F(s_{i+1}) - F(s_i)| < \epsilon, \tag{2.9}$$

for all $J_i = (s_i, s_{i+1}]$, s.t. $\sum_i (s_{i+1} - s_i) < \delta$.

It will be clear from the analysis below of IFS-measures $\mu$ (with gaps) that the condition of absolute continuity will not be satisfied for the cumulative distribution functions computed from these measures.

**2.1. Applications to iterated function system measures.** The theory of iterated function system (IFS) measures is extensive. IFS measures arise in diverse applications, geometric analysis, fractal harmonic analysis, chaotic dynamics and more. Here we wish to cite the following papers of most direct relevance for our current discussion, [**JT19a, HJW19, BJ19, BJ18, JT18, BF21, MP21, WLLZ20**].

Definition 2.3. Let $(X, \mathscr{B}_X)$ be a measurable space, $N \in \mathbb{N}$, and let $\{\sigma_i\}_{i=1}^N$ be a system of continuous endomorphisms $\sigma_i : X \to X$. Let $\{p_i\}_{i=1}^N$, $p_i > 0$, $\sum_{i=1}^N p_i = 1$ be fixed.

A Borel measure $\mu$ on $X$ is said to be a iterate function system (IFS) measure w.r.t. the data iff (Def) the following identity holds:

$$\mu = \sum_{i=1}^N p_i\, \mu \circ \sigma_i^{-1} \tag{2.10}$$

on the Borel $\sigma$-algebra $\mathscr{B}_X$.

We now turn to an explicit realization of the IFS-measure from e.q. (2.10).

Lemma 2.4. *Let $X$, $N$, $\{\sigma_i\}_{i=1}^N$, $\{p_i\}_{i=1}^N$ be as above. Consider the infinite product*

$$\Omega_N := \prod_{\mathbb{N}_0} \{1, 2, \cdots, N\}, \tag{2.11}$$

*and suppose, for all $\omega \in \Omega_N$, the intersection below is a singleton, i.e.,*

$$\bigcap_{\substack{n=1 \\ \omega|_n = \{i_1, \cdots, i_n\}}}^{\infty} \sigma_{i_n} \sigma_{i_{n-1}} \cdots \sigma_{i_1}(X) = \{x(\omega)\}; \tag{2.12}$$

*then there is an associated IFS measure $\mu$ constructed from the infinite product*

$$\pi := p \times p \times p \times \cdots \text{ on } \Omega_N. \tag{2.13}$$

*See Figure* 2.1.

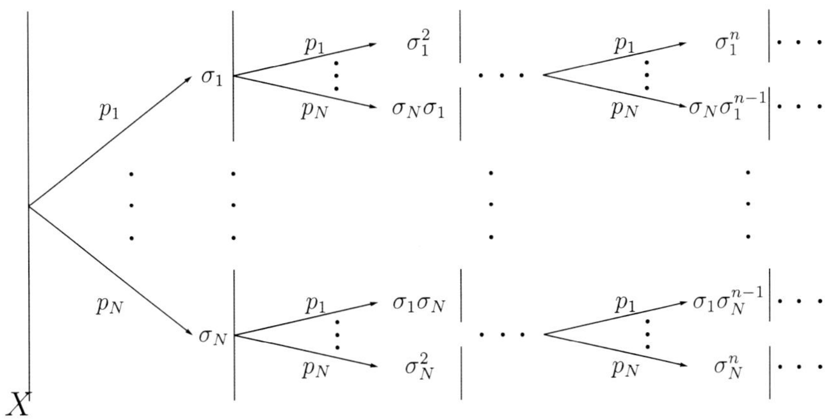


FIGURE 2.1. IFS as a Markov process.

PROOF. With (2.12), we define the random variable $W^{(IFS)}(\omega) := x(\omega)$, $\omega \in \Omega_N$. If $(i_1, \cdots, i_n) \in \prod_1^n \{1, \cdots, N\}$, then the measure $\pi$ is specified on cylinder sets as follows:

$$\pi([i_1, \cdots, i_n]) = p_{i_1} p_{i_2} \cdots p_{i_n}. \tag{2.14}$$

The measure $\pi$ is then defined on $\Omega_N$ via Kolomogorov's consistency extension theorem, see [**Hid80**, **Kol83**, **Kol77**]. Let

$$W : \Omega_N \longrightarrow X \tag{2.15}$$

be the random variable specified by the condition in (2.12), and set

$$\mu := \pi \circ W^{-1}. \tag{2.16}$$

One checks that $\mu$ will then satisfy the IFS condition in (2.10). □

COROLLARY 2.5. *In this corollary we fix a system $\{\sigma_i\}_{i=1}^N$ of endomorphisms, and we consider the IFS measure $\mu$ as it depends on choice of probability weights $p = (p_i)_{i=1}^N$, $\sum p_i = 1$. Set $\mu^{(p)}$ = the solution to* (2.10)*; (see also* (2.15)*). Then if $p \neq q$ (i.e., $\exists i$ such that $p_i \neq q_i$) then the two measures $\mu^{(p)}$ and $\mu^{(q)}$ are mutually singular.*

PROOF. The result is immediate from (2.16) and Kakutani's dichotomy theorem for infinite product measures, applied to (2.13). By Kakutani [**Kak43**], the two measures $\times_{\mathbb{N}} p$ and $\times_{\mathbb{N}} q$ are mutually singular; and by (2.16), so are the two IFS measures $\mu^{(p)}$ and $\mu^{(q)}$. □

### 2.2. Gaussian processes.

DEFINITION 2.6 (Generalized Brownian motion). For earlier related literature, see e.g., [**AJ15, AJL18, AJ21, JT22b**]. Let $(S, \mathscr{B}_S, \mu)$ be a measure space, where $S$ is a set, $\mathscr{B}_S$ a specified $\sigma$-algebra of subsets of $S$; and $\mu$ a fixed $\sigma$-finite positive measure defined on $\mathscr{B}_S$. Set $\mathscr{B}_S(fin) := \{A \in \mathscr{B}_S : \mu(A) < \infty\}$, and

$$K^{(\mu)}(A, B) := \mu(A \cap B), \quad \forall (A, B) \in \mathscr{B}_S(fin) \times \mathscr{B}_S(fin). \tag{2.17}$$

We shall denote by $W^{(\mu)}$ the centered Gaussian process with covariance kernel $K^{(\mu)}$, i.e., $W^{(\mu)}$ is indexed by $\mathscr{B}_S(fin)$ and realized in a probability space $(\Omega, \mathscr{C}, \mathbb{P})$ with $\mathbb{E}(\cdot) = \int \cdot \cdot d\mathbb{P}$, such that

$$\mathbb{E}\left(W_A^{(\mu)}\right) = 0, \text{ and} \tag{2.18}$$

$$\mathbb{E}\left(W_A^{(\mu)} W_A^{(\mu)}\right) = K^{(\mu)}(A, B), \quad \forall A, B \in \mathscr{B}_S(fin). \tag{2.19}$$

When $\mu$ is fixed, there is a variety of choices for probability spaces $(\Omega, \mathscr{C}, \mathbb{P})$ which realizes the corresponding centered Gaussian process $W^{(\mu)}$. Note that, if the process is indexed by time, i.e., by $\mathbb{R}$, then, for the purpose of Monte Carlo simulations, it is convenient to fix $\Omega$ to be functions on $\mathbb{R}$. But this means that $\Omega$ is fixed, and so the probability path-space measure $\mathbb{P}$ from $(\Omega, \mathscr{C}, \mathbb{P})$ will depend on $\mu$. On occasion we shall write $\mathbb{P}^{(\mu)}$ to stress this fact. For a detailed discussion of the general case of the Gaussian process $W^{(\mu)}$, see Section 3 below, especially (3.5).

REMARK 2.7. Three examples of such Gaussian process $W^{(\mu)}$ are discussed in Section 3 below.

**2.3. Skew symmetric vs skew-adjoint.** Note $\lambda_1 \longleftrightarrow \mu$ when $\lambda_1 = dx$ is standard Lebesgue measure restricted to an interval. We work with $[0, 1]$, but for some calculations the choice of $[0, 2\pi]$ is simpler. We make use $L^2(\mu)$ and $T_\mu (= \nabla_\mu)$, then we get ONBs

$$\left\{e^{i(n+\theta)x}\right\}_{n\in\mathbb{Z}} \text{ in } L^2([0, 2\pi], \lambda_1), \quad \left\{e^{i(n+\theta)g(\cdot)}\right\}_{n\in\mathbb{Z}} \text{ in } L^2(\mu), \tag{2.20}$$

with a choice of boundary condition

$$f(2\pi) = \alpha f(0), \quad \alpha = e^{i\theta} \text{ fixed.} \tag{2.21}$$

Let $\xi \in \mathbb{R}$, apply to $x \to e^{i\xi x} = e_\xi(x)$ and $e_\xi \circ g = e^{i\xi \cdot g(x)}$:

$$\nabla_\mu (e_\xi \circ g) = i\xi e_\xi \circ g. \tag{2.22}$$

Now put $\xi$ as in (2.20), $\{\theta + n\}_{n\in\mathbb{Z}}$, $\xi_n = \theta + n$, then

$$\underbrace{\sum_n c_n e_{\xi_n}}_{\text{ONB in } L^2(J,\lambda_1)} \longleftrightarrow \underbrace{\sum_n c_n e_{\xi_n} \circ g}_{\text{ONB in } L^2(J,\mu)}. \tag{2.23}$$

EXAMPLE 2.8. Let $\psi(t) = e^{-\frac{u^2 \mu([0,t])}{2}}$, i.e.,

$$g_\mu(t) = \mu([0, t]) \longrightarrow e^{-\frac{u^2 x}{2}}.$$

Then we have

$$(\nabla_\mu \psi)(t) = \psi' \circ g_\mu(t) = -\frac{u^2}{2} e^{-\frac{u^2 g_\mu(t)}{2}}.$$

**2.4. More about $\mu \circ g^{-1} = \lambda_1$.** For simplicity put $J = [0, 2\pi]$ and consider the periodic boundary condition

$$f(2\pi) = f(0) \tag{2.24}$$

in both $L^2(\mu)$ and $L^2(\lambda_1)$. Then we get skew-adjoint realizations of the following two operators:

$$\nabla_\mu \quad \text{skew-adjoint in } L^2(J, \mu) \tag{2.25}$$

$$\nabla_\lambda \quad \text{skew-adjoint in } L^2(J, \lambda_1) \tag{2.26}$$

We set $V_\mu f = f \circ g$, where

$$g(x) = g_\mu(x) = \mu([0, x]) \tag{2.27}$$

so the question is: When can we say

$$V_\mu : L^2(\lambda_1) \to L^2(\mu) \tag{2.28}$$

is an isometric isomorphism? We show that

$$L^2(\lambda_1) \overset{V_\mu}{\simeq} L^2(\mu) \Longleftrightarrow \mu \circ g^{-1} = \lambda_1.$$

Assumptions:

(1) $J = [0, 2\pi]$;
(2) periodic boundary condition as in (2.24);
(3) $\mu(J) = \lambda(J) = 1$.

REMARK 2.9. Direct verification of

$$\mu \circ g^{-1} = \lambda_1 \tag{2.29}$$

for the special case of Cantor measure on $[0, 1]$. Here, $\lambda_1 = dx$,

$$g(x) = \mu([0, x]), \tag{2.30}$$

$\mu$ is the $\frac{1}{3}$ Cantor measure, so

$$\mu = \frac{1}{2}\left(\mu \circ \tau_0^{-1} + \mu \circ \tau_1^{-1}\right), \tag{2.31}$$

and

$$\tau_0(x) = \frac{x}{3}, \quad \tau_1(x) = \frac{x+2}{3}. \tag{2.32}$$

In order to verify (2.29), i.e., $\mu \circ g^{-1} = \lambda_1$ we need to check that

$$\int \varphi \, d\mu \circ g^{-1} = \frac{1}{2}\left(\int \varphi\left(\frac{x}{2}\right) d\mu \circ g^{-1} + \int \varphi\left(\frac{x+1}{2}\right) d\mu \circ g^{-1}\right) \tag{2.33}$$

since for the Lebesgue measure $\lambda_1$,

$$\int \varphi \, d\lambda_1 = \frac{1}{2}\left(\int \varphi\left(\frac{x}{2}\right) d\lambda_1 + \int \varphi\left(\frac{x+1}{2}\right) d\lambda_1\right). \tag{2.34}$$

Note, $\lambda_1$ is the unique measure fixed by

$$\lambda_1 = \frac{1}{2}\left(\lambda_1 \circ \sigma_0^{-1} + \lambda_1 \circ \sigma_1^{-1}\right)$$

where

$$\sigma_0(x) = \frac{x}{2}, \quad \sigma_1(x) = \frac{x+1}{2}. \tag{2.35}$$

See e.g., [**Hut81**]. The main theorem in Hutchinson's paper is uniqueness (up to scale) of solutions to the IFS equation (as acting on measures). There is a proof that uses the Banach fixed point theorem, applied to spaces of measures with Hausdorff metric.

So the question is whether

$$\frac{1}{2}\left(\int \varphi\circ\sigma_0\, d\mu\circ g^{-1}+\int \varphi\circ\sigma_1\, d\mu\circ g^{-1}\right)=\int \varphi\, d\mu\circ g^{-1}$$
$$\Updownarrow$$
$$\frac{1}{2}\left(\int \varphi\circ\sigma_0\circ g\, d\mu+\int \varphi\circ\sigma_1\circ g\, d\mu\right)=\int \varphi\, d\mu\circ g^{-1}?$$

Alternatively, we may use the Leibniz rule (see e.g., [**BR87**, **Jor04**]) from above to verify (2.33):

In the right-hand side of (2.33), we have

$$\frac{1}{2}\int_0^1 \varphi\left(\frac{g(x)}{2}\right)d\mu=\Phi\left(\frac{1}{2}\right)-\Phi(0)$$

and

$$\frac{1}{2}\int_0^1 \varphi\left(\frac{g(x)+1}{2}\right)d\mu=\Phi\left(\frac{g(1)+1}{2}\right)-\Phi\left(\frac{1}{2}\right)=\Phi(1)-\Phi\left(\frac{1}{2}\right),$$

where $\Phi'=\varphi$. Therefore,

$$\begin{aligned}\mathrm{RHS}_{(2.33)}&=\Phi\left(\frac{1}{2}\right)-\Phi(0)+\Phi(1)-\Phi\left(\frac{1}{2}\right)\\&=\Phi(1)-\Phi(0)\\&=\int_0^1\varphi\, d\lambda_1=\mathrm{LHS}_{(2.33)}.\end{aligned}$$

While the argument above (Remark 2.9) is restricted to the transformation-rule for the Cantor 1/3 measure, in fact, it applies to the general case of positive Borel measures $\mu$, assumed finite and supported in a finite interval. Further note that, for the sake of proof-details, we may take the interval to be $[0,1]$ by the usual rescaling consideration. In conclusion, the transformation rule for general $\mu$ , as opposed to just for the Cantor 1/3 measure.

### 2.5. $V_\mu$ maps $L^2(\lambda)$ onto $L^2(\mu)$.

Let $\mu$ be as before: Borel measure, positive, non-atomic, and set $g=g_\mu$, $g(x)=\mu([0,x])$. We set $V_\mu(\varphi)=\varphi\circ g_\mu$. Now compare the two processes $W_t^{(\mu)}$ (generalized Brownian motion) and $B_t$ (standard Brownian motion), i.e., centered Gaussian processes with

$$\mathbb{E}\left(W_s^{(\mu)}W_t^{(\mu)}\right)=\mu([0,s\wedge t]),\tag{2.36}$$

$$\mathbb{E}(B_sB_t)=s\wedge t.\tag{2.37}$$

PROPOSITION 2.10 (Transformation of the corresponding Ito-integrals). *We have the following:*

$$\int(\psi\circ g)\,dW_t^{(\mu)}\underset{=}{(dist)}\int\psi\, dB_t.\tag{2.38}$$

PROOF. Since both sides in (2.38) are centered Gaussian, we only need to prove that they have the same variance quadratic form. Compute:

$$\begin{aligned}\mathbb{E}\left(\left|\int \psi\circ g\, dW^{(\mu)}\right|^{2}\right) &\underset{\text{Ito for } W^{(\mu)}}{=} \int \left|\psi\circ g\right|^{2} d\mu \\ &= \int \left|\psi\right|^{2} d\left(\mu\circ g^{-1}\right) \\ &\underset{\text{since } \mu\circ g^{-1}=\lambda_1}{=} \int \left|\psi\right|^{2} d\lambda_1 \\ &\underset{\text{Ito for } B_t}{=} \mathbb{E}\left(\left|\int \psi\, dB_t\right|^{2}\right).\end{aligned}$$

($\lambda_1 = dt =$ Lebesgue measure). □

Since $\mu\circ g^{-1} = \lambda_1$ we have conditional measures $\{\rho_y\}_{y\in J}$ subject to the partition $\{g^{-1}(\{y\})\}_{y\in J}$. By the disintegration theorem, applied to $\mu$, we therefore get the representation:

$$\mu(\cdot) = \int_J \rho_y(\cdot)\, dy. \tag{2.39}$$

For details regarding (2.39), we refer readers to the literature, e.g., [**BJ18**, ch2. pg 13-21] and [**Roh49**, **Roh52**, **Roh64a**, **Roh64b**]. In more detail, for each $y$, the measure $\rho_y$ is obtained by conditioning on $g^{-1}(\{y\})$; and

$$\mu(A) = \int_J \rho_y(A)\, dy, \quad A\in\mathscr{B}_1,$$

where $dy = \lambda_1(dy)$ is the standard Lebesgue measure.

THEOREM 2.11. *Let $(J,\mathscr{B}_1,\mu)$ be as above, i.e., $\mu$ is a fixed Borel measure, positive and non-atomic. Let $V_\mu : L^2(\lambda_1)\to L^2(\mu)$ be the isometry given by $g = g_\mu$, $g(x) = \mu([0,x])$, and*

$$V_\mu\psi = \psi\circ g, \tag{2.40}$$

*i.e., composition. Then the adjoint $V_\mu^* : L^2(\mu)\to L^2(\lambda_1)$ is as follows: For $f\in L^2(\mu)$, we have*

$$\left(V_\mu^* f\right)(y) = \int_{g^{-1}(\{y\})} f(\cdot)\, d\rho_y(\cdot). \tag{2.41}$$

PROOF. With the setting as in (2.39)–(2.40) above, we now get the following identity, valid for all $\psi\in L^2(\lambda_1)$, and $f\in L^2(\mu)$:

$$\begin{aligned}\int \overline{(\psi\circ g)} f\, d\mu &\underset{\text{by (2.39)}}{=} \int_J\int_{g^{-1}(\{y\})} \overline{(\psi\circ g)} f\, d\rho_y(\cdot)\, dy \\ &= \int_J \overline{\psi(y)} \int_{g^{-1}(\{y\})} f(\cdot)\, d\rho_y(\cdot)\, dy, \text{ and}\end{aligned}$$

the result (2.41) follows. □

COROLLARY 2.12. *With $\mu$ and $V_\mu$ specified as above, we then have:*

$$ker\left(V_\mu^*\right) = 0. \tag{2.42}$$

PROOF. Assume, for $f \in L^2(\mu)$, that we have $\left(V_\mu^* f\right)(\cdot) = 0$ a.e. $\lambda_1$. Recall $V_\mu^* : L^2(\mu) \to L^2(\lambda_1)$ is a (*a priori*) co-isometry.

The disintegration measures $\rho_y$ in (2.39) are obtained from the disintegration of $\mu$ as follows: $\mu(\cdot \mid g = y)$. Since $\left\{g^{-1}(\{y\})\right\}_{y \in J}$ is a measurable partition in $(J, \mathscr{B}_1, \mu)$, we get that for a.a. $y$ (w.r.t. $\lambda_1$)

$$f = 0 \text{ on } g^{-1}(\{y\}) \text{ a.e. w.r.t. } \mu \tag{2.43}$$

and therefore $f = 0$ on $J$ w.r.t. $\mu$. □

COROLLARY 2.13. $V_\mu\left(L^2(\lambda)\right) = L^2(\mu)$.

## 3. Results for Reproducing Kernel Hilbert Spaces (RKHS) associated to general measure spaces

**3.1. Gaussian Fields.** In this section we establish a close connection between generalized Gaussian fields, and associated Ito-integrals, on the one hand, and the class of positive definite kernels considered here, on the other. This section is motivated in part by a number of earlier works, for example [**AD93, BP17, EMESO17, Jr68, JT16, JT17, LMP09, MSF16, SBM07, SZ09, KS93, AL08, Tak11, AJ12**].

Fix a sigma-finite measure space $(X, \mathscr{B}, \mu)$, and a positive definite kernel $K$ on $\mathscr{B}_{fin} \times \mathscr{B}_{fin}$. We further assume that $K$ has an $L^2(\mu)$ factorization. We saw that there is an unbounded selfadjoint operator $S$, with dense domain $\mathscr{D}_{fin}$ in $L^2(\mu)$, such that we may take $\{k_A\}_{A \in \mathscr{B}_{fin}}$ to have the form $k_A = S(\chi_A)$, where $\chi_A$ is the indicator function. Hence, we may take as core domain for the operator $S$,

$$\mathscr{D}_{fin} := span\left\{\chi_A \,;\, A \in \mathscr{B}_{fin}\right\}. \tag{3.1}$$

It is a core for the full domain of $S$. Here, by "span", we mean all *finite* linear combinations.

3.1.1. *The path-space* $\Omega$. Let $\dot{\mathbb{R}}$ denote the one-point compactification of $\mathbb{R}$, and set

$$\Omega := \dot{\mathbb{R}}^{\mathscr{B}_{fin}}, \tag{3.2}$$

$\mathscr{C} :=$ the cylinder sigma-algebra of subsets of $\Omega$; and $\mathbb{P}$ be the Gaussian probability measure on $\Omega$, defined on $\mathscr{C}$, and indexed by $K$.

In details, if $F = \{A_i\}_1^n$ is a finite system, $A_i \in \mathscr{B}_{fin}$, set

$$\mathscr{C}_F := \prod_1^n A_i \times \prod_{F^c} \dot{\mathbb{R}}, \tag{3.3}$$

(a cylinder subset); and let $\mathscr{C}$ be the sigma-algebra of all subsets of $\Omega$ which is generated by the cylinder sets.

To construct $\mathbb{P}$ as a probability measure, and defined on $\mathscr{C}$, we first specify its finite-dimensional joint distributions,

$$\begin{aligned}\mathbb{P}(\cdot \mid \mathscr{C}_F) :=\ &\text{the Gaussian on } \mathbb{R}^n \text{ which has 0 mean, and}\\ &\text{covariance matrix } \left(K(A_i, A_j)\right)_{i,j=1}^n.\end{aligned} \tag{3.4}$$

By Kolmogorov's consistency property (see e.g., [**Hid80, Kol83, Kol77**]), we then get a unique probability measure $\mathbb{P}$ on $(\Omega, \mathscr{B})$ which satisfies (3.4).

For $A \in \mathscr{B}_{fin}$, and $\omega \in \Omega$ (see (3.2)), set

$$W_A(\omega) := \omega(A). \tag{3.5}$$

Then, by construction, the expectation $\mathbb{E}$, defined from $\mathbb{P}$, satisfies $\mathbb{E}(W_A) = 0$,

$$\mathbb{E}(W_A W_B) = K(A, B). \tag{3.6}$$

Specifically, each $W_A$ is Gaussian with distribution $N(0, K(A, A))$; and the joint distribution of $(W_{A_1}, W_{A_2}, \cdots, W_{A_n})$ is the Gaussian from (3.4).

Now, let

$$\begin{aligned} \mathscr{A} := \{A_i\} &= \text{a countable partition of } X, A_i \in \mathscr{B}_{fin} \\ A_i \cap A_j &= \emptyset, i \neq j, \end{aligned} \tag{3.7}$$

and let $\mathscr{C}_{\mathscr{A}} :=$ the sigma-subalgebra of $\mathscr{C}$ which is generated by $\{W_{A_i}\}$, $A_i \in \mathscr{A}$. Introducing conditional expectations, we get an inductive system of Ito-integrals, indexed by the set of partitions $\mathscr{A}$ (as above) where we use the usual ordering of partitions $\mathscr{A} \leq \mathscr{A}'$ given by refinement: If $\varphi = \sum_i \alpha_i \chi_{A_i}$,

$$\mathbb{E}\left(\left|\int \varphi dW_x\right|^2 \mid \mathscr{C}_{\mathscr{A}}\right) = \sum_i \sum_j \alpha_i \alpha_j K(A_i, A_j) \tag{3.8}$$

where the "|" stands for conditional expectation.

Passing to the limit, over the set of all partitions, we get a necessary and sufficient condition for when the Ito-integral $\int_X \varphi(x)\, dW_x$ is well defined, and is in $L^2(\Omega, \mathscr{C}, \mathbb{P})$, i.e., when $\mathbb{E}\left(\left|\int \varphi dW_x\right|^2\right) < \infty$.

3.1.2. *$L^2(\mu)$-factorizations.* We now introduce a particular $L^2(\mu)$ factorization,

$$K(A, B) = \langle k_A, k_B \rangle_{L^2(\mu)}, \quad A, B \in \mathscr{B}_{fin}. \tag{3.9}$$

We introduce the selfadjoint operator $S$, so

$$k_A = S(\chi_A). \tag{3.10}$$

For applications, it is not a serious restriction to further assume that $S$ has a bounded inverse (in $L^2(\mu)$).

THEOREM 3.1. *Let $(X, \mathscr{B}, \mu)$, $(\Omega, \mathscr{C}, \mathbb{P})$, and $\{W_A\}_{A \in \mathscr{B}_{fin}}$ be as above. We shall consider the corresponding Ito-integral $\int_X \varphi(x)\, dW_x$ for deterministic functions $\varphi$ and $\psi$ on $(X, \mathscr{B})$. We have:*

(1) *$\int \varphi dW \in L^2(\Omega, \mathscr{C}, \mathbb{P})$ iff $S(\varphi) \in L^2(\mu)$, i.e., iff $\varphi$ is in the domain of $S$.*
(2) *When (1) holds, then*

$$\mathbb{E}\left(\left|\int \varphi dW\right|^2\right) = \|S(\varphi)\|^2_{L^2(\mu)}, \tag{3.11}$$

*and*

$$\mathbb{E}\left(\left(\int \varphi dW\right)\left(\int \psi dW\right)\right) = \langle S\varphi, S\psi \rangle_{L^2(\mu)}. \tag{3.12}$$

PROOF. In view of the discussion before the statement of the theorem, it is enough to prove (3.12). We shall do this by first establishing the formula when a partition is fixed. The general case will then follow from the inductive limit argument outlined above.

In detail: Since the Ito-integral is defined as an inductive limit on partitions as specified in (3.8), to prove (3.12), it is enough to fix a partition $\mathscr{A} = \{A_i\}$, and functions $\varphi = \sum_i \alpha_i \chi_{A_i}$, $\psi = \sum_j \beta_j \chi_{A_j}$; then

$$
\begin{aligned}
\mathbb{E}\left(\left(\int \varphi dW\right)\left(\int \psi dW\right)\right) &= \sum_i \sum_j \alpha_i \beta_j \mathbb{E}\left(W_{A_i} W_{A_j}\right) \\
&\underset{(\text{by } (3.6))}{=} \sum_i \sum_j \alpha_i \beta_j K\left(A_i, A_j\right) \\
&\underset{(\text{by } (3.10))}{=} \sum_i \sum_j \alpha_i \beta_j \left\langle S\left(\chi_{A_i}\right), S\left(\chi_{A_j}\right)\right\rangle_{L^2(\mu)} \\
&= \left\langle S\left(\sum_i \alpha_i \chi_{A_i}\right), S\left(\sum_j \beta_j \chi_{A_j}\right)\right\rangle_{L^2(\mu)} \\
&= \langle S\varphi, S\psi\rangle_{L^2(\mu)}
\end{aligned}
$$

which is the desired conclusion (3.12) when the partition $\mathscr{A}$ is fixed.

The general result follows from use of the ordering of partitions, and the corresponding Kolmogorov inductive limit over all partitions; see (3.7). For more details, we refer to the cited literature on standard considerations for generalized Ito limits. □

COROLLARY 3.2 (Monotonicity). *Let the setting be as in the statement of Theorem* 3.1. *We consider the limit over partitions* $\mathscr{A}$ *as in* (3.7)*:*

$$
\mathbb{E}\left(\left|\int \varphi dW\right|^2 \mid \mathscr{C}_{\mathscr{A}}\right), \tag{3.13}
$$

*so limit over* $\mathscr{A}$. *Here* $\mathscr{C}_{\mathscr{A}} :=$ *(the sigma-algebra generated by* $\{W_{A_i}\}$ *for* $A_i \in \mathscr{A}$*); and we take monotone limit over refinements, written* $\mathscr{A} \leq \mathscr{A}'$. *Hence, for* $\mathscr{A} \leq \mathscr{A}'$, *we have:*

$$
\mathbb{E}\left(\left|\int \varphi dW\right|^2 \mid \mathscr{C}_{\mathscr{A}}\right) \leq \mathbb{E}\left(\left|\int \varphi dW\right|^2 \mid \mathscr{C}_{\mathscr{A}'}\right), \tag{3.14}
$$

*and*

$$
\sup_{\mathscr{A}} \mathbb{E}\left(\left|\int \varphi dW\right|^2 \mid \mathscr{C}_{\mathscr{A}}\right) = \mathbb{E}\left(\left|\int \varphi dW\right|^2\right) \tag{3.15}
$$

*where the supreme in* (3.15) *over all partitions is finite iff* $\varphi \in$ *domain* $(S)$.

PROOF. From the construction, we get

$$
\mathbb{E}\left(\cdot\cdot \mid \mathscr{C}_{\mathscr{A}}\right) \leq \mathbb{E}\left(\cdot\cdot \mid \mathscr{C}_{\mathscr{A}'}\right)
$$

in the ordering of projections in $L^2(\Omega, \mathscr{C}, \mathbb{P})$. Equivalently, when $\mathscr{A} \leq \mathscr{A}'$, we have:

$$
\mathbb{E}\left(\left(F \mid \mathscr{C}_{\mathscr{A}'}\right) \mid \mathscr{C}_{\mathscr{A}}\right) = \mathbb{E}\left(F \mid \mathscr{C}_{\mathscr{A}}\right),
$$

and the assertions in (3.14)–(3.15) follow immediately from this. □

**3.2. Krein-Feller diffusion.** We now turn to a detailed discussion of diffusion, and its connection to the Krein-Feller operator. As before, our starting point is a fixed measure $\mu$ supported on the real line, and we let $K_F$ be the corresponding Krein-Feller operator. The measure $\mu$ under consideration is assumed to be defined on the Borel-sigma algebra, and further assumed positive, sigma-finite, and non-atomic. The measure $\mu$ may be singular, and the main novelty in our analysis of the diffusion semigroup is for the singular case; including the case of IFS-measures. We first introduce the *centered Gaussian process* $W^{(\mu)}$ having $\mu$ as its quadratic variation. We then note that Ito's lemma applies to $W^{(\mu)}$.

Fix a measure space $(J, \mathscr{B}, \mu)$ with $J = [0, \infty)$. We introduce the corresponding positive definite kernel:

$$K_{\mu}(x, y) = \mu([0, x \wedge y]). \tag{3.16}$$

Note that if $\mu = \lambda = dx =$ the Lebesgue measure then

$$K_{\lambda}(x, y) = x \wedge y, \tag{3.17}$$

the usual *covariance function* for Brownian motion.

Starting with $\mu$ (assumed non-atomic) and the RKHS $\mathscr{H}(\mu)$, we arrive at a generalized Brownian motion $W_x^{(\mu)}$; i.e., a Gaussian process with

$$\mathbb{E}\left(W_x^{(\mu)}\right) = 0, \tag{3.18}$$

and

$$\mathbb{E}\left(W_x^{(\mu)} W_y^{(\mu)}\right) = K_{\mu}(x, y), \quad \forall x, y \in [0, \infty). \tag{3.19}$$

A detailed description of $\{W_x^{(\mu)}\}$ and its Ito-calculus is contained in many relevant papers, and books; see e.g., [**JT20, JT19b, AJL17, AJ15, AJ12, AJL11, IM74, IM63, HØUZ10, JT21**].

Here we shall need the following: Let $\varphi \in C^2$, and consider the corresponding diffusion semigroup (depending on $\mu$):

$$u(t, x) := \mathbb{E}\left(\varphi\left(W_t^{(\mu)}\right) \mid W_0^{(\mu)} = x\right) \tag{3.20}$$

where $\mathbb{E}$ in (3.20) refers to the expectation $\mathbb{E}(\cdot) = \int_{\Omega}(\cdot)\, d\mathbb{P}$ for the probability space associated to (3.16). Then $\mathbb{E}(\cdots \mid W_0^{(\mu)} = x)$ in (3.20) refers to conditioning with all paths $\omega$ s.t. $\omega(0) = x$. Here we also use the representation

$$W_t^{(\mu)}(\omega) = \omega(t), \quad \forall \omega \in \Omega \tag{3.21}$$

for the Gaussian process $\{W_t^{(\mu)}\}$. Since the Gaussian process $W^{(\mu)}$ has independent increments, it follows that eq. (3.20) defines a semigroup (see e.g., [**IM74, Phi61, Kas12, KS16**]), and we shall call it the *Markov semigroup*. Its properties and its infinitesimal generator will be identified in Lemma 3.3 below, and in the subsequent discussion.

The Gaussian process from (3.19) and (3.21) is often called a generalized Brownian motion, or a Gaussian field. The associated Ito-integral is also used in the proof of Lemma 3.3 below.

We now introduce the Krein-Feller operator

$$K_F := \frac{\partial^2}{\partial\mu\partial x} \tag{3.22}$$

(see above.) In our discussion of (3.20), we shall consider $K_F$ as acting in the $x$-variable.

LEMMA 3.3. *The diffusion* (3.20) *is generated by the following generalized heat equation:*

$$\frac{\partial u}{\partial t} = \frac{1}{2} K_F u, \quad u(0,x) = \varphi(x) \tag{3.23}$$

*where $\varphi$ is a fixed continuous function.*

REMARK 3.4. The special case of (3.23) corresponding to $\mu = \lambda =$ Lebesgue measure is

$$\frac{\partial u}{\partial t} = \frac{1}{2} \frac{\partial^2 u}{\partial x^2}, \quad u(0,x) = \varphi(x). \tag{3.24}$$

PROOF. We shall refer to the literature, e.g., [**JT20, JT19b, AJL17, AJ15, JT21**]. Suffice it to say that *the Ito-lemma* for the Gaussian process $\{W_t^{(\mu)}\}$ is a key tool; together with the following fact for the quadratic variation:

$$\left(dW_t^{(\mu)}\right)^2 = \mu(dt). \tag{3.25}$$

□

## 4. Dual pairs

Below we show that every positive non-atomic Borel measure (see Lemma 2.1) gives rise to a naturally associated dual pair of operators in the sense of (4.1), and we discuss selfadjoint extensions of $K_F$.

Set $J = [0,1]$. Fix $\mu$, $\sigma$-finite and non-atomic. Let $\lambda = dx =$ Lebesgue measure, and consider the following diagram

$$\mathscr{L}^2(\mu) \underset{D=-d/dx}{\overset{T_\mu}{\rightleftarrows}} L^2(\mu) \tag{4.1}$$

with

$$\mathscr{L}^2(\mu) := \overline{dom(T_\mu)}^{L^2(\lambda)}.$$

Let $K_F$ be as in (3.22), then

$$K_F \subseteq T_\mu T_\mu^*, \tag{4.2}$$

both are operators in $L^2(\mu)$. But $T_\mu T_\mu^*$ is *selfadjoint* by general theory (see e.g., [**DS88**]), and restriction of symmetric is symmetric.

THEOREM 4.1. *Consider a fixed positive non-atomic Borel measure $\mu$ on $J = [0,b]$, $b < \infty$; and define an operator $A$ on $L^2(\mu) = L^2(J,\mu)$ as follows:*

*For $\varphi \in L^2(\mu) \cap C$, set*

$$(A\varphi)(x) = \int_0^x \left( \int_0^y \varphi(s)\, \mu(ds) \right) dy, \tag{4.3}$$

*then*

$$K_F A\varphi = \varphi. \tag{4.4}$$

PROOF. Since $L^2(\mu) \subset L^1(\mu)$ the function $y \longmapsto \int_0^y \varphi(s)\,\mu(ds)$ is continuous, and so $x \longmapsto (A\varphi)(x)$ is $C^1$ (one time differentiable with $(A\varphi)' \in C(J)$.) Hence, for the LHS of (4.4) we have

$$K_F A : \varphi \longmapsto \nabla_\mu \frac{d}{dx} A\varphi = \nabla_\mu \int_0^x \varphi(s)\,\mu(ds) = \varphi(x),$$

which is the desired conclusion (4.4). □

COROLLARY 4.2. *Let the measure $\mu$ be specified as above, and set*

$$(A\varphi)(x) = \int_0^x \left( \int_0^y \varphi(s)\,\mu(ds) \right) dy$$

*(see (4.3)), defines a compact integral operator in $L^2(\mu)$.*

*Then $A = A_\mu$ is bounded and compact with triangular integral kernel*

$$a(x,s) = \chi_{[0,x]}(s)(x-s). \tag{4.5}$$

PROOF. Without loss of generality we may work with $b = 1$ and real Hilbert space. An easy application of Schwarz to $L^2(\mu)$ shows that $A : L^2(\mu) \to L^2(\mu)$ is bounded. We have

$$\langle A\varphi, \psi \rangle_{L^2(\mu)} = \int_0^1 \int_0^x (x-s)\,\varphi(s)\,\psi(x)\,\mu(ds)\,\mu(dx). \tag{4.6}$$

The asserted symmetry follows from this, or equivalently,

$$(A\varphi)(x) = \int_0^1 a(x,s)\,\varphi(s)\,\mu(ds) \tag{4.7}$$

where $a(x,s) = \chi_{[0,x]}(s)(x-s)$, see Figure 4.1.

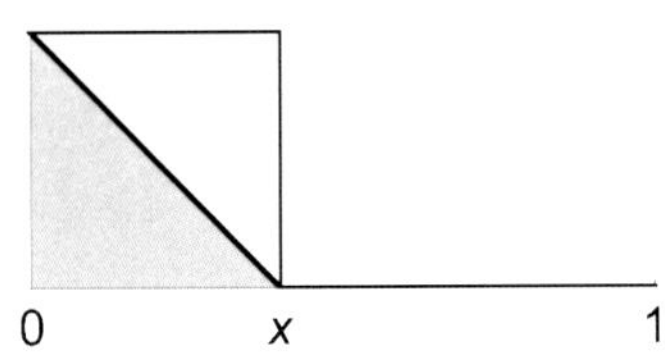


FIGURE 4.1. The kernel $a(x,s) = \chi_{[0,x]}(s)(x-s) = [x-s]_+ = \max(0, x-s)$.

To see that the operator $A$ of (4.3) is compact as an operator in $L^2(\mu)$, we make use of (4.6) and (4.7) as follows: We recall the fact the compact operators are the norm-closure of finite rank operators. We then create such norm-limits of finite rank operators with the use of the kernel (4.5), and a choice of a filter of partitions $P$ with disjoint Borel subsets of the support of $\mu$. For each such partition $P$, we form rank-one operators from the corresponding indicator functions from pairs of sets $B, B'$ in $P$, and we then form the associated span of the rank-one operators $|\chi_B\rangle\langle\chi_{B'}|$ by evaluation of (4.5) with sample points chosen from the partition sets. (We use Dirac's terminology $|\cdot\rangle\langle\cdot|$ for rank-one operators.) The Borel sets $B$, making up partitions, are chosen with $\mu(B)$ finite, so the corresponding indicator functions are in $L^2(\mu)$. For a fixed partition, we then form pairs of such indicator functions, and the corresponding rank-one operators. Since $\mu$ is chosen non-atomic, the partition-limit refinements can be constructed such that the limit of the corresponding numbers $\mu(B)$ is zero. □

The following result gives a spectral decomposition for the Krein-Feller operator $K_F$.

COROLLARY 4.3. *Let $J = [0, 1]$, and $\mu$ be a non-atomic positive Borel measure on $J$. Set $g(x) = \mu([0, x])$.*

*Consider $K_F = \frac{d}{d\mu}\frac{d}{dx}$ as a selfadjoint operator in $L^2(\mu)$ with Neumann boundary condition, i.e.,*

$$dom(K_F) = \left\{ \psi(x) = \psi(0) + \int_0^x \left( \int_0^y f d\mu \right) dx : f \in L^2(\mu),\ \psi'(0) = \psi'(1) = 0 \right\}.$$

*Let $V : L^2(\mu) \to L^2(\mu)$ be the integral operator defined as*

$$Vf(x) = \int_0^1 H(x, t) f(t)\, dt \tag{4.8}$$

*where*

$$H(x, t) = \begin{cases} (g(x) - 1) g(t) & t \leq x, \\ (g(t) - 1) g(x) & x \geq t. \end{cases} \tag{4.9}$$

*Then $V$ is compact and selfadjoint, and we have the following eigenvalue correspondence:*

$$c \in sp(K_F) \Longleftrightarrow c^{-1} \in sp(V). \tag{4.10}$$

PROOF. Selfadjointness of $V$ follows from (4.9), and the argument for compactness is the same as that of Corollary 4.2.

Let $\psi(x) = \psi(0) + \int_0^x \left( \int_0^t f d\mu \right) dt \in dom(K_F)$, where $f \in L^2(\mu)$, so that $K_F\psi(x) = f(x)$. Assume that

$$K_F\psi = c\psi,$$

for some constant $c < 0$. Then,

$$\begin{gathered} K_F\psi = c\psi \\ \Updownarrow \\ f(x) = c \int_0^x \left( \int_0^t f d\mu \right) dt \\ \Downarrow \\ f'(x) = c \int_0^x f d\mu. \end{gathered}$$

The last line can be written as follows:

Set $g(x) := \mu([0,x])$, then

$$\begin{aligned} f'(x) &= c\int_0^x f d\mu \\ &= c\int_0^x f dg \\ &= c\left(f(x)g(x) - \int_0^x g(t) f'(t)\,dt\right) \\ &= c\left(\left(f(0) + \int_0^x f(s)\,ds\right)g(x) - \int_0^x g(t) f'(t)\,dt\right) \\ &= cf(0)g(x) + c\int_0^x (g(x) - g(t)) f'(t)\,dt. \end{aligned} \tag{4.11}$$

Since $f' = c\psi'$, the boundary condition $\psi'(1) = 0$ and (4.11) imply that

$$cf(0) + c\int_0^1 (1 - g(t)) f'(t)\,dt = 0. \tag{4.12}$$

Substitute (4.12) into (4.11), then

$$\begin{aligned} f'(x) &= c\int_0^x (g(x) - g(t)) f'(t)\,dt + cf(0)g(x) \\ &= c\left(\int_0^x (g(x) - g(t)) f'(t)\,dt - g(x)\int_0^1 (1 - g(t)) f'(t)\,dt\right) \\ &= c\left(\int_0^x (g(x) - 1) g(t) f'(t)\,dt + \int_x^1 g(x)(g(t) - 1) f'(t)\,dt\right) \\ &= c\int_0^1 H(x,t) f'(t)\,dt \end{aligned}$$

with $H$ as defined in (4.9), and the assertion (4.10) follows. □

## 5. An application

In this section we introduce a notion of implied probability distributions. From the general Black-Scholes-Merton theory of option prices [**BSS23**], one studies how choices of numeraires, and the assumption of completeness of markets lead to explicit formulas for such standard options as calls and puts. With this one obtains realization of associated martingales. Here we turn the tables, addressing the question of inducing probability distributions from option prices [**WW97**, **WHD95**]. More precisely we present a duality between probability measures and call option prices $C(t,T,r,q,K)$, with the variable terminology as usual, $t$ calendar time, $T$ strike time, $K$ strike price, and $\sigma$ volatility. Our analysis begins with our consideration of the payoff at maturity as a $K$ ReLu function [**JLL23**, **HS23**]. And we obtain the probability measures as a double partial derivative of $C(t,T,r,q,K)$ in the $K$ variable.

**5.1. Singular calculus of ReLu functions.** While the framework of functions discussed is wider than that of options, we have been motivated by the latter. An option to buy is denoted call, written $C$, and to sell, put, written $P$. Both refer to a future time, called strike or expiration, denoted $T$. The asset that is bought or sold is referred to as an underlying, and its value $S(t)$ is a stochastic process [**DR23, GKLH22**]. The options are traded at calendar times between 0 and $T$. A call gives the holder and option (but not an obligation) to sell at $\$K$ at time $T$. A put gives the holder an option to buy at $\$K$ at time $T$. Hence graph of the pay-off functions for these options take the form of hockey sticks (see Fig 5.1 below); they are examples of ReLu functions from neural networks.

Considering distributional derivatives, also termed as weak derivatives:

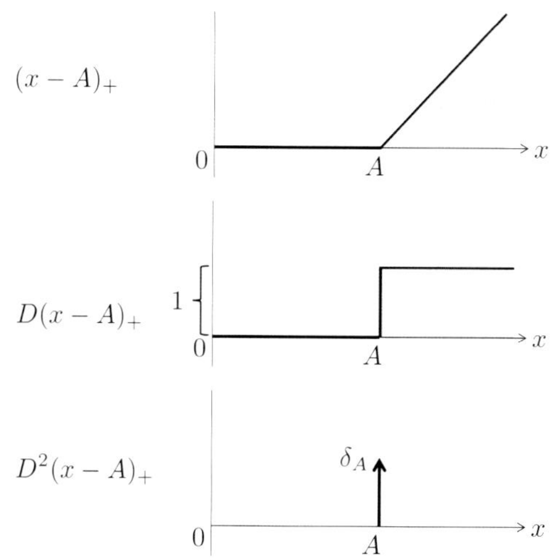


FIGURE 5.1. A ReLu function: its first and its second distributional derivative.

In our Figure 5.1 illustration, use is made of weak derivative, also referred to as derivative in the sense of Laurent Schwartz distributions. It will be used also in subsequent proofs involving stochastic calculus elsewhere in our present section. For the benefit of readers, we offer pointers for details in the references [**Sch88, Sch67**].

A financial option $C$ is an instrument whose value depends on an underlying risky asset $S$, for example a stock. Additionally, the pricing of an option $C$ depends on a strike time $T$ and a strike price $K$, as well as the other parameters which determine the risky asset $S$, e.g., rates and volatility. This is summarized in (5.1) below, with $\mathbb{E}$ representing a conditional expectation [**HS18, OKT17**]. For our present option $C$, the payoff at $T$ is given as a ReLu function $(S-K)_+$, but the nature of the random variable $S(T)$ is unknown. Nonetheless, there are a diversity of tools for pricing the option $C(t)$ at values of time $t < T$. See e.g., [**Bly14, WHD95, WW97**]. Existing techniques in the literature, typically start with stochastic calculus techniques [**ESH22**], based on arriving on probabilities from risk-neutral Ito calculus. Our present approach goes in the other direction, we start with a function $C$ as in (5.1) and we then derive the implied probability which, when used on the RHS in (5.1), will yield the function $C$. In our analysis the strike price $K$ will play a central role, and it will be convenient for us to indicate this fact with the subscript notation $C_K$.

Furthermore, based on Blyth's work [**Bly14**], we have:

$$C_K = \mathbb{E}\left((S-K)_+\right). \tag{5.1}$$

S. Blyth has introduced specific conditions which integrate distribution theory with Malliavin calculus, further detailed in Denis R. Bell's book [**Bel06**].

The option price formula (5.1) is a simplification. There is also a discount factor, called a zero-coupon bond $Z(t,T)$ [**CEJP22**]. Zero-coupon bonds depend on discount rates, yield curves. They trade at discounts and offer full face value (par) profits at maturity, i.e., $Z(T,T)=1$. For our initial discussion, $Z(t,T)$ does not play a role, but we shall resume a further discuss on of $Z(t,T)$ in section 5.4 below.

Consider a probability space $(\Omega,\mathscr{F},\mathbb{P})$ with expectation $\mathbb{E}$, and let $S:\Omega\to\mathbb{R}_{\geq 0}$ be a random variable with distribution $\mu$ i.e.,

$$\mu := \mathbb{P}\circ S^{-1}. \tag{5.2}$$

Then consider the function of $K$, $0\leq K<\infty$, defined by

$$C_K : K \longmapsto \mathbb{E}\left((S-K)_+\right) \tag{5.3}$$

where

$$(S-K)_+ = \max(0, S-K), \tag{5.4}$$

and $\mathbb{E}$ denotes a conditional expectation.

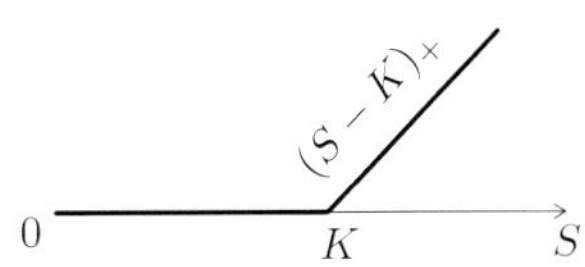


FIGURE 5.2. The function which represent pay-off for call options.

REMARK 5.1. Note that the probability densities (distributions) which are determined via the implied method (see also Theorem 5.2) are conditional distributions. More specifically, when measurements are made at calendar time $t<T$ (the strike time), then the $K$-implied method will only yield the distribution of $S(T)$ conditioned by $S(t)$. Moreover, the implied method will not determine the stochastic process $S$. Now in the other direction, with Black-Scholes, we input a Log-normal process for $S$. And so when the implied method is applied, we will recover log-normal distributions.

**5.2. Implied Distributions.** From the definition of options as sketched in sec 5.1 above, they involve values of the underlying at expiration $T$. But the option has a market value, say $C(t)$ at every calendar time $t$ between 0 and $T$. Hence it may be evaluated as a $t$-conditional expectation of $S(T)$ and our object is to identify the corresponding implied probabilities.

Then there is a Schwartz/Malliavin (see e.g., [**Mal97, Sch75, Sch64**]) notion of derivative, and we arrive at the following:

THEOREM 5.2. $\frac{\partial}{\partial K}C_K = \operatorname{dist}(S)=\mu$, *i.e., $\mu$ is determined by*

$$\mathbb{E}(\varphi(S)) = \int_0^\infty \varphi(x)\,\mu(dx) \underset{def}{\Longleftrightarrow} \mu = dist(S). \tag{5.5}$$

*Moreover, let $g\in C^2$ and $K\in[0,\infty]$, then*

$$g(x) = g(0) + g'(0)\,x + \int (x-K)_+\, g''(K)\,dK. \tag{5.6}$$

*With the substitution* $S \longleftrightarrow x$, *an application of* $\mathbb{E}$ *to both sides of the above equation yields:*

$$\mathbb{E}\left(g\left(S\right)\right)=g\left(0\right)+g'\left(0\right)S+\int_0^\infty C_K\left(S\right)g''\left(K\right)dK. \tag{5.7}$$

PROOF. For clarify, we shall consider measures on $(\mathbb{R},\mathscr{B})$, $\mathscr{B}$ denoting the Borel $\sigma$-algebra, as linear functionals with $\langle\cdot,\cdot\rangle$ terminology $\langle\varphi,\mu\rangle:=\int\varphi\left(x\right)\mu\left(dx\right)$ where $\varphi$ is a Borel function on $\mathbb{R}$.

Let $S$ denote a random variable, or more generally a stochastic process, and let $\mu$ denote its probability distribution (p.d.f.); we then have

$$\mathbb{E}\left(\left(S-K\right)_+\right)=\int_{\mathbb{R}}\left(x-K\right)_+\mu\left(dx\right),$$

and more generally

$$\mathbb{E}\left(\left(\varphi\left(S\right)-K\right)_+\right)=\int_{\mathbb{R}}\left(\varphi\left(x\right)-K\right)_+\mu\left(dx\right).$$

Introduce the convolution product $*$, we arrive at the following:

$$\begin{aligned}\left\langle\varphi\left(\cdot\right),\int_K\left(\cdot-K\right)_+d\mu\left(\cdot\right)\right\rangle&=\int_K^\infty\varphi\left(x\right)\left(x-K\right)_+\mu\left(dx\right)\\&=\left\langle\varphi,\left(\left(x-\cdot\right)_+*\mu\right)\left(K\right)\right\rangle\\&=\left\langle\varphi*\left(\cdot-K\right)_+,\mu\right\rangle.\end{aligned}$$

As a result,

$$\begin{aligned}\frac{\partial}{\partial K}\underbrace{\left\langle\varphi\left(\cdot\right),\int_K\left(\cdot-K\right)_+d\mu\left(\cdot\right)\right\rangle}_{\mathbb{E}\left(\left(S-K\right)_+\right)}&=\left\langle\frac{\partial}{\partial K}\varphi*\left(\cdot-K\right)_+,\mu\left(\cdot\right)\right\rangle\\&=\left\langle\varphi*\underbrace{\frac{\partial}{\partial K}\left(\cdot-K\right)_+}_{\delta_0},\mu\left(\cdot\right)\right\rangle\\&=\left\langle\varphi*\delta_0,\mu\right\rangle=\left\langle\varphi,\mu\right\rangle\end{aligned}$$

which is the desired conclusion; see the statement of Theorem 5.2. □

We shall consider the following more general framework:

THEOREM 5.3. *Setting* $C_K:=\mathbb{E}\left(\left(S-K\right)_+\right)$ *as a function of* $K\in[0,\infty]$, *where* $S$ *is a random variable with distribution* $\mu$.

(1) *It holds that*

$$dist\left(S\right)\ll\nu\Longleftrightarrow\frac{\partial^2}{\partial\nu\partial K}C_K\neq0. \tag{5.8}$$

*In the case* $\operatorname{dist}\left(S\right)\ll\nu$, *then*

$$\mu=dist\left(S\right)=\left(\frac{d\mu}{d\nu}\right)d\nu, \tag{5.9}$$

*where* $d\mu/d\nu$ *is the Radon-Nikodym derivative (or, more generally, the Krein-Feller derivative; see* [**MF23, JT23, AN06**]*).*

(2) *There is a set of mutually singular measures* $\nu$ *such that*

$$\mu \in \sum^{\oplus} \mathscr{H}(\nu), \quad \frac{d\mu}{d\nu} \in \mathscr{H}(\nu). \tag{5.10}$$

PROOF. For part (1), one checks that

$$\begin{aligned}
\mathbb{E}\left[(S-K)_{+}\right] &= \int (x-K)_{+}\,\mu(dx) \\
&\Downarrow \\
\frac{\partial}{\partial K}\mathbb{E}\left[(S-K)_{+}\right] &= \int \frac{d}{dk}(x-K)_{+}\,\mu(dx) \\
&= -\int \chi_{(-\infty,x)}(K)\,\mu(dx) \\
&= -\int \chi_{(K,\infty)}(x)\,\mu(dx) \\
&= \mu((-\infty,K)) - 1
\end{aligned} \tag{5.11}$$

Therefore, $\mu \ll \nu$ if and only if

$$\frac{\partial^2}{\partial\nu\partial K}\mathbb{E}\left[(S-K)_{+}\right] = \frac{d\mu}{d\nu} \neq 0.$$

For part (2) and the universal Hilbert space ($\sigma$-densities), we refer to [**Nel69**, **BJ18**, **AJ15**, **Jr14**]. □

REMARK 5.4. Let $\mu = \mathrm{dist}(S) =$ the distribution of the random variable

$$\Omega \xrightarrow{S} \mathbb{R}_{\geq 0}. \tag{5.12}$$

That is, $\mu = \mathbb{P}\circ S^{-1}$, where $\mathbb{P}$ is the probability measure from the probability space $(\Omega, \mathscr{F}, \mathbb{P})$. Since $\mu$ is a Borel probability measure on $(\mathbb{R}, \mathscr{B})$, it is also the Stieljes measure of the Cauchy distribution, i.e.,

$$\begin{aligned}
F(K) = F_{\mu}(K) &= \mu([0,K]) = \mathbb{E}\left(\chi_{[0,K]}\right) \\
&= \mathbb{P}(S \leq K)\,(= \mathbb{P}(\{\omega; S(\omega) \leq K\})).
\end{aligned} \tag{5.13}$$

We write $\mu = dF/d\lambda$ (terminology from standard measure theory), and so $\mu$ is the unique Borel measure with

$$\mu((x_0, x_1]) := F(x_1) - F(x_0), \quad \forall x_0, x_1 \in \mathbb{R}_{\geq 0}, x_0 \leq x_1.$$

Indeed, every Borel probability measure $\mu$ is of the form $\mu = dF$ (Stieljes measure). Now start with $S$ on $(\Omega, \mathscr{F}, \mathbb{P})$ and set

$$C_K := \mathbb{E}\left((S-K)_{+}\right), \quad \forall K \in [0,\infty) \tag{5.14}$$

then from the distribution derivative $\frac{\partial}{\partial K}$ apply to (5.14) we get (5.11) (see Figure 5.1).

**5.3. Option prices as stochastic integrals.** Below we take a closer look at the way option prices arise as functions of $K$, the strike price.

By utilizing $C_K : x \to (x-K)_{+}$, we generalize the ReLu function. If $S$ is a real-valued random variable, then $C_K(S) := (S-K)_{+}$ is defined in Blyth's work (and in the literature more broadly) as a general function. Essentially, all options can be generalized from the set $\{C_K(S)\}$ options where $K \in (0,\infty]$.

There is a relation between Taylor's formula and our analysis, expressed as follows:

$$g(x) = g(0) + g'(0)\,x + \int_0^\infty (x-K)_+\, g''(K)\, dK \tag{5.15}$$

We introduce a random variable $S$ representing the price of a risky asset (e.g., stocks). With the substitution $S \leftrightarrow x$ in (5.15), then

$$g(S) = g(0) + g'(0)\,S + \int_0^\infty (S-K)_+\, g''(K)\, dK. \tag{5.16}$$

Taking expectation, the right-hand side of (5.16) becomes:

$$\mathbb{E}(g(S)) = g(0) + g'(0)\,\mathbb{E}(S) + \int_0^\infty C_K(S)\, g''(K)\, dK. \tag{5.17}$$

Note that the distribution of $S$, $\text{dist}(S) = \mu$, may possibly be singular, meaning it is not of the form $\mu(dx) = f_S(x)\,dx$.

In the discussion involving (5.15)–(5.17), we further refine these expressions using calculus. Essentially, the general option prices $\mathbb{E}(g(S))$ can be computed using the ReLu functions leading to $C_K(S)$:

$$g(x) \to g(S) \to \mathbb{E}(g(S)). \tag{5.18}$$

Especially, fix $x \geq 0$, and consider the formula $(x-K)_+$ as a function of $K$

$$(x-K)_+ = \begin{cases} x-K & \text{if } x \geq K \\ 0 & 0 \text{ otherwise} \end{cases}$$

The derivatives with respect to $K$ are given as in Figure 5.3.

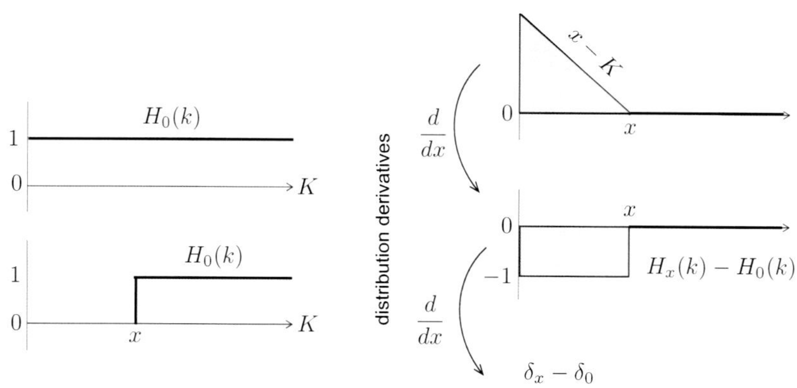


FIGURE 5.3. Analysis of functions representing pay-off for a put option.

**Summary: Big picture.** Let $g \in C^2$. Introducing a ReLu system, $x \to (x-K)_+$or equivalently, $K \to (x-K)_+$, we have (5.15). So, we verify (5.15) in the form

$$g(x) - g(0) - g'(0)\,x = \int_0^\infty (x-K)_+\, g''(K)\, dK \tag{5.19}$$

in two steps:

(1) Verify the equivalence of the two derivatives $\frac{d}{dx}$ and $\left(\frac{d}{dx}\right)^2$

(2) Ensure the boundary conditions align, as shown in Figure 5.3:

$$\frac{d}{dx}\mathrm{LHS}_{(5.19)} = g'(x) - g'(0)$$

$$\frac{d}{dx}\mathrm{RHS}_{(5.19)} = \frac{d}{dx}\int_0^\infty (x-K)_+ g''(K)\,dK$$

$$= \int_0^x g''(K)\,dK = g'(x) - g'(0) = \mathrm{LHS}_{(5.19)}.$$

We can derive general options $\mathbb{E}(g(S))$ for $g \in C^2$, for spread options. Specifically, for a plain vanilla European option:

$$C_K(S) = \mathbb{E}((S-K))_+ .$$

This is governed by the formula:

$$\mathbb{E}(g(S)) = g(0) + g'(0)\,\mathbb{E}(S) + \int_0^\infty C_K(S)\,g''(K)\,dK.$$

Note $S$ is determined by its distribution, $\mathrm{dist}(S) = \mathbb{P} \circ S^{-1}$. We seek to derive $\mathrm{dist}(S)$ from $\frac{\partial}{\partial K}C_K(S)$, and $\left(\frac{\partial}{\partial K}\right)^2 C_K(S)$. The approximations for these derivatives are given by:

$$\frac{\partial}{\partial K}C_K \sim \lambda\left(C_K - C_{K+\frac{1}{\lambda}}\right)$$

$$\left(\frac{\partial}{\partial K}\right)^2 C_K \sim \lambda^2\left(C_{K-\frac{1}{\lambda}} - 2C_K + C_{K+\frac{1}{\lambda}}\right)$$

Thus, as $\lambda \to \infty$, we have

$$\frac{C_K - C_{K+\frac{1}{\lambda}}}{\frac{1}{\lambda}} \xrightarrow[\lambda\to\infty]{} -\frac{\partial}{\partial K}C_K$$

$$\text{Butterfly} \xrightarrow[\lambda\to\infty]{} \left(\frac{\partial}{\partial K}\right)^2 C_K$$

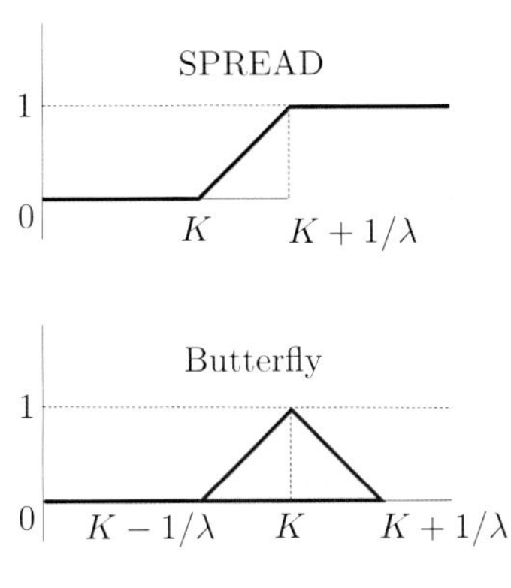


FIGURE 5.4. The options which represent (finite element) difference operations, computed for first (spread), and for second order (butterfly).

**5.4. Butterfly Options.** In Section 5.3, the finite element approximation of the double derivative with respect to $K$ uses double differences of the ReLu functions. One sees that their graphs resemble butterflies. Hence, these are referred to as butterfly options.

**Pricing vanilla European call options.** The $C_K(t,T)$ price is denoted in dollars and depends on time:

$$C_K(t,T) = \underbrace{Z(t,T)}_{\text{zero coupon bound}} \mathbb{E}_A\left((S_T - K)_+ \mid \underbrace{S_t}_{\text{conditioning}}\right)$$

or

$$C_K(t,T) = e^{-r(T-t)}\mathbb{E}_*\left((S_T - K)_+ \,|S_t\right)$$

A vanilla call option provides the right to buy a stock at a future time $T$. At expiration $T$, one may choose to exercise the option or not. Exercising the option is beneficial if $S_T - K > 0$ (otherwise not), leading to a payment of $(S_T - K)_+$.

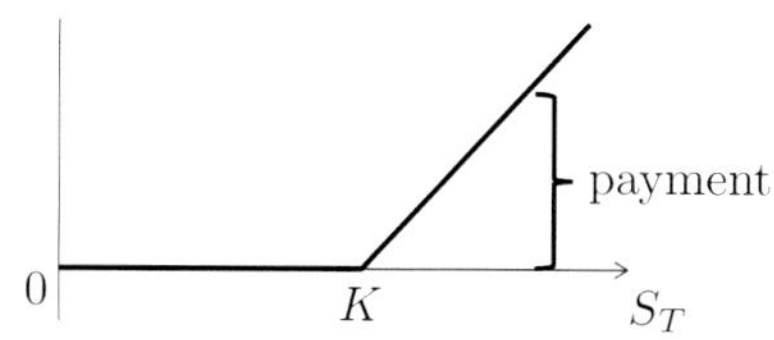


FIGURE 5.5. A $K$-diagram.

When the distribution of $S_T$ is known or assumed, we have:

$$C_K(t,T) = Z(h,T)\, S_T \mid S_t(x)\, dx.$$

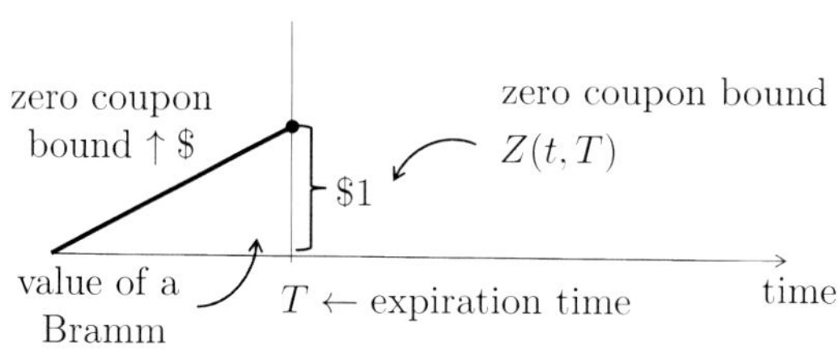


FIGURE 5.6. Zero-coupon bonds.

At expiration, $Z(t,T)$ represents a zero-coupon bond:

$$Z(t,T) = e^{-r(T-t)}$$

where $r$ is the risk-free bank interest rate.

The option $C_K(t,T)$ grants the right to purchase the stock at a predetermined price $K$, termed the strike price, at the future time $T$. The option is exercised if $S_T > K$. The probability $Prob^*(S_T > K)$ depends on the distribution of $S_T$, and $C_K(t,T)$ relies on $Prob^*(S_T \mid \mathscr{F}_{S_t})$, where $S_t$ is the known stock price at calendar time $t$.

## References


[AD93] Daniel Alpay and Harry Dym, *On a new class of structured reproducing kernel spaces*, J. Funct. Anal. **111** (1993), no. 1, 1–28, DOI 10.1006/jfan.1993.1001. MR1200633

[AJ12] Daniel Alpay and Palle E. T. Jorgensen, *Stochastic processes induced by singular operators*, Numer. Funct. Anal. Optim. **33** (2012), no. 7-9, 708–735, DOI 10.1080/01630563.2012.682132. MR2966130

[AJ15] Daniel Alpay and Palle Jorgensen, *Spectral theory for Gaussian processes: reproducing kernels, boundaries, and $L^2$-wavelet generators with fractional scales*, Numer. Funct. Anal. Optim. **36** (2015), no. 10, 1239–1285, DOI 10.1080/01630563.2015.1062777. MR3402823

[AJ21] Daniel Alpay and Palle E. T. Jorgensen, *New characterizations of reproducing kernel Hilbert spaces and applications to metric geometry*, Opuscula Math. **41** (2021), no. 3, 283–300, DOI 10.7494/opmath.2021.41.3.283. MR4302453

[AJ22] Daniel Alpay and Palle Jorgensen, *Finitely additive functions in measure theory and applications*, arXiv e-prints (2022), `arXiv:2205.07462`.

[AJL11] Daniel Alpay, Palle Jorgensen, and David Levanony, *A class of Gaussian processes with fractional spectral measures*, J. Funct. Anal. **261** (2011), no. 2, 507–541, DOI 10.1016/j.jfa.2011.03.012. MR2793121

[AJL17] Daniel Alpay, Palle Jorgensen, and David Levanony, *On the equivalence of probability spaces*, J. Theoret. Probab. **30** (2017), no. 3, 813–841, DOI 10.1007/s10959-016-0667-7. MR3687240

[AJL18] Daniel Alpay, Palle Jorgensen, and Izchak Lewkowicz, *W-Markov measures, transfer operators, wavelets and multiresolutions*, Frames and harmonic analysis, Contemp. Math., vol. 706, Amer. Math. Soc., [Providence], RI, [2018] ©2018, pp. 293–343, DOI 10.1090/conm/706/14219. MR3796644

[AL08] Antoine Ayache and Werner Linde, *Approximation of Gaussian random fields: general results and optimal wavelet representation of the Lévy fractional motion*, J. Theoret. Probab. **21** (2008), no. 1, 69–96, DOI 10.1007/s10959-007-0101-2. MR2384473

[AN06] Sergio Albeverio and Leonid Nizhnik, *A Schrödinger operator with a $\delta'$-interaction on a Cantor set and Krein-Feller operators*, Math. Nachr. **279** (2006), no. 5-6, 467–476, DOI 10.1002/mana.200310371. MR2213587

[Bel06] Denis R. Bell, *The Malliavin calculus*, Dover Publications, Inc., Mineola, NY, 2006. Reprint of the 1987 edition. MR2250060

[BF21] Simon Baker and Michael Farmer, *Quantitative recurrence properties for self-conformal sets*, Proc. Amer. Math. Soc. **149** (2021), no. 3, 1127–1138, DOI 10.1090/proc/15285. MR4211868

[BJ18] Sergey Bezuglyi and Palle E. T. Jorgensen, *Transfer operators, endomorphisms, and measurable partitions*, Lecture Notes in Mathematics, vol. 2217, Springer, Cham, 2018, DOI 10.1007/978-3-319-92417-5. MR3793614

[BJ19] Sergey Bezuglyi and Palle E. T. Jorgensen, *Graph Laplace and Markov operators on a measure space*, Linear systems, signal processing and hypercomplex analysis, Oper. Theory Adv. Appl., vol. 275, Birkhäuser/Springer, Cham, 2019, pp. 67–138. MR3968033

[Bly14] Stephen Blyth, *An introduction to quantitative finance*, Oxford University Press, Oxford, 2014. MR3115237

[BP17] Christopher J. Bishop and Yuval Peres, *Fractals in probability and analysis*, Cambridge Studies in Advanced Mathematics, vol. 162, Cambridge University Press, Cambridge, 2017, DOI 10.1017/9781316460238. MR3616046

[BR87] Ola Bratteli and Derek W. Robinson, *Operator algebras and quantum statistical mechanics. 1: $C^*$- and $W^*$-algebras, symmetry groups, decomposition of states*, 2nd ed., Texts and Monographs in Physics, Springer-Verlag, New York, 1987, DOI 10.1007/978-3-662-02520-8. MR887100

[BSS23] L. N. Bakirova, M. A. Shurygina, and V. V. Shurygin Jr., *Symmetries of the Black-Scholes-Merton equation for European options*, Lobachevskii J. Math. **44** (2023), no. 4, 1256–1263. MR4621597

[CEJP22] Rafael Company, Vera N. Egorova, Lucas Jódar, and Jorge Peris, *A front-fixing method for American option pricing on zero-coupon bond under the Hull and White model*, Math. Methods Appl. Sci. **45** (2022), no. 6, 3334–3344, DOI 10.1002/mma.7505. MR4395658

[CP68] Michael G. Crandall and Ralph S. Phillips, *On the extension problem for dissipative operators*, J. Functional Analysis **2** (1968), 147–176, DOI 10.1016/0022-1236(68)90015-3. MR231220

[DL17] Qi-Rong Deng and Ka-Sing Lau, *Structure of the class of iterated function systems that generate the same self-similar set*, J. Fractal Geom. **4** (2017), no. 1, 43–71, DOI 10.4171/JFG/44. MR3631376

[DR23] Amaury Durand and François Roueff, *Weakly stationary stochastic processes valued in a separable Hilbert space: Gramian-Cramér representations and applications*, ESAIM Probab. Stat. **27** (2023), 776–809, DOI 10.1051/ps/2023014. MR4634740

[DS88] Nelson Dunford and Jacob T. Schwartz, *Linear operators. Part II: Spectral theory. Selfadjoint operators in Hilbert space*, Wiley Classics Library, John Wiley & Sons, Inc., New York, 1988. With the assistance of William G. Bade and Robert G. Bartle; Reprint of the 1963 original; A Wiley-Interscience Publication. MR1009163

[EMESO17] Mohamed El Machkouri, Khalifa Es-Sebaiy, and Idir Ouassou, *On local linear regression for strongly mixing random fields*, J. Multivariate Anal. **156** (2017), 103–115, DOI 10.1016/j.jmva.2017.02.002. MR3624688

[ESH22] A. M. A. El-Sayed and H. H. G. Hashem, *Stochastic Itô-differential and integral of fractional-orders*, J. Fract. Calc. Appl. **13** (2022), no. 2, 251–258. MR4435796

[Fre11] Uta Freiberg, *Refinement of the spectral asymptotics of generalized Krein Feller operators*, Forum Math. **23** (2011), no. 2, 427–445, DOI 10.1515/FORM.2011.017. MR2787628

[GKLH22] Battulga Gankhuu, Jacob Kleinow, Altangerel Lkhamsuren, and Andreas Horsch, *Dividends and compound Poisson processes: a new stochastic stock price model*, Int. J. Theor. Appl. Finance **25** (2022), no. 3, Paper No. 2250014, 36, DOI 10.1142/S0219024922500145. MR4446297

[GL20] Qingsong Gu and Ka-Sing Lau, *Dirichlet forms and critical exponents on fractals*, Trans. Amer. Math. Soc. **373** (2020), no. 3, 1619–1652, DOI 10.1090/tran/8004. MR4068276

[GLQ19] Qingsong Gu, Ka-Sing Lau, and Hua Qiu, *On a recursive construction of Dirichlet form on the Sierpiński gasket*, J. Math. Anal. Appl. **474** (2019), no. 1, 674–692, DOI 10.1016/j.jmaa.2019.01.070. MR3912922

[Hid80] Takeyuki Hida, *Brownian motion*, Applications of Mathematics, vol. 11, Springer-Verlag, New York-Berlin, 1980. Translated from the Japanese by the author and T. P. Speed. MR562914

[HJW19] John E. Herr, Palle E. T. Jorgensen, and Eric S. Weber, *Positive matrices in the Hardy space with prescribed boundary representations via the Kaczmarz algorithm*, J. Anal. Math. **138** (2019), no. 1, 209–234, DOI 10.1007/s11854-019-0026-6. MR3996038

[HØUZ10] Helge Holden, Bernt Øksendal, Jan Ubøe, and Tusheng Zhang, *Stochastic partial differential equations*, 2nd ed., Universitext, Springer, New York, 2010. A modeling, white noise functional approach, DOI 10.1007/978-0-387-89488-1. MR2571742

[HS18] Wei He and Yeneng Sun, *Conditional expectation of correspondences and economic applications*, Econom. Theory **66** (2018), no. 2, 265–299, DOI 10.1007/s00199-017-1067-7. MR3836200

[HS23] Christoph Hertrich and Leon Sering, *ReLU neural networks of polynomial size for exact maximum flow computation*, Integer programming and combinatorial optimization, Lecture Notes in Comput. Sci., vol. 13904, Springer, Cham, [2023] ©2023, pp. 187–202, DOI 10.1007/978-3-031-32726-1_14. MR4606832

[Hut81] John E. Hutchinson, *Fractals and self-similarity*, Indiana Univ. Math. J. **30** (1981), no. 5, 713–747, DOI 10.1512/iumj.1981.30.30055. MR625600

[IM63] K. Itô and H. P. McKean Jr., *Brownian motions on a half line*, Illinois J. Math. **7** (1963), 181–231. MR154338

[IM74] Kiyosi Itô and Henry P. McKean Jr., *Diffusion processes and their sample paths*, Die Grundlehren der mathematischen Wissenschaften, Band 125, Springer-Verlag, Berlin-New York, 1974. Second printing, corrected. MR345224

[JLL23] Yuling Jiao, Yanming Lai, Xiliang Lu, Fengru Wang, Jerry Zhijian Yang, and Yuanyuan Yang, *Deep neural networks with ReLU-sine-exponential activations break curse of dimensionality in approximation on Hölder class*, SIAM J. Math. Anal. **55** (2023), no. 4, 3635–3649, DOI 10.1137/21M144431X. MR4629853

[Jor04] Palle E. T. Jorgensen, *The existence problem for dynamics of dissipative systems in quantum probability*, J. Math. Phys. **45** (2004), no. 9, 3605–3619, DOI 10.1063/1.1777401. MR2081648

[Jr68] Ole Groth Jørsboe, *Equivalence or singularity of Gaussian measures on function spaces*, Various Publications Series, No. 4, Aarhus Universitet, Matematisk Institut, Aarhus, 1968. MR277027

[Jr14] Palle E. T. Jørgensen, *A universal envelope for Gaussian processes and their kernels*, J. Appl. Math. Comput. **44** (2014), no. 1-2, 1–38, DOI 10.1007/s12190-013-0678-9. MR3147727

[JT16] Palle Jorgensen and Feng Tian, *Graph Laplacians and discrete reproducing kernel Hilbert spaces from restrictions*, Stoch. Anal. Appl. **34** (2016), no. 4, 722–747, DOI 10.1080/07362994.2016.1170613. MR3507188

[JT17] Palle Jorgensen and Feng Tian, *Non-commutative analysis*, World Scientific Publishing Co. Pte. Ltd., Hackensack, NJ, 2017. With a foreword by Wayne Polyzou, DOI 10.1142/10317. MR3642406

[JT18] Palle Jorgensen and Feng Tian, *Metric duality between positive definite kernels and boundary processes*, Int. J. Appl. Comput. Math. **4** (2018), no. 1, Paper No. 3, 13, DOI 10.1007/s40819-017-0434-1. MR3736758

[JT19a] Palle Jorgensen and Feng Tian, *Dynamical properties of endomorphisms, multiresolutions, similarity and orthogonality relations*, Discrete Contin. Dyn. Syst. Ser. S **12** (2019), no. 8, 2307–2348, DOI 10.3934/dcdss.2019146. MR4025985

[JT19b] Palle Jorgensen and Feng Tian, *Realizations and factorizations of positive definite kernels*, J. Theoret. Probab. **32** (2019), no. 4, 1925–1942, DOI 10.1007/s10959-018-0868-3. MR4020693

[JT20] Palle Jorgensen and James Tian, *Sampling with positive definite kernels and an associated dichotomy*, Adv. Theor. Math. Phys. **24** (2020), no. 1, 125–154, DOI 10.4310/ATMP.2020.v24.n1.a4. MR4106884

[JT21] Palle Jorgensen and James Tian, *Infinite-dimensional analysis*, WORLD SCIENTIFIC, 2021.

[JT22a] Palle E. T. Jorgensen and James Tian, *Dual pairs of operators, harmonic analysis of singular nonatomic measures and Krein-Feller diffusion*, J. Operator Theory **89** (2023), no. 1, 205–248, DOI 10.1007/s43670-023-00051-z. MR4567343

[JT22b] Palle Jorgensen and James Tian, *Reproducing kernels and choices of associated feature spaces, in the form of $L^2$-spaces*, J. Math. Anal. Appl. **505** (2022), no. 2, Paper No. 125535, 31, DOI 10.1016/j.jmaa.2021.125535. MR4295177

[JT23] Palle E. T. Jorgensen and James Tian, *Dual pairs of operators, harmonic analysis of singular nonatomic measures and Krein-Feller diffusion*, J. Operator Theory **89** (2023), no. 1, 205–248, DOI 10.1007/s43670-023-00051-z. MR4567343

[Kak43] Shizuo Kakutani, *Notes on infinite product measure spaces. II*, Proc. Imp. Acad. Tokyo **19** (1943), 184–188. MR14404

[Kas12] Yuji Kasahara, *Asymptotic behavior of the transition density of an ergodic linear diffusion*, Publ. Res. Inst. Math. Sci. **48** (2012), no. 3, 565–578, DOI 10.2977/PRIMS/81. MR2973393

[KLLW21] Shi-Lei Kong, Ka-Sing Lau, Jun Jason Luo, and Xiang-Yang Wang, *Hyperbolic graphs induced by iterations and applications in fractals*, Analysis and partial differential equations on manifolds, fractals and graphs, Adv. Anal. Geom., vol. 3, De Gruyter, Berlin, 2021, pp. 143–181, DOI 10.1515/9783110700763-006. MR4320089

[KLW17] Shi-Lei Kong, Ka-Sing Lau, and Ting-Kam Leonard Wong, *Random walks and induced Dirichlet forms on self-similar sets*, Adv. Math. **320** (2017), 1099–1134, DOI 10.1016/j.aim.2017.09.029. MR3709130

[Kol77] A. Kolmogoroff, *Grundbegriffe der Wahrscheinlichkeitsrechnung* (German), Springer-Verlag, Berlin-New York, 1977. Reprint of the 1933 original. MR494348

[Kol83] A. N. Kolmogorov, *On logical foundations of probability theory*, Probability theory and mathematical statistics (Tbilisi, 1982), Lecture Notes in Math., vol. 1021, Springer, Berlin, 1983, pp. 1–5, DOI 10.1007/BFb0072897. MR735967

[KS93] T. V. Kurien and Jayaram Sethuraman, *Singularities in Gaussian random fields*, J. Theoret. Probab. **6** (1993), no. 1, 89–99, DOI 10.1007/BF01046770. MR1201059

[KS16] B. I. Kopytko and R. V. Shevchuk, *On Feller semigroups associated with one-dimensional diffusion processes with membranes*, Theory Stoch. Process. **21** (2016), no. 1, 31–44. MR3571410

[LMP09] Sneh Lata, Meghna Mittal, and Vern I. Paulsen, *An operator algebraic proof of Agler's factorization theorem*, Proc. Amer. Math. Soc. **137** (2009), no. 11, 3741–3748, DOI 10.1090/S0002-9939-09-09928-6. MR2529882

[Mal97] Paul Malliavin, *Stochastic analysis*, Grundlehren der mathematischen Wissenschaften [Fundamental Principles of Mathematical Sciences], vol. 313, Springer-Verlag, Berlin, 1997, DOI 10.1007/978-3-642-15074-6. MR1450093

[MF23] Max Menzel and Uta Freiberg, *Fractal transformation of Krein-Feller operators* (English, with English and Ukrainian summaries), J. Math. Phys. Anal. Geom. **19** (2023), no. 2, 482–502. MR4633992

[Min20] Lenon Alexander Minorics, *Spectral asymptotics for Krein-Feller operators with respect to $V$-variable Cantor measures*, Forum Math. **32** (2020), no. 1, 121–138, DOI 10.1515/forum-2018-0188. MR4048458

[MP21] Simone Marsiglio and Fabio Privileggi, *Three dimensional fractal attractors in a green transition economic growth model*, Commun. Nonlinear Sci. Numer. Simul. **93** (2021), Paper No. 105509, 13, DOI 10.1016/j.cnsns.2020.105509. MR4149147

[MSF16] Krikamol Muandet, Bharath Sriperumbudur, Kenji Fukumizu, Arthur Gretton, and Bernhard Schölkopf, *Kernel mean shrinkage estimators*, J. Mach. Learn. Res. **17** (2016), Paper No. 48, 41. MR3504608

[Nel69] Edward Nelson, *Topics in dynamics. I: Flows*, Mathematical Notes, Princeton University Press, Princeton, NJ; University of Tokyo Press, Tokyo, 1969. MR282379

[OKT17] Sergio Ortobelli, Noureddine Kouaissah, and Tomáš Tichý, *On the impact of conditional expectation estimators in portfolio theory*, Comput. Manag. Sci. **14** (2017), no. 4, 535–557, DOI 10.1007/s10287-017-0282-9. MR3717273

[Phi61] R. S. Phillips, *On the integration of the diffusion equation with boundary conditions*, Trans. Amer. Math. Soc. **98** (1961), 62–84, DOI 10.2307/1993513. MR126625

[Roh49] V. A. Rohlin, *On the fundamental ideas of measure theory* (Russian), Mat. Sbornik N.S. **25(67)** (1949), 107–150. MR30584

[Roh52] V. A. Rohlin, *On the fundamental ideas of measure theory*, Amer. Math. Soc. Translation **1952** (1952), no. 71, 55. MR0047744

[Roh64a] V. A. Rohlin, *Exact endomorphism of a Lebesgue space*, Magyar Tud. Akad. Mat. Fiz. Oszt. Közl. **14** (1964), 443–474. MR0228654

[Roh64b] V. A. Rohlin, *Metric properties of endomorphisms of compact commutative groups* (Russian), Izv. Akad. Nauk SSSR Ser. Mat. **28** (1964), 867–874. MR168697

[SBM07] Roger B. Sidje, Kevin Burrage, and Shev Macnamara, *Inexact uniformization method for computing transient distributions of Markov chains*, SIAM J. Sci. Comput. **29** (2007), no. 6, 2562–2580, DOI 10.1137/060662629. MR2357627

[Sch64] Laurent Schwartz, *Sous-espaces hilbertiens d'espaces vectoriels topologiques et noyaux associés (noyaux reproduisants)* (French), J. Analyse Math. **13** (1964), 115–256, DOI 10.1007/BF02786620. MR179587

[Sch67] Laurent Schwartz, *Analyse mathématique. II*, Hermann, Paris, 1967. MR226973

[Sch75] L. Schwartz, *Fonctions mesurables et *-scalairement mesurables, mesures banachiques majorées, martingales banachiques, et propriété de Radon-Nikodým* (French), Séminaire Maurey-Schwartz 1974–1975: Espaces $L^p$, applications radonifiantes et géométrie des espaces de Banach, École Polytech., Paris, 1975, pp. Exp. No. IV, 18 pp. (erratum, pp. 1–2). MR448075

[Sch88] Laurent Schwartz, *Calcul infinitésimal stochastique* (French), Analyse mathématique et applications, Gauthier-Villars, Montrouge, 1988, pp. 445–463. MR956973

[SZ09] Steve Smale and Ding-Xuan Zhou, *Geometry on probability spaces*, Constr. Approx. **30** (2009), no. 3, 311–323, DOI 10.1007/s00365-009-9070-2. MR2558684

[Tak11] Masayoshi Takeda, *A large deviation principle for symmetric Markov processes with Feynman-Kac functional*, J. Theoret. Probab. **24** (2011), no. 4, 1097–1129, DOI 10.1007/s10959-010-0324-5. MR2851247

[WHD95] Paul Wilmott, Sam Howison, and Jeff Dewynne, *The mathematics of financial derivatives: A student introduction*, Cambridge University Press, Cambridge, 1995, DOI 10.1017/CBO9780511812545. MR1357666

[WLLZ20] Huoyun Wang, Xing Liao, Qing Liu, and Peng Zeng, *Topological entropy pairs for an iterated function system*, J. Math. Anal. Appl. **488** (2020), no. 2, 124076, 18, DOI 10.1016/j.jmaa.2020.124076. MR4080661

[WW97] A. E. Whalley and P. Wilmott, *An asymptotic analysis of an optimal hedging model for option pricing with transaction costs*, Math. Finance **7** (1997), no. 3, 307–324, DOI 10.1111/1467-9965.00034. MR1459062

Department of Mathematics, The University of Iowa, Iowa City, Iowa 52242-1419
*Email address*: palle-jorgensen@uiowa.edu

Mathematical Reviews, 535 W William St, Ann Arbor, Michigan 48103
*Email address*: jft@ams.org

Contemporary Mathematics
Volume **839**, 2026
https://doi.org/10.1090/conm/839/16733

# Generalised fractal interpolation functions with partial self-similarity using Rakotch contraction mappings

Gwang-Jin O and Vasileios Drakopoulos

*This paper is dedicated to our advisors.*

Abstract. Natural geometric structures often do not exhibit exact self-similarity, making them difficult to model mathematically. Previous research has addressed this by using fractal interpolation functions with partial self-similarity, typically constructed through Banach contraction mappings. However, these approaches are limited in their ability to represent the full complexity of natural forms. A novel method for constructing generalised fractal interpolation functions with partial self-similarity, utilising Rakotch contraction mappings, is presented. By relaxing the contraction requirements, our approach offers greater flexibility in modelling intricate geometries. We outline the construction process in detail and include illustrative examples to highlight how this method more effectively captures the complexity of natural patterns compared to traditional fractal interpolation techniques.

## 1. Introduction

A fractal interpolation function, or FIF for short, that interpolates a given set of points is a continuous fractal function whose graph is the attractor of an iterated function system, or IFS for short. FIFs are non-differentiable everywhere [**5**]. FIFs are useful for modelling experimental datasets with more complicated geometric objects found in nature. For example, the theory of FIFs has been applied to signal processing [**24**], [**25**], [**45**], [**46**], as well as modelling coastlines and shapes; see [**18**], [**19**], [**20**] and the references therein.

In [**2**], [**3**], Barnsley first introduced the concept of FIFs, which has since been studied by many researchers. Several types of FIFs have been constructed in various ways. For example, in [**3**], [**4**], the author constructed hidden variable FIFs, while in [**7**], [**21**], vector-valued FIFs were developed. Additionally, bilinear FIFs were introduced in [**6**], spline FIFs and Hermite FIFs were constructed in [**10**], [**12**], and in [**13**], [**14**], [**30**], respectively.

Many important properties of FIFs, such as similarity dimension, box-counting dimension, smoothness, stability, convexity preservation, sensitivity, shape preservation, monotonicity preservation, and approximation properties, have also been

2020 *Mathematics Subject Classification.* Primary 28A80, 54H25.

*Key words and phrases.* Fractals, interpolation, fractal interpolation functions; fixed points; self- similarity.

studied in [**2**], [**6**], [**12**], [**15**], [**26**], [**27**], [**28**], [**29**], [**30**], [**31**], [**33**], [**43**], [**44**]. See also the books [**3**], [**22**], [**23**], and others. Nevertheless, the authors of previous papers provided a construction of fractal interpolation functions using Banach contraction mappings.

Among the numerous geometric objects found in nature, very few exhibit strict self-similarity. Although studies have been conducted on constructing fractal interpolation functions with partial self-similarity using Banach contraction mappings, the scope of existing results is limited in the present study compared to the variety of geometric objects found in nature.

The purpose of this article is to present a new fractal interpolation method that better describes more natural geometric objects than previous methods. The structure of the article is as follows. Section 2 outlines the mathematical framework for constructing fractal interpolation functions using continuous functions on a metric space. In Section 3, introduces the generalized FIF construction with partial self-similarity using Rakotch contraction mappings. In Section 4, we provide a corresponding illustrative example.

## 2. Fractal interpolation functions

Let $(Y, d_Y)$ be a complete metric space. Let the set of data points $\Delta = \{(x_i, y_i) \in \mathbb{R} \times Y : i = 0, 1, \ldots, N\}$ be given, where $N > 1$ and $x_0 < x_1 < \cdots < x_N$. To avoid trivial interpolation, we assume that not all data points in $\Delta$ are collinear. Let $K = \mathbb{R} \times Y$ and $\theta > 0$. Define a metric $d$ on $K$ by $d((x, y), (x^*, y^*)) = |x - x^*| + \theta\, d_Y(y, y^*)$ for all points $(x, y)$ and $(x^*, y^*)$ in $K$. Then, the space $K$ is a complete metric space with respect to this distance $d$. For each $k = 1, 2, \ldots N$, define $l_k \colon \mathbb{R} \to \mathbb{R}$ as

$$l_k(x) = a_k x + b_k, \tag{2.1}$$

where

$$a_k = \frac{x_k - x_{k-1}}{x_N - x_0}, \quad b_k = \frac{x_N x_{k-1} - x_0 x_k}{x_N - x_0}.$$

Note that $l_k([x_0, x_N]) = [x_{k-1}, x_k]$. Let $0 \le s < 1$ and $c > 0$, and let $F_k \colon K \to Y$ be a function that satisfies the following three end-point conditions:

$$d_Y(F_k(x, y), F_k(x^*, y)) \le c\,|x - x^*| \quad \text{for all} \quad x, x^* \in \mathbb{R}, \quad y \in Y,$$

$$d_Y(F_k(x, y), F_k(x^*, y^*)) \le s\, d_Y(y, y^*) \quad \text{for all} \quad x \in \mathbb{R}, \quad y, y^* \in Y, \tag{2.2}$$

$$F_k(x_0, y_0) = y_{k-1}, \quad F_k(x_N, y_N) = y_k. \tag{2.3}$$

For each $k = 1, 2, \ldots, N$, define $w_k \colon K \to K$ as

$$w_k(x, y) = (l_k(x), F_k(x, y)) \quad \text{for all} \quad (x, y) \in K.$$

If we set $\theta = \frac{1-\alpha}{2c}$, where $\alpha = \max\{a_1, a_2, \ldots, a_N\}$, then, by Theorems 4.1 and 4.2 in Chapter 6 of [**3**], all the $w_k$ are contraction mappings on $K$, and there is a unique attractor $G \subseteq K$ of the IFS $\{K; w_1, w_2, \ldots, w_N\}$. The set $G$ is the graph of a continuous function $f$ such that $f(x_i) = y_i$ for all $i = 0, 1, \ldots, N$. This function $f$ is called a FIF that interpolates all the given data points in $\Delta$.

In [**2**], Barnsley considered the case where $Y = [a, b]$, $d_Y(y, y^*) = |y - y^*|$, $K = [x_0, x_N] \times [a, b]$, and $d((x, y), (x^*, y^*)) = \max\{|x - x^*|, |y - y^*|\}$ for $(x, y), (x^*, y^*) \in K$. A FIF is constructed by the IFS $\{X; w_1, w_2, \ldots, w_N\}$ in Theorem 1 of [**2**] under the conditions that, for each $k = 1, 2, \ldots, N$, $l_k \colon [x_0, x_N] \to [x_{k-1}, x_k]$ is a contractive homeomorphism such that $l_k(x_0) = x_{k-1}$, $l_k(x_N) = x_k$, and $F_k\colon K \to [a, b]$ is a continuous function with Conditions (2.2) and (2.3) for some $0 \le s < 1$.

If we define the function $l_k$ as in (2.1) and $F_k(x, y) = c_k x + \alpha_k y + e_k$ for $k \in \{1, 2, \dots, N\}$, then each $w_k$ is an affine map. Here, we call $\alpha_k$ the vertical scaling factor of the mapping $w_k$. If $|\alpha_k| < 1$ for all $k = 1, 2, \dots, N$, then each $w_k$ is a contractive mapping on $K$, and the obtained FIFs are called linear, or affine, FIFs (see Example 1 of [**2**]). In Example 2 of [**2**], a class of $F_k$ is considered, where $F_k(x, y) = \alpha_k y + p_k(x)$ and $p_k(x) = u(l_k(x) - \alpha_k b(x))$. Here, $u$ and $b$ are continuous functions satisfying $u(x_i) = y_i$ for all $i = 0, 1, \dots, N$, and the end-point equation $b(x_0) = y_0$, $b(x_N) = y_N$. The FIF obtained by such an IFS is called an $\alpha$-fractal function obtained with $u$, denoted by $u^\alpha$, where $\alpha = (\alpha_1, \alpha_2 \dots, \alpha_N)$ is a vector. The operator $F^\alpha\colon u \mapsto u^\alpha$ is called an *$\alpha$-fractal operator.* Here, $l$ is a bounded linear operator on the space of all continuous functions defined on $[x_0, x_N]$.

The development of many properties of $F^\alpha$ delineates a theory referred to as fractal interpolation theory. See also [**26**], [**27**], [**29**], [**42**]. By using variable vertical scaling parameters, Wang and Yu [**44**] obtained FIFs with more flexibility in a general sense. Let $[x_0, x_N] = [0, 1]$ and $F_k(x, y) = \alpha_k(x)y + p_k(x)$ for each $k \in \{1, 2, \dots, N\}$. Here, the functions $\alpha_k$ and $q_k$ are Lipschitz continuous functions defined on $[0, 1]$ that satisfy $\sup_{x\in[0,1]} |\alpha_k(x)| < 1$, and $p_k(0) = y_{k-1} - \alpha_k(0)y_0$, $p_k(1) = y_k - \alpha_k(1)y_N$.

In the case where $p_k(x) = u(l_k(x)) - \alpha_{k(x)}b(x)$, where $u$ is the piecewise linear interpolation function through the set of points $\Delta$ and $b$ is the linear function through the points $(x_0, y_0)$ and $(x_N, y_N)$, the stability and sensitivity of FIFs were investigated. Various types of FIFs have been constructed in the literature by considering different forms of $F_k$, with $l_k$ typically assumed to be a contraction homeomorphism from $[x_0, x_N]$ to $[x_{k-1}, x_k]$ such that $l_k(x_0) = x_{k-1}$ and $l_k(x_N) = x_k$.

In [**4**], a recurrent structure for IFSs is introduced, and recurrent FIFs are obtained by such IFSs. We briefly describe the method for constructing recurrent FIFs. For more details, see [**4**] and [**8**]. Consider the set of points $\Delta$ with $Y = \mathbb{R}$ and $[x_0, x_N] = [0, 1]$. For $k = 1, 2, \dots, N$, let $I_k = [x_{k-1}, x_k]$ and $I_k^* = [x_l, x_m]$, where $x_k - x_{k-1} < x_m - x_l$. Define $K = [0, 1] \times \mathbb{R}$ and define $w_k$ to be an affine map on $K$ such that $w_k(x_l, y_l) = (x_{k-1}, y_{k-1})$ and $w_k(x_m, y_m) = (x_k, y_k)$ (or $w_k(x_l, y_l) = (x_k, y_k)$ and $w_k(x_m, y_m) = (x_{k-1}, y_{k-1})$). Define a metric on $K$ so that each $w_k$ is a contraction mapping on $K$. The recurrent structure is given by the connection matrix $C = [C_{ij}]$, where $C_{ij} = 1$ if $I_j \subset I_i^*$ and $C_{ij} = 0$ otherwise. The authors define a map $W\colon H^N \to H^N$ by $(W(A))_i = \bigcup_{j:[C_{ij}=1} F_i(A_j)$, where $H$ denotes the set of all non-empty compact subsets of $K$, and $F_i$ is the set map $F_i(S) = \{w_i(x, y) : (x, y) \in S\}$ for $S \in H$. It is proven in [**4**] that $W$ has a unique fixed point $A = (A_1, \dots, A_N) \in H^N$ and $G = \bigcap_{i=1}^N A_i$ is the graph of a continuous function on $[0, 1]$. This function is called a *recurrent FIF.*

In [**2**] and [**3**], the construction of FIFs assumes that the domains of the functions $l_k$ and $F_k$ are $I$ and $I \times Y$, respectively, where $I = [x_0, x_N]$. We can easily see that the graph on $I$ and the subgraph on $I_k$ of a linear FIF have obvious self-similarity. Practically, the graph of some natural objects has weaker local self-similarity but not global self-similarity. Various types of FIFs with weak self-similarity have been developed with more flexibility and diversity by choosing different types of $F_k$. In the present work, we consider a new approach to establish

FIFs with locally self-similar graphs. We follow the process in [**2**], [**3**], and [**44**] to construct FIFs, and based on the idea of recurrent FIFs, we consider the domains of $l_k$ and $F_k$ to be $J_k$ and $J_k \times Y$, respectively. Here, $J_k = [x_{j(k)}, x_{m(k)}]$ is a subinterval of $I$ that depends on $k$.

Additionally, let the constant vertical scaling factors $\alpha_k$ be replaced by functions $\alpha_k(x)$. Suppose that the length of $J_k$ is not larger than the length of $I_k$, and $l_k : J_k \to I_k$ is not a contraction mapping. Then we see that a FIF obtained using this method exhibits self-similarity between its graph on the subinterval $J_k$ and the subinterval $I_k$.

Throughout this article, we use the following symbols and spaces. Let $(Y, d_Y)$ be a complete metric space, and let $\Delta = \{(x_k, y_k) \in \mathbb{R} \times Y : k = 0, 1, \dots, N\}$ be a given set of data points, where $N > 1$ and $x_0 < x_1 < \cdots < x_N$. To avoid trivial interpolation, we assume that no three data points in $\Delta$ are collinear. Let $I = [x_0, x_N]$. We define the set of all continuous functions on $I$ as $C[I, Y]$, and $C_\Delta[I, Y]$ denotes the subset of functions in $C[I, Y]$ that pass through all data points in $\Delta$.

We define the graph of $f \in C[I, Y]$ as $G(f) = \{(x, f(x)) : x \in I\}$. The collection $\{G(f) : f \in C[I, Y]\}$ is denoted as $\mathbf{G}[I, Y]$, and $\{G(f) : f \in C_\Delta[I, Y]\}$ is denoted by $\mathbf{G}_\Delta[I, Y]$. For $A \in \mathbf{G}[I, Y]$, we denote the unique function in $C[I, Y]$ such that $A = G(f_{[A]})$ by $f_{[A]}$. We define the sup-metric $d_\infty$ on $C[I, Y]$ by

$$d_\infty(f, g) = \max_{x \in I} d_Y(f(x), g(x)) \quad \text{for } f, g \in C[I, Y].$$

It is well known that $(C[I, Y], d_\infty)$ is a complete metric space. Since $C_\Delta[I, Y]$ is a closed subspace of $(C[I, Y], d_\infty)$, the space $(C_\Delta[I, Y], d_\infty)$ is also complete. For $A, B \in \mathbf{G}[I, Y]$, we define the distance $d_{\mathbf{G}}(A, B)$ as

$$d_{\mathbf{G}}(A, B) = d_\infty(f_{[A]}, f_{[B]}),$$

and it is easy to see that $(\mathbf{G}[I, Y], d_{\mathbf{G}})$ and $(\mathbf{G}_\Delta[I, Y], d_{\mathbf{G}})$ are both complete metric spaces.

For $A \in \mathbf{G}_\Delta[I, Y]$ and $k \in \{1, 2, \dots, N\}$, we define $A_{J_k} = A \cap (J_k \times Y)$. Then we have

$$W_k(A_{J_k}) = \{(l_k(x), F_k(x, f_{[A]}(x))) : x \in J_k\}.$$

Due to Assumption 1) in [**17**], where $l_k : J_k \to I_k$ is a homeomorphism, $W_k(A_{J_k})$ can be written as

$$W_k(A_{J_k}) = \{(x, F_k(l_k^{-1}(x), f_{[A]}(l_k^{-1}(x)))) : x \in I_k\}.$$

Thus, if we define the function $h_k(x)$ as

$$h_k(x) - F_k(l_k^{-1}(x), f_{[A]}(l_k^{-1}(x))), \quad x \in I_k,$$

then $W_k(A_{J_k})$ is the graph of the function $h_k\colon I_k \to Y$. Since $F_k$, $l_k^{-1}$, and $f_{[A]}$ are all continuous functions, $h_k$ is continuous on $I_k$. By the relations $l_k^{-1}(x_{k-1}) = x_{j(k)}$, $l_k^{-1}(x_k) = x_{m(k)}$, and $f_{[A]} \in C_\Delta[I, Y]$, we obtain $h_k(x_{k-1}) = y_{k-1}$ and $h_k(x_k) = y_k$.

Let

$$\mathbf{W}(A) = \bigcup_{k=1}^{N} W_k(A_{J_k}).$$

Then $\mathbf{W}(A)$ is the graph of the function $h\colon I \to Y$ defined by $h(x) = h_k(x)$ for $x \in I_k$. Furthermore, since each function $h_k$ is continuous on $I_k$ and $h(x_i) = y_i$, we have $h \in C_\Delta[I, Y]$ for every $i \in \{0, 1, \cdots, N\}$.

Define a mapping $T\colon C_\Delta[I,Y] \to C_\Delta[I,Y]$ by $T(h) = f_{[\mathbf{W}(G(h))]}$. That is, for $h \in C_\Delta[I,Y]$,

$$T(h)(x) = F_k(l_k^{-1}(x), h(l_k^{-1}(x))), \quad x \in I_k,\ k = 1, 2, \ldots, N.$$

In [**17**], the author constructed fractal interpolation functions with partial self-similar graphs using the class of simplest Banach contraction mappings.

THEOREM 2.1. *(see Theorem 2.2 on p. 914 of* [**17**]*)* $\mathbf{W}$ *is a contraction mapping on* $(\mathbf{G}_\Delta[I,Y], d_\mathbf{G})$.

THEOREM 2.2. *(see Theorem 2.3 on p. 915 of* [**17**]*)* $T$ *is a contraction mapping on* $C_\Delta[I,Y]$, *and* $f_{[\mathbf{G}]}$ *is the fixed point of* $T$ *in* $C_\Delta[I,Y]$.

## 3. Generalised fractal interpolation functions with partial self-similarity

In this Section, we present a method of constructing generalized fractal interpolation functions with partial self-similarity using Rakotch contraction mappings.

**3.1. Generalised fractal interpolation functions with self-similarity.** In this subsection, we recall the well known typical contractive mappings (see [**1**], [**9**], [**34**], [**35**]), fixed point theorems (see [**1**], [**9**], [**34**]) and generalized fractal interpolation functions with self-similarity using Rakotch contraction mappings (see [**16**], [**36**]), which are the key object of discussion in this paper.

DEFINITION 3.1. (see [**35**]) Let $(X,d)$ be a complete metric space with distance $d$, $f$ be a mapping $X$ into itself.

(1) (Banach [**1**]) There exists a number $\alpha$, $0 \le \alpha < 1$, such that, for each $x, y \in X$,

$$d(f(x), f(y)) \le \alpha d(x,y)$$

(2) (Rakotch [**34**]) There exists a monotone decreasing function $\alpha : (0, +\infty) \to [0,1)$ such that, for each $x, y \in X$, $x \ne y$,

$$d(f(x), f(y)) \le \alpha d(x,y)$$

By definitions of Banach and Rakotch, it is obvious that (1) $\Rightarrow$ (2). Now we give some well-known fixed point theorems.

THEOREM 3.2. *(see* [**1**], [**34**], [**9**]*) Let* $X$ *be a complete metric space.*

1. *Let* $f\colon X \to X$, $0 \le \alpha < 1$ *and*

$$d(f(x), f(y)) \le \alpha d(x,y), \ \forall x, y \in X.$$

*Then* $f$ *has a unique fixed point in* $X$.

2. *Let* $f\colon X \to X$ *and*

$$d(f(x), f(y)) \le \alpha d(x,y), \ \forall x, y \in X, \ x \ne y,$$

*where* $\alpha$ *is a monotone decreasing function such that* $\alpha\colon (0,+\infty) \to [0,1)$.

*Then* $f$ *has a unique fixed point in* $X$.

REMARK 3.3. 1) There is another convenient definition for Rakotch contraction mappings (see [**34**], p.463, Corollary, see [**35**], p.257, see [**36**], see [**37**], see [**40**], see [**39**], p.100). Let $(X,d)$ be a metric space and $f\colon X \to X$. Let $\varphi\colon (0,+\infty) \to (0,+\infty)$ be a function satisfying for all $x, y \in X$,

$$d(f(x), f(y)) \le \varphi(d(x,y)),$$

where $\varphi$ is an increasing function and for every $t > 0$, $\varphi(t) < t$ and a function $\frac{\varphi(t)}{t}$ is a decreasing function, then $f$ is called Rakotch contraction mapping.

2) There is a Rakotch contraction mapping that is not a Banach contraction mapping. To show that (2) $\nRightarrow$ (1), define $\varphi : (0, +\infty) \to [0, 1)$ by $\varphi(d) = \frac{1}{d+1}$ (see Example in p.262 of [**35**]).

Let $f(x) = \frac{1}{x+1}$, $0 \le x \le 1$. Then $f\colon [0,1] \to [0,1]$ and has a fixed point at $\frac{\sqrt{5}-1}{2}$.

For any fixed $\alpha \in (0,1)$, choose $y < -1 + \frac{1}{\alpha}$ and $y \in (0, \frac{\sqrt{5}-1}{2})$. Then

$$d(f(0), f(y)) = \frac{y}{y+1} > \alpha y = \alpha d(0, y),$$

and it does not satisfy (1).

On the other hand, for $0 \le x < y < 1$,

$$d(f(x), f(y)) = \frac{y-x}{(x+1)(y+1)} \le \frac{y-x}{y-x+1} = \varphi(d(x,y)),$$

and $f$ satisfies (2).

THEOREM 3.4. *(Existence and uniqueness of fractals, see Theorem 2 in p.218 of* [**3**], [**37**], [**38**]*) Let $(X, d)$ be a complete metric space and for all $n = 1, 2, \ldots, N$, $w_n\colon X \to X$. Let $\{X; w_1, w_2, \ldots, w_N\}$ be an IFS consisting of Banach contraction mappings (or Rakotch contraction mappings). Then there exist a unique fractal generated by this IFS.*

We introduce generalised fractal interpolation functions with self-similarity using Rakotch contraction mappings with the following assumptions.

Let $\{(x_i, y_i) \in \mathbb{R}^2 : i \in \{0, 1, \ldots, N\}\}$ be a data set, where $N \ge 2$, $-\infty < x_0 < x_1 < x_2 < \cdots < x_N < +\infty$ and $y_i \in \mathbb{R}$. Let for $(x, y) \in [x_0, x_N] \times \mathbb{R}$,

$$w_n \begin{pmatrix} x \\ y \end{pmatrix} := \begin{pmatrix} l_n(x) \\ F_n(x,y) \end{pmatrix},$$

where

$$l_n(x) = u_n x + v_n,$$
$$u_n := \frac{x_n - x_{n-1}}{x_N - x_0},$$
$$v_n := x_{n-1} - \frac{x_n - x_{n-1}}{x_N - x_0} x_0,$$

and $F_n$ is a continuous function such that $F_n\colon [x_0, x_N] \times \mathbb{R} \to \mathbb{R}$ and every $w_n$ satisfies the following conditions

$$w_n \begin{pmatrix} x_0 \\ y_0 \end{pmatrix} = \begin{pmatrix} x_{n-1} \\ y_{n-1} \end{pmatrix}, \quad w_n \begin{pmatrix} x_N \\ y_N \end{pmatrix} = \begin{pmatrix} x_n \\ y_n \end{pmatrix}.$$

The following theorem gives the generative principle of one variable FIFs by using Banach contraction (or Rakotch contraction) mappings.

THEOREM 3.5. *(see* [**2**], [**3**], [**16**], [**36**]*) If for every $n \in \{1, 2, \ldots, N\}$, a mapping $F_n(\cdot, y)$ satisfies Banach contraction (or Rakotch contraction) condition, and $F_n(x, \cdot)$ satisfies Lipschitz condition (*[**36**]*) (or Hölder condition (*[**16**]*)), then there is a continuous function that passes through the given data, and the graph of this continuous function is an attractor of the IFS defined above.*

REMARK 3.6. If $F_n(\cdot, y)$ is a special Banach contraction mapping, and $F_n(x, \cdot)$ is a special linear Lipschitz function, then a continuous function that passes through the given data is Barnsley's affine FIF (see [**3**]).

**3.2. Generalised fractal interpolation functions with partial self-similarity.** In this subsection, we present a method of constructing generalized fractal interpolation functions with partial self-similarity using Rakotch contraction mappings.

Let be $I_k = [x_{k-1}, x_k]$ and $J_k = [x_{j(k)}, x_{m(k)}]$ for $k = 1, 2, \ldots, N$, where $j$ and $m$ are two functions from the set $\{1, 2, \ldots, N\}$ to the set $\{0, 1, \ldots, N\}$ and $j(k) < m(k)$ for each $k = 1, 2, \ldots, N$.

We introduce the following two functions $l_k$ and $F_k$ for $k = 1, \cdots, N$.

(1) $l_k \colon J_k \to I_k$ is a homeomorphism such that $l_k(x_{j(k)}) = x_{k-1}$ and $l_k(x_{m(k)}) = x_k$.

(2) $F_k \colon J_k \times Y \to Y$ is a continuous function such that $F_k(x_{j(k)}, y_{j(k)}) = y_{k-1}$ and $F_k(x_{m(k)}, y_{m(k)}) = y_k$.

For each $k$, suppose that the condition

$$d_Y(F_k(x, y), F_k(x, y^*)) \leq \varphi_k(d_Y(y, y^*)) \quad for \;\; x \in J_k \;\; y, y^* \in Y \tag{3.1}$$

holds, where for $k = 1, \cdots, N$, each function $\varphi_k : (0, +\infty) \to (0, +\infty)$ is the non-decreasing function such that $\varphi_k(x) < x$ for $x > 0$ and $x \to \frac{\varphi_k(x)}{x}$ is non-increasing, that is, suppose that the function $F_k(x, \cdot)$ is the Rakotch contraction mapping with respect to the second variable $y$.

Define $W_k : J_k \times Y \to I_k \times Y$ by

$$W_k(x, y) = (l_k(x), F_k(x, y)), x \in J_k, y \in Y. \tag{3.2}$$

Note that the mapping $W_k(x_{j(k)}, y_{j(k)})$ satisfies the following end point equations.

$$W_k(x_{j(k)}, y_{j(k)}) = (x_{k-1}, y_{k-1}), \quad W_k(x_{m(k)}, y_{m(k)}) = (x_k, y_k).$$

LEMMA 3.7. *For* $A \in \mathbf{G}_\Delta[I, Y]$, *we obtain* $\mathbf{W}(A) \in \mathbf{G}_\Delta[I, Y]$ *and*

$$f_{[\mathbf{W}(A)]}(x) = F_k(l_k^{-1}(x), f_{[A]}(l_k^{-1}(x))), x \in I_k, k = 1, \cdots, N. \tag{3.3}$$

THEOREM 3.8. $\mathbf{W}$ *is a contractive mapping on* $(\mathbf{G}_\Delta[I, Y], d_\mathbf{G})$.

PROOF OF THEOREM 2.8. From the (3.3) and (3.1), we can see that, for $A_1, A_2 \in \mathbf{G}_\Delta[I, Y]$,

$$d_\mathbf{G}(\mathbf{W}(A_1), \mathbf{W}(A_2)) = \max_{x \in I} d_\mathbf{Y}(f_{[\mathbf{W}(A_1)]}(x), f_{[\mathbf{W}(A_2)]}(x))$$

$$= \max_{k=1,\cdots N} \max_{x \in I_k} d_Y(F_k(l_k^{-1}(x), f_{[A_1]}(l_k^{-1}(x))), F_k(l_k^{-1}(x), f_{[A_2]}(l_k^{-1}(x))))$$

$$= \max_{k=1,\cdots,N} \max_{x^* \in J_k} d_Y(F_k(x^*, f_{[A_1]}(x^*)), F_k(x^*, f_{[A_2]}(x^*)))$$

$$\leq \max_{k=1,\cdots,N} \max_{x^* \in J_k} \varphi_k(d_Y(f_{[A_1]}(x^*), f_{[A_2]}(x^*))) \leq \varphi(\max_{x^* \in J} d_Y(f_{[A_1]}(x^*), f_{[A_2]}(x^*))), \tag{3.4}$$

where $\varphi = \max_{k=1,\cdots,N} \{\varphi_1, \cdots, \varphi_N\}$ and $J = \bigcap_{k=1}^N J_k$. Since $J \subseteq I$, we have

$$d_\mathbf{G}(\mathbf{W}(A_1), \mathbf{W}(A_2)) \leq \varphi(\max_{x^* \in I} d_Y(f_{[A_1]}(x^*), f_{[A_2]}(x^*))) = \varphi(d_\mathbf{G}(A_1, A_2)) \tag{3.5}$$

and therefore, $\mathbf{W}$ is a Rakotch contractive mapping on $(\mathbf{G}_\Delta[I, Y], d_\mathbf{G})$. □

Define a mapping $T\colon C_\Delta[I,Y]\to C_\Delta[I,Y]$ by $T(h)=f_{[\mathbf{W}(G(h))]}$. That is, for $h\in C_\Delta[I,Y]$, $T(h)(x)=F_k(l_k^{-1}(x),h(l_k^{-1}(x))), x\in I_k, k=1,\cdots,N$. We denote the fixed point of $\mathbf{W}$ in $\mathbf{G}_\Delta[I,Y]$ by $\mathbf{G}$. The function $f_{[\mathbf{G}]}\in C_\Delta[I,Y]$ is called a FIF on $I$ that interpolates the given set of points $\Delta$.

THEOREM 3.9. *The mapping $T$ is a contractive on $C_\Delta[I,Y]$ and $f_{[\mathbf{G}]}$ is the fixed point of the mapping $T$ in $\in C_\Delta[I,Y]$.*

PROOF OF THEOREM 2.9. From (3.1) and a similar proof of Theorem 2.8, we can see that

$$d_\infty(T(h_1),T(h_2))\le\varphi(d_\infty(h_1,h_2))\quad for\quad h_1,h_2\in C_\Delta[I,Y]. \tag{3.6}$$

Hence the mapping $T$ is a Rakotch contraction mapping on $C_\Delta[I,Y]$. Let $\mathbf{G}$ denote the fixed point of $\mathbf{W}$ in $\mathbf{G}_\Delta[I,Y]$. Because of $\mathbf{G}=\mathbf{W}(\mathbf{G})$, from the (3.3), we can see that $f_{[\mathbf{G}]}$ satisfies the functional equation

$$f_{[\mathbf{G}]}(x)=f_{[\mathbf{W}(\mathbf{G})]}(x)=F_k(l_k^{-1}(x),f_{[\mathbf{G}]}(l_k^{-1}(x))),x\in I_k,k=1,\cdots,N. \tag{3.7}$$

So we have $f_{[\mathbf{G}]}(x)=T(f_{[\mathbf{G}]})(x)$ for $x\in I$ by the definition of $T$ and (3.7) and therefore, we obtain that $f_{[\mathbf{G}]}$ is the fixed point of $T$ in $C_\Delta[I,Y]$. □

## 4. Example

The following example can be used to compare FIFs established by our approach with the results obtained in the earlier works. To see the difference, we consider the following nonlinear FIFs.

Let $l_k$ and $F_k$ be of the forms $l_k(x)=a_kx+b_k$ and $F_k(x,y)=c_kx+\alpha_k(y)+e_k$, respectively. Suppose that $l_k$ satisfies (1) and $F_k$ satisfies (2). We now use the following Rakotch contraction mappings.

$$\alpha_1(y)=0.8y,\quad \alpha_2(y)=\tfrac{y}{1+y},\quad \alpha_3(y)=\tfrac{1}{1+y},$$
$$\alpha_4(y)=0.5y,\quad \alpha_5(y)=0.3y,\quad \alpha_6(y)=\tfrac{y}{\sqrt[3]{1+y^3}},$$
$$\alpha_7(y)=\tfrac{2y}{2y+3},\quad \alpha_8(y)=\tfrac{y}{y+2},\quad \alpha_9(y)=0.7y,$$

$a_k,b_k,c_k$ and $e_k$ are obtained using the formulae

$$a_k=\frac{x_k-x_{k-1}}{x_N-x_0},\quad b_k=\frac{x_Nx_{k-1}-x_0x_k}{x_N-x_0},$$
$$c_k=\frac{y_k-y_{k-1}}{x_N-x_0}-\frac{\alpha_k(y_N)-\alpha_k(y_0)}{x_N-x_0},\quad e_k=\frac{x_Ny_0-x_0y_N}{x_N-x_0}-\frac{x_N\alpha_k(y_N)-x_0\alpha_k(y_0)}{x_N-x_0}.$$

From the figures obtained in the following example, one may easily see the effect of choices of each $J_k$ on the fractal interpolation curves.

EXAMPLE 4.1. Let $\Delta=\{(0,7),(1,6),(2,7),(3,5),(4,10),(5,0),(6,10),(7,8),(8,4),(9,2)\}$.

(1) In Figure 1, all the subintervals are $J_k=I=[0,9]$.

(2) In Figure 2, the subintervals are $J_1=J_2=J_3=[0,3]$, $J_4=J_5=J_6=[3,6]$ and $J_7=J_8=J_9=[6,9]$.

(3) In Figure 3, the subintervals are $J_1=J_2=J_3=[0,3]$, $J_4=[1,4]$, $J_5=[2,5]$, $J_6=[3,6]$, $J_7=[4,7]$, $J_8=[5,8]$, $J_9=[6,9]$.

(4) For the subintervals $J_1=J_2=J_3=[0,3]$, $J_4=[1,4]$, $J_5=[2,5]$, $J_6=[3,6]$, $J_7=[4,7]$, $J_8=[5,8]$, $J_9=[6,9]$, the graph of the FIF is given in Figure 4.

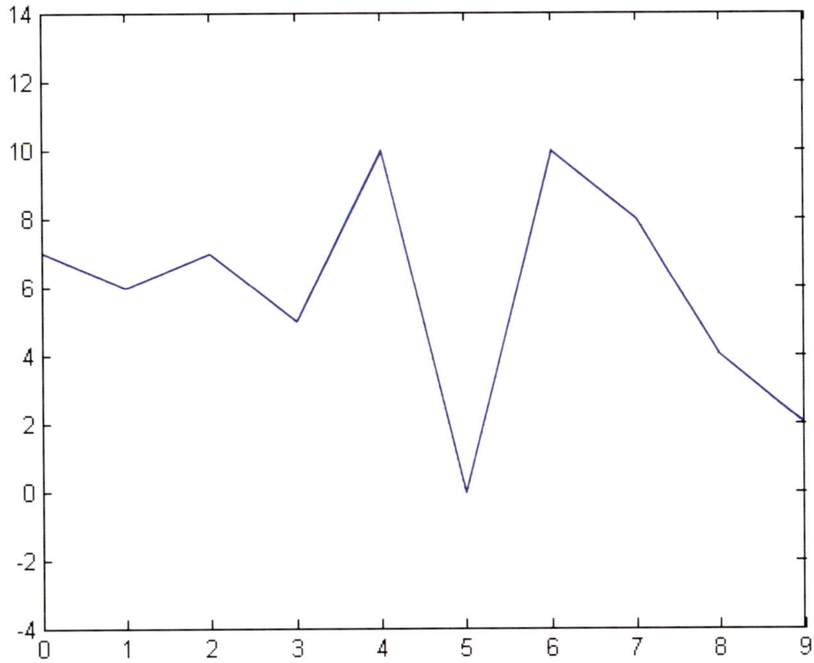


FIGURE 1. The graph of a piecewise linear interpolation function that interpolates a given set of points

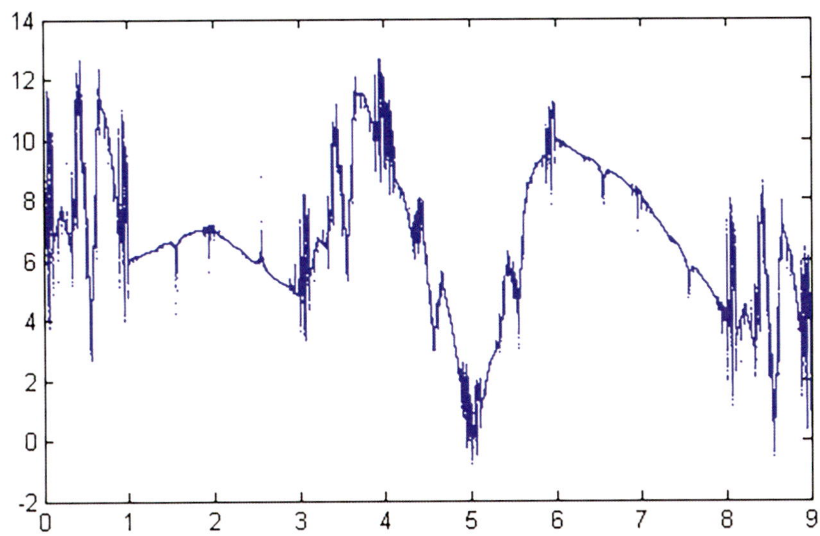


FIGURE 2. The graph of the linear FIF by choosing $J_k = I = [0, 9]$ for each $k = 1, 2, \ldots 9$.

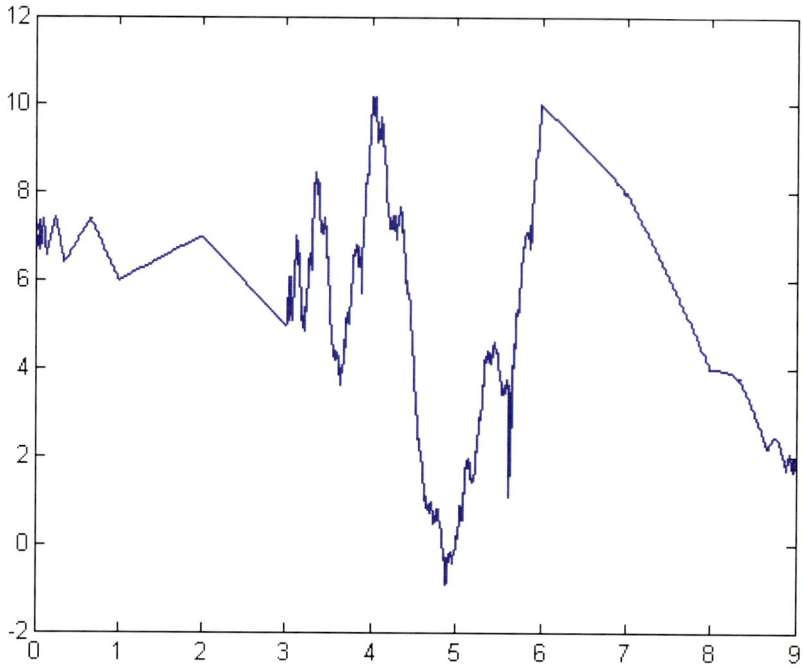


FIGURE 3. The graph of FIF that is constructed by choosing $J_1 = J_2 = J_3 = [0, 3]$, $J_4 = J_5 = J_6 = [3, 6]$ and $J_7 = J_8 = J_9 = [6, 9]$

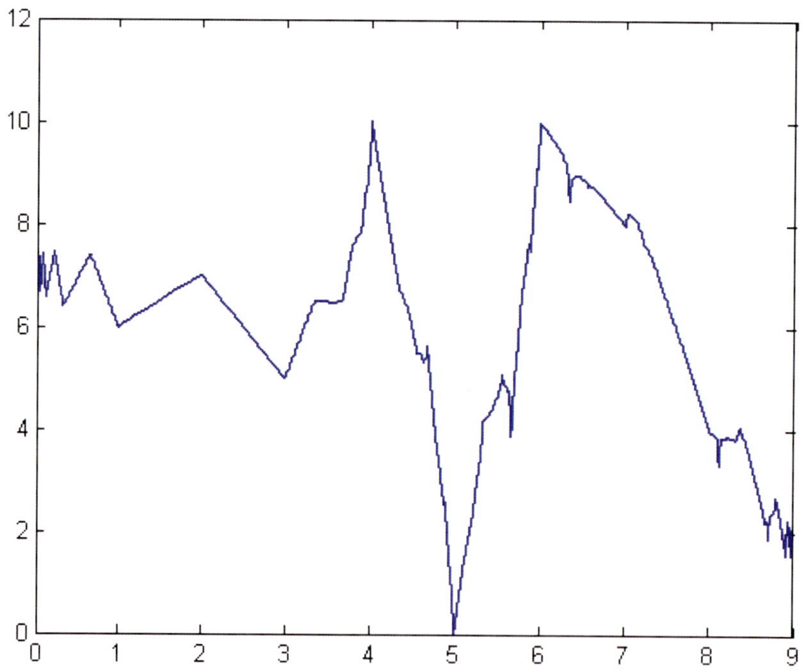


FIGURE 4. The graph of FIF that is constructed by choosing $J_1 = J_2 = J_3 = [0, 3]$, $J_4 = [1, 4]$, $J_5 = [2, 5]$, $J_6 = [3, 6]$, $J_7 = [4, 7]$, $J_8 = [5, 8]$, $J_9 = [6, 9]$

Note that if, for all $k = 1, 2, \ldots, 9$, the vertical scaling factor is $\alpha_k = 0.5$, then the resulting graphs of the fractal interpolation functions coincide with those presented in [**17**].

## References

[1] Stefan Banach, *Sur les opérations dans les ensembles abstraits et leur application aux équations intégrales* (French), Fund. Math. **3** (1922), 133–181, DOI 10.4064/fm-3-1-133-181. MR3949898

[2] Michael F. Barnsley, *Fractal functions and interpolation*, Constr. Approx. **2** (1986), no. 4, 303–329, DOI 10.1007/BF01893434. MR892158

[3] Michael Barnsley, *Fractals everywhere*, Academic Press, Inc., Boston, MA, 1988. MR977274

[4] Michael F. Barnsley, John H. Elton, and Douglas P. Hardin, *Recurrent iterated function systems*, Constr. Approx. **5** (1989), no. 1, 3–31, DOI 10.1007/BF01889596. Fractal approximation. MR982722

[5] Michael F. Barnsley and Andrew Vince, *Fractal continuation*, Constr. Approx. **38** (2013), no. 2, 311–337, DOI 10.1007/s00365-013-9205-3. MR3097049

[6] Michael F. Barnsley and Peter R. Massopust, *Bilinear fractal interpolation and box dimension*, J. Approx. Theory **192** (2015), 362–378, DOI 10.1016/j.jat.2014.10.014. MR3313490

[7] P. Bouboulis and L. Dalla, *Hidden variable vector valued fractal interpolation functions*, Fractals **13** (2005), no. 3, 227–232, DOI 10.1142/S0218348X05002854. MR2166281

[8] P. Bouboulis and L. Dalla, *Fractal interpolation surfaces derived from fractal interpolation functions*, J. Math. Anal. Appl. **336** (2007), no. 2, 919–936, DOI 10.1016/j.jmaa.2007.01.112. MR2352989

[9] V. W. Bryant, *A remark on a fixed-point theorem for iterated mappings*, Amer. Math. Monthly **75** (1968), 399–400, DOI 10.2307/2313440. MR226621

[10] A. K. B. Chand and G. P. Kapoor, *Generalized cubic spline fractal interpolation functions*, SIAM J. Numer. Anal. **44** (2006), no. 2, 655–676, DOI 10.1137/040611070. MR2218964

[11] A. K. B. Chand and G. P. Kapoor, *Spline coalescence hidden variable fractal interpolation functions*, J. Appl. Math., posted on 2006, Art. ID 36829, 17, DOI 10.1155/JAM/2006/36829. MR2268485

[12] A. K. B. Chand and G. P. Kapoor, *Stability of affine coalescence hidden variable fractal interpolation functions*, Nonlinear Anal. **68** (2008), no. 12, 3757–3770, DOI 10.1016/j.na.2007.04.017. MR2416082

[13] A. K. B. Chand and M. A. Navascués, *Generalized Hermite fractal interpolation*, Rev. R. Acad. Cienc. Exactas Fís. Quím. Nat. Zaragoza (2) **64** (2009), 107–120. MR2676244

[14] A. K. B. Chand and P. Viswanathan, *A constructive approach to cubic Hermite fractal interpolation function and its constrained aspects*, BIT **53** (2013), no. 4, 841–865, DOI 10.1007/s10543-013-0442-4. MR3129835

[15] A. K. B. Chand, N. Vijender, and M. A. Navascués, *Shape preservation of scientific data through rational fractal splines*, Calcolo **51** (2014), no. 2, 329–362, DOI 10.1007/s10092-013-0088-2. MR3210161

[16] Vasileios Drakopoulos and Song-Il Ri, *Hölder continuous fractal interpolation functions*, New perspectives on nonlinear dynamics and complexity, Nonlinear Syst. Complex., vol. 35, Springer, Cham, [2023] ©2023, pp. 141–152, DOI 10.1007/978-3-030-97328-5_9. MR4487140

[17] Dah-Chin Luor, *Fractal interpolation functions with partial self similarity*, J. Math. Anal. Appl. **464** (2018), no. 1, 911–923, DOI 10.1016/j.jmaa.2018.04.041. MR3794121

[18] P. Manousopoulos, V. Drakopoulos, T. Theoharis, Fractal active shape models, in: Computer Analysis of Images and Patterns, Springer, 2007, pp. 645-652.

[19] Polychronis Manousopoulos, Vassileios Drakopoulos, and Theoharis Theoharis, *Curve fitting by fractal interpolation*, Transactions on computational science. I, Lecture Notes in Comput. Sci., vol. 4750, Springer, Berlin, 2008, pp. 85–103, DOI 10.1007/978-3-540-79299-4_4. MR2454259

[20] Polychronis Manousopoulos, Vassileios Drakopoulos, and Theoharis Theoharis, *Parameter identification of 1D recurrent fractal interpolation functions with applications to imaging and signal processing*, J. Math. Imaging Vision **40** (2011), no. 2, 162–170, DOI 10.1007/s10851-010-0253-z. MR2782124

[21] Peter R. Massopust, *Vector-valued fractal interpolation functions and their box dimension*, Aequationes Math. **42** (1991), no. 1, 1–22, DOI 10.1007/BF01818475. MR1112180

[22] Peter R. Massopust, *Fractal functions, fractal surfaces, and wavelets*, Academic Press, Inc., San Diego, CA, 1994. MR1313502

[23] Peter Massopust, *Interpolation and approximation with splines and fractals*, Oxford University Press, Oxford, 2010. MR2723033

[24] D.S. Mazel, Representation of discrete sequences with three-dimensional iterated function systems, IEEE Trans. Signal Process. 42 (1994) 3269-3271.

[25] D.S. Mazel, M.H. Hayes, Using iterated function systems to model discrete sequences, IEEE Trans. Signal Process. 40(1992) 1724-1734.

[26] M. A. Navascues, *Fractal trigonometric approximation*, Electron. Trans. Numer. Anal. **20** (2005), 64–74. MR2149505

[27] M. A. Navascués, *A fractal approximation to periodicity*, Fractals **14** (2006), no. 4, 315–325, DOI 10.1142/S0218348X06003283. MR2282780

[28] M. A. Navascués, *Fractal approximation*, Complex Anal. Oper. Theory **4** (2010), no. 4, 953–974, DOI 10.1007/s11785-009-0033-1. MR2735314

[29] M. A. Navascués, *Fractal bases of $\mathcal{L}^p$ spaces*, Fractals **20** (2012), no. 2, 141–148, DOI 10.1142/S0218348X12500132. MR2950202

[30] M. A. Navascués and A. K. B. Chand, *Fundamental sets of fractal functions*, Acta Appl. Math. **100** (2008), no. 3, 247–261, DOI 10.1007/s10440-007-9182-2. MR2369080

[31] M. A. Navascués and M. V. Sebastián, *Generalization of Hermite functions by fractal interpolation*, J. Approx. Theory **131** (2004), no. 1, 19–29, DOI 10.1016/j.jat.2004.09.001. MR2103831

[32] M. A. Navascués and M. V. Sebastián, *Smooth fractal interpolation*, J. Inequal. Appl., posted on 2006, Art. ID 78734, 20, DOI 10.1155/JIA/2006/78734. MR2253425

[33] Srijanani Anurag Prasad, *Node insertion in coalescence fractal interpolation function*, Chaos Solitons Fractals **49** (2013), 16–20, DOI 10.1016/j.chaos.2013.02.002. MR3042092

[34] E. Rakotch, *A note on contractive mappings*, Proc. Amer. Math. Soc. **13** (1962), 459–465, DOI 10.2307/2034961. MR148046

[35] B. E. Rhoades, *A comparison of various definitions of contractive mappings*, Trans. Amer. Math. Soc. **226** (1977), 257–290, DOI 10.2307/1997954. MR433430

[36] Songil Ri, *A new nonlinear fractal interpolation function*, Fractals **25** (2017), no. 6, 1750063, 12, DOI 10.1142/S0218348X17500633. MR3725302

[37] Nicolae-Adrian Secelean, *Generalized iterated function systems on the space $l^\infty(X)$*, J. Math. Anal. Appl. **410** (2014), no. 2, 847–858, DOI 10.1016/j.jmaa.2013.09.007. MR3111871

[38] Nicolae-Adrian Secelean, *Generalized $F$-iterated function systems on product of metric spaces*, J. Fixed Point Theory Appl. **17** (2015), no. 3, 575–595, DOI 10.1007/s11784-015-0235-2. MR3411806

[39] Filip Strobin, *Attractors of generalized IFSs that are not attractors of IFSs*, J. Math. Anal. Appl. **422** (2015), no. 1, 99–108, DOI 10.1016/j.jmaa.2014.08.029. MR3263451

[40] Filip Strobin and Jarosław Swaczyna, *On a certain generalisation of the iterated function system*, Bull. Aust. Math. Soc. **87** (2013), no. 1, 37–54, DOI 10.1017/S0004972712000500. MR3011940

[41] P. Viswanathan and A. K. B. Chand, *A fractal procedure for monotonicity preserving interpolation*, Appl. Math. Comput. **247** (2014), 190–204, DOI 10.1016/j.amc.2014.06.090. MR3270831

[42] Puthan Veedu Viswanathan and Arya Kumar Bedabrata Chand, *$\alpha$-fractal rational splines for constrained interpolation*, Electron. Trans. Numer. Anal. **41** (2014), 420–442. MR3284416

[43] P. Viswanathan and A. K. B. Chand, *Fractal rational functions and their approximation properties*, J. Approx. Theory **185** (2014), 31–50, DOI 10.1016/j.jat.2014.05.013. MR3233064

[44] P. Viswanathan, A. K. B. Chand, and M. A. Navascués, *Fractal perturbation preserving fundamental shapes: bounds on the scale factors*, J. Math. Anal. Appl. **419** (2014), no. 2, 804–817, DOI 10.1016/j.jmaa.2014.05.019. MR3225407

[45] Hong-Yong Wang and Jia-Shan Yu, *Fractal interpolation functions with variable parameters and their analytical properties*, J. Approx. Theory **175** (2013), 1–18, DOI 10.1016/j.jat.2013.07.008. MR3101056

[46] Ming-Yue Zhai, Juan Luis Fernández-Martínez, and James W. Rector, *A new fractal interpolation algorithm and its applications to self-affine signal reconstruction*, Fractals **19** (2011), no. 3, 355–365, DOI 10.1142/S0218348X11005427. MR2826746

Department of Mathematics, University of Science, Pyongyang, Democratic People's Republic of Korea

*Current address*: Department of Mathematics and Statistics, Case Western Reserve University, Cleveland, Ohio 43403

Department of Computer Science and Biomedical Informatics, University of Thessaly, Lamia, Greece

*Email address*: `vdrakop@uth.gr`

FSC
www.fsc.org
MIX
Papier | Fördert
gute Waldnutzung
FSC® C083411

Kolibri 360 GmbH
Ferdinand-Jühlke-Straße 7
99095 Erfurt, Deutschland
produktsicherheit@kolibri360.de